# Embedded Systems and Wireless Technology

Editors
Dr. Raul Aquino Santos
*University of Colima*
*Colima, Mexico*

and

Dr. Arthur Edwards Block
*University of Colima*
*Colima, Mexico*

CRC Press
Taylor & Francis Group
Boca Raton London New York

CRC Press is an imprint of the
Taylor & Francis Group, an **informa** business
A SCIENCE PUBLISHERS BOOK

CRC Press
Taylor & Francis Group
6000 Broken Sound Parkway NW, Suite 300
Boca Raton, FL 33487-2742



Printed in the United States of America on acid-free paper
Version Date: 20120327

International Standard Book Number: 978-1-57808-803-4 (Hardback)



*CIP Data will be provided on request*

**Visit the Taylor & Francis Web site at**
**http://www.taylorandfrancis.com**

**CRC Press Web site at**
**http://www.crcpress.com**

**Science Publishers Web site at**
**http://www.scipub.net**

# Preface

The first mention most people over 50 years of age had of the term "sensors" was in Gene Roddenberry's Star Trek television series, which first aired in the United States in 1966. This groundbreaking science fiction program almost predictably used this term when Captain Kirk would ask Spock or some lesser member of the Enterprise's crew to give him sensor readings of a planet's surface as the starship arrived at its destination many light years in space. As children or teenagers, the concept of a sensor became clear. It was some kind of gadget that would give us information about a place without us having to be there. That sounded cool and we ate it up.

People were fascinated by watching their first visual example of how ubiquitous computing worked as they saw doors automatically open. They also saw how ambient conditions in crew quarters and the starship, in general, were set by voice command and sensors then controlled the ambient conditions according to the parameters set by the crewmembers. And we can never forget how Leonard "Bones" McCoy could monitor and control many basic functions to keep his patients alive and well. Ubiquitous computing and wireless communications made all of these science fiction things possible.

Back then, people did not know that much of that future, hundreds of years away, would be reality within 50 years. Today, millions of Americans and others living around the world fondly remember this series and some are almost religiously fanatical about the view of a humanity that no longer was at war with each other and used technology to further human growth and evolution at a time when the world lived under the ever-present threat of nuclear annihilation.

This series, almost 50 years ago, presented the subject of sensors, wireless communications and ubiquitous computing as far away future technological developments. Little did people know that the National Aeronautic and Space Administration (NASA) and the United States Department of Defense were developing sensors for space exploration and military use at that very moment.

Sensors, although they did exist in a more primitive form, suffered from four major disadvantages:

First, they only functioned for very short periods. They consumed too much energy for them to be useful for any prolonged period. In the 1960's, the computing capacity of a mainframe computer was necessary to run the algorithms now used to save the energy required to give sensor networks the extended lifetime they require to carry out many important tasks. Today, sensors require little batter power or can even be powered by alternative energy sources, particularly solar.

Secondly, sensors were much larger than they are now. Their size did not make them practical for many uses. However, the size of sensors has been reduced greatly and can be used in many applications that were unthought-of of 50 years ago. This miniaturization has led to a great expansion of sensor technology embedded in an ever-greater number of applications. A sensor the size of a shoe box weighing 5 pounds can hardly be worn by a human throughout the day to monitor a patient's heartbeat. The size of sensors has become so small that many now weigh mere grams and fit into a thimble.

Thirdly, the processing power of sensors has increase from almost nothing to what entire computers had in the late 1980's. Rodenburry's future vision has James T. Kirk employing embedded systems using the smallest of crystals to sense, process, and control an entire spaceship without his even having to command. Today's microprocessors have more computing capacity than man's first space flight did in the late 1950's. Finally, sensors transferring information through wireless communications, untethered by "old fashioned" cables, can now go "where no man has gone before" -or better said, should never ever go- due to health and safety considerations.

Today, optical sensors relay information wirelessly to actuators that open doors automatically for people and intelligent houses now permit persons to control ambient conditions such as lighting and temperature automatically. Nowadays, security applications use motion and heat sensors, among others, to determine if there is unauthorized entry into a home or business. Ubiquitous computing, based on sensors and wireless communications, also provide benefits to persons in the area of health care, where sensors, attached to the individual or embedded in the individual's clothing can remotely send vital real-time information regarding a patient's heart beat,body temperature, etc.

Although still in its initial stages, sensors in embedded systems, communicated wirelessly among themselves and to what we presently call the Web, will evolve from being a novelty, to becoming prevalent, to becoming indispensable. As a result, our lives will become increasingly dependent on embedded systems and digital information technology that is ubiquitously embedded in our environment. A decade ago ubiquitous computing was barely become a viable option to traditional technologies. Today, ten years later, more than 98% of processors used today are in embedded systems that are no longer visible to their owner. This serves as evidence of how technologies are evolving at a staggering rate and evidences the evolution from times when processors were located almost exclusively in personal computers.

Gene Roddenbury's future vision is quickly becoming a reality. However, what he envisioned as a reality in several hundred years is quickly becoming a reality in the early 21st century. For persons having lived the past half decade, the advances have been nothing short of miraculous. Using Star Trek as a reference of what the future might be in a few hundred years motivated many current scientists to look beyond what is and seek a future where technology makes life longer and more satisfying. This book is dedicated to the many persons who in the past and in the future will develop embedded systems and bring back the crew of the Enterprise to a 21st century earth that is very similar to the one from which they come.

# Contents

# List of Contributors

**Agis, Carlos Gómez:** Universidad Autónoma de Baja California, Km. 103 Carretera Tijuana-Ensenada. Ensenada, B.C. México. Código Postal 22860. E-mail: cgomez@uabc.edu.mx

**Al-Huneidi, Ahmad:** Hasselt University, Belgium. Campus Diepenbeek, Agoralaan Gebouw D, BE 3590 Diepenbeek. E-mail: ahmad.alhuneidi@ uhasselt.be

**Angeles, Alfonso:** Centro de Investigación y Desarrollo de Tecnología Digital—IPN, Av. del Parque No. 131O, Mesa de Otay, Tijuana, Baja California, México. C.P. 22510. E-mail: aangeles@ipn.mx

**Ávila, Humberto Cervantes De:** Facultad de Ingeniería, Arquitectura y Diseño (FIAD). México, E-mail: hcervantes@uabc.edu.mx

**Azizi, Abdelmalek:** Department of Mathematics and Computer Science, FSO, University Mohamed Ist, Morocco. Academy Hassan II of Sciences and Technology, Oujda, Morocco. E-mail: abdelmalekazizi@yahoo.fr

**Azizi, Mostafa:** Department of Computer Engineering, ESTO, University Mohamed Ist, Oujda, Morocco. E-mail: azizi.mos@gmail.com

**Baldivia, Margarita Glenda Mayoral:** University of Colima, College of Telematics, Avenue Universidad 333, Colima, 28040, Mexico. E-mail: mglenda@ucol.mx

**Barriba, Itzel:** Universidad Autónoma de Baja California, Facultad de Ciencias Químicas e Ingeniería, Calzada Universidad 14418, Parque Industrial Internacional Tijuana, B.C., C.P. 22390, México. E-mail: itzel. barriba}@uabc.edu.mx

**Cardenas, Omar Alvarez:** University of Colima, College of Telematics, Avenue Universidad 333, Colima, 28040, Mexico. E-mail: xe1aom@ ucol.mx

**Carrillo, Jesus Rafael:** University of Colima, College of Telematics, Avenue Universidad 333, Colima, 28040, Mexico. E-mail: jesuscarrillo8@gmail. com

**Castanheira, F.M.:** Ph.D. Student, Electromechanical Engineering Department, University of Beira Interior, Covilhã, Portugal. E-mail: D953@ubi.pt.

**Deáky, Bogdan-Alexandru:** AMTS Research Department, "Transilvania" University of Bra'ov, Address: Str. MihaiViteazul Nr. 5, 500174, Bra'ov, Romania. E-mail: bogdan_deaky@xu.unitbv.ro

**Elboukhari, Mohamed:** Department of Mathematics and Computer Science, FSO, University Mohamed Ist, Morocco. E-mail: elboukharimohamed@ gmail.com

**Folea, Silviu:** Automation Department, Technical University of Cluj-Napoca, Romania. E-mail: Silviu.Folea@aut.utcluj.ro

**García, Jesús Luna:** Technische Universität Darmstadt, Germany. E-mail: jlunagar@gmail.com

**Garcia-Ruiz, Miguel A.:** Algoma University, Dept. of Computer Science and Mathematics, 1520 Queen Street East Sault Ste. Marie, Ontario, P6A 2G4, Canada. E-mail: miguel.garcia@algomau.ca

**Gonzalez-Gutierrez, M.:** College of Telematics, University of Colima, Mexico. E-mail: maythegonzalez@ucol.mx

**Hernández-Capuchin, Ismael:** Universidad Autónoma de Baja California, Km. 103 Carretera Tijuana-Ensenada. Ensenada, B.C. México. Código Postal 22860. E-mail: ihernandez@uabc.edu.mx

**Hipólito, Juan Iván Nieto:** Autonomous University of Baja California, Km. 103 Carretera Tijuana-Ensenada. Ensenada, B.C. México. Código Postal 22860. E-mail: jnieto@uabc.edu.mx

**Hulea, Mihai:** Automation Department, Technical University of Cluj-Napoca, Romania. E-mail: Mihai.Hulea@aut.utcluj.ro

**Izquierdo, Victor:** GPPI Telecomunicaciones S.A. de R.L. de C.V., Blvd. Díaz Ordaz 1460 Interior A, Colonia Reynoso, La Mesa, Tijuana, Baja California C.P. 22106, México. E-mail: vizquierdo@ gppitelecom.com

**Janković, Radomir:** Ph.D., Union University School of Computing, 6 Knez Mihailova Street, 11000 Belgrade, Serbia. E-mail (school): rjankovic@raf.edu.rs

**Juárez-Ramírez, Reyes:** Universidad Autónoma de Baja California, Facultad de Ciencias Químicas e Ingeniería, Calzada Universidad 14418, Parque Industrial Internacional Tijuana, B.C., C.P. 22390, México. E-mail: reyesjua@uabc.edu.mx

**León, María de los Ángeles Cosio:** M.Sc.,AutonomousUniversity of Baja California, Km. 103 Carretera Tijuana-Ensenada. Ensenada, B.C. México. Código Postal 22860. E-mail: mary.cosio@gmail.com

**Licea, Guillermo:** Universidad Autónoma de Baja California, Facultad de Ciencias Químicas e Ingeniería, Calzada Universidad 14418, Parque Industrial Internacional Tijuana, B.C., C.P. 22390, México. E-mail: glicea@uabc.edu.mx

**Martínez-Rosas, Miguel E.:** Facultad de Ingeniería, Arquitectura y Diseño (FIAD). México. E-mail:emartine@uabc.edu.mx

**Martínez-Rosas, Miguel Enrique:** Universidad Autónoma de Baja California, Km. 103 Carretera Tijuana-Ensenada. Ensenada, B.C. México. Código Postal 22860. E-mail: hmartine@uabc.edu.mx

**Martínez-Sandoval, Jesús R.:** Facultad de Ingeniería, Arquitectura y Diseño (FIAD). México. E-mail: rmartinez33@uabc.edu.mx

**Murillo-Bracamontes, Eduardo A.:** Facultad de Ingeniería, Arquitectura y Diseño (FIAD). Mexico, Centro de Ingeniería y Tecnología (CITEC), Universidad Autónoma de Baja California (UABC). E-mail: emurillo@uabc.edu.mx

**Princen, Tom:** Hasselt University, Belgium. Campus Diepenbeek, Agoralaan Gebouw D, BE 3590 Diepenbeek . E-mail:tprincen@gmail.com

**Reyes, Horacio Luis Martínez:** Universidad Autónoma de Baja California, Km. 103 Carretera Tijuana-Ensenada. Ensenada, B.C. México. Código Postal 22860. E-mail: hmartine@uabc.edu.mx

**Ribeiro, B.J.F.:** Ph.D., Assistant Professor, Electromechanical Engineering Department, University of Beira Interior, Covilhã, Portugal. E-mail: e-mail: bruno@ubi.pt

**Santo, Espírito:** Ph.D., Assistant Professor, Electromechanical Engineering Department, University of Beira Interior, Covilhã, Portugal. E-mail: aes@ubi.pt

**Schreurs, Jeanne:** Hasselt University, Belgium. Campus Diepenbeek, Agoralaan Gebouw D, BE 3590 Diepenbeek. E-mail: Jeanne.SCHREURS@uhasselt.be

**Sen, Jaydip:** Innovation Lab, Tata Consultancy Services Ltd., Bengal Intelligent Park, Kolkata 700091, India. E-mail: Jaydip.sen@acm.org

**Serna-Olvera, Jetzabel M.:** Universitat Politécnica de Catalunya, Spain. E-mail: jetzabel@ac.upc.edu

**Toledo, Cora Beatriz Excelente:** LaboratorioNacional de InformáticaAvanzada (LANIA) A. C., Research Centre and Innovation, Rébsamen 80, Esq. Circuito Presidentes, Col. Centro, C.P. 91000, Xalapa, Veracruz, México. E-mail: cora@lania.mx

**Trinidad, Marco Antonio López:** Laboratorio Nacional de Informática Avanzada (LANIA) A.C., Research Centre and Innovation, Rébsamen 80, Esq. Circuito Presidentes, Col. Centro, C.P. 91000, Xalapa, Veracruz, México. E-mail: mlopez@lania.mx

**Ursuţiu, Doru:** Automation, Electronics and Computers Department, "Transilvania" University of Brasov, Romania. E-mail: udoru@unitbv.ro

**Valle, Maurizio:** Universitàdegli Studi di Genova, MicroelectronicsGroup, Viaall' Opera Pia 11a, 16145, Genoa, Italy. E-mail: maurizio.valle@unige.it

**Velasco, Manuel M. Miranda:** Facultad de Ingeniería, Arquitectura y Diseño (FIAD). México. E-mail: mmiranda@uabc.edu.mx

**Velasco, Manuel Moisés Miranda:** Universidad Autónoma de Baja California, Km. 103 Carretera Tijuana-Ensenada. Ensenada, B.C. México. Código Postal 22860. E-mail: mmiranda@uabc.edu.mx

**Zoican, Sorin:** Professor, Electronics, Telecommunications and Information Technology, Politehnica University of Bucharest, 1-3 IuliuManiu, Bucharest, sect 6, 061071, Romania. E-mail: sorin@elcom.pub.ro

# Introduction

The growing presence of ubiquitous computing and ambient intelligence in products and services is presently creating huge opportunities in a variety of different areas. Embedded systems already play an important role not only in consumer electronics, but in many important and safety-critical systems. As a consequence, there is growing scientific interest in the conceptual and practical tools necessary to develop embedded systems. These systems are becoming increasingly more popular in a wide range of value-added applications. As a result, our lives are becoming increasingly dependent on embedded systems and digital information technology that is ubiquitously embedded in our environment. A decade ago ubiquitous computing was barely become a viable option to traditional technologies. Today, ten years later, more than 98% of processors used today are in embedded systems that are no longer visible to their owner. This serves as evidence of how technologies are evolving at a staggering rate and evidences the evolution from times when processors were located almost exclusively in personal computers. As society witnessed the size of personal computers diminish while their computational capacity increased, the same has occurred with sensor technologies, which have undergone the same process. This miniaturization has led to a great expansion of sensor technology embedded in an ever-greater number of applications. This process has led to the integration of embedded systems in not only safety-critical applications such as intelligent smart houses, controllers for automotive devices, railways, aircrafts, aerospace, health care and medical devices, but also in communications technologies, mobile systems, environmental monitoring and control systems, mobile phones, PDA's, DVD players, cameras, etc. All of these technologies have wide-ranging impacts on society in the areas of security, privacy, the workplace and the home. (Schoitsch 2003); (Butazzo

2006); (Hollabaugh 2006); (Hallinan 2007); (Aquino et al., 2008); (Aquino et al., 2009).

In smart homes, most electronic and electrical devices (including computer-based systems) are connected to each other in an integrated network to provide safety, entertainment and comfort. However, this indoor integrated network is connected to the outside world via wireless computer and sensor networks. These networks possess limited processing capacity and are connected to the Internet, where remote diagnostics can be carried out, leading to real-time intervention. The potential of embedded systems ranges from the simplicity of sharing digital media to the coordination of a variety of complex joint actions carried out between collections of networked devices. Examples of embedded systems are becoming increasingly common in energy efficient homes as well as technologies to support independent living and augmented lifestyles (entertainment, comfort, safety, and communications).

Smart homes, for example, enable their owners to create ambient intelligence by deconstructing conventional appliances and applications into a network of sophisticated services and applications which can be recombined in various ways. Most intelligent homes are still at a pre-commercial deployment stage, but because they offer much value to conventional homes, their incorporation into homes will be sooner than later. The main obstacles to their actual application in real-life settings are the complex and highly variable interactions between various independent and interdependent systems, as well as between humans and systems that provide functions to make their lives safer, easier or more enjoyable. One important aspect of this evolution is the rise of some unique issues, including robustness, energy consumption, intercommunication and integration of the different embedded systems, among others. In addition to technical considerations, embedded applications tend to be extremely cost-sensitive because they are more often than not extremely high volume devices in very competitive markets.

On the other hand, the continuous monitoring and analysis of vital signs is the key to detecting potential health risks in otherwise healthy-looking patients. There are presently several projects around the world that aim to monitor people's health. The authors in (Falck et al., 2006) describe the BASUMA project, which focuses on developing a robust and energy efficient platform for human wireless body sensor networks to provide at-home monitoring of chronically ill patients. The initial goals of the BASUMA project are to improve the treatment of obstructive pulmonary disease and provide support for female breast cancer patients undergoing chemotherapy. In (Krco 2003), the author describes how to implement a personal sensor network to monitor patients and help provide health care.

This project combines several intelligent sensors and an integrated control node that functions in conjunction with a Bluetooth network. In (Zhao and Cui 2005), the authors present a system based on wireless sensor network technology. This project describes an architecture composed of medical sensors incorporated around the human body using the Zigbee standard. The WHAM-Bios project in (Ming-Hui et al., 2005) proposes telemedicine applications to provide real-time emergency medical services. The WHAM-Bios project is based on a device the authors call "Human Body Gateway," where the sensor nodes provide the information needed to produce instantaneous monitoring results. Real-time monitoring requires algorithms that facilitate contention-free communication in order to reduce the power needed to transmit data.

The long-term health effect of the electromagnetic fields caused by in-home monitoring has been a cause of concern. The author in (Poon et al., 2006) focuses on the safety of body sensor networks and wireless communications in close and constant proximity to humans.

In other studies (Knight et al., 2005); (Teller and Stivoric 2004), the objective is to incorporate technologies into clothes or common accessories (for example, watches, bracelets, etc.) to measure, register and transmit different physiological parameters, including heart rate, body temperature, and movement. The authors in (Mamykina et al., 2006) describe a prototype that monitors diabetic patients and the authors in (Schwiebert et al., 2001) describe another prototype of a retinal prosthesis, based on embedded implanted intelligent sensors.

There are various projects using 802.15.4 and Zigbee to transfer patient information. The authors in (Marco et al., 2006) present ZUPS. This is an ultrasound-based position system that provides mul-cell coverage. The system uses Zig-Bee and ultrasound to measure distances between mobile devices carrying tags and beacons with known locations; however, it uses proximity and multi-lateration localization methods simultaneously. This combination reduces the infrastructure needed for the ultrasound system and also provides accurate information, even at very short distances, enabling the system to provide guidance and spatial orientation training inside buildings for the elderly and people with disabilities.

Projects presenting plural layer architecture have also been developed. In (Zhou et al., 2007) a three-layer network structure for a pervasive medical supervision system is proposed. The first layer is the medical sensor network providing information of oximetry, heart rate, and blood pressure, as well contextual data such as temperature and the patient's video/picture in emergency situations. This network is configured in a star topology with a gateway node. The second layer provides reliable transmission. The data is transmitted to where nearest wireless node emplaced in the house and the transmission is relayed to a PC with an internet connection inside the house. The third layer of the system is responsible for the aggregation of physiological data in a remote medical center for analysis and provides feedback to the patient using a mobile phone, PDA, or web services.

Some projects developed aim to considerably reduce the power consumption of the network in long-term monitoring applications where there is no need to alert persons about a threatening condition or event. SATIRE (Ganti et al., 2006) is designed to identify a user's activity based on accelerometers and global positioning systems (GPS). The system uses SHIMMER motes and accelerometers to sense this data and then logs it to local flash. These data are opportunistically transmitted using a low-power radio when the SHIMMER node is within communication range with the base station, thus allowing the use of an aggressive DutyCycle strategy to extend battery lifetime. The data is then processed offline to characterize the user's activity patterns.

In the medical application field, there have been some projects focusing directly in ECG measuring and processing. In (Chen et al., 2008) a variable control system is proposed to optimize an ECG measure resolution to save power. This control system permits users to set the ECG Signal-to-Noise-Ratio to achieve this.

CodeBlue (Shnayder et al., 2005) is a hardware and software platform developed at Harvard University. The network architecture is based on publish/subscribe routing framework. The sensors do not publish data at an arbitrary rate; instead, they filter the data locally. Multihop routing can be used when the subscribers and publishers are not within a single hop radio range. The publishers and subscribers are mobile, so position information has to be available to define routing paths. This information is obtained using a localization system called MoteTrack (Lorincz and Welsh, 2005).

With the increasing popularity of mobile ECG measurement, data transmission to a remote place for processing and diagnosis(such as a Medical Center) gains ever-greater importance. In (Chung et al., 2007), the authors describe a cardiac healthcare system that can use WLAN and CDMA technologies to transmit data. When the ECG sensor detects a WLAN, it transmits data using that path; otherwise, a cell phone with a prototype wireless dongle that is capable of performing a simple electrocardiogram diagnosis algorithm is used. Afterwards, the data is then transmitted only when an abnormality is detected.

Mobile ECG (Kailanto et al., 2008) is an ECG measurement and analysis system which uses a smart mobile phone as a base station. The ECG is sent by means of Bluetooth technology. The mobile phone analyzes the received data and if any abnormality is detected, the ECG data is sent to a server for further analysis by healthcare professionals.

The development of embedded applications actually is entering into new domains because of the more readily availability of new high-speed processors that provide greater processing power and lower power consumption at a lower cost. As a result of this new generation of hardware, there is an increasing interest in enabling multiple applications to share a single processor and memory. To facilitate this type of architecture, the execution time and memory space of each application must be protected from other applications in the system.

Partitioning operating systems represents the future of embedded systems, which have evolved to meet a wide variety of functions required for ubiquitous computing and ambient intelligence to become more widespread, particularly where predictability and reliability are the main concerns. In a partitioned operating system, memory is divided between statically allocated partitions in a fixed manner. The idea is to take a processor and make it serve the functions of several processors by completely isolating the subsystems. This concept is commonly referred to as virtualization. The importance of virtualization in the area of embedded computing is currently emerging. The industry is making rapid advances in the areas of system virtualization. Hypervisor is the term that refers to the software layer that provides this virtualization. This software layer, in conjunction with hardware mechanisms, allows a single microprocessor to run several independent (both spatially and temporally isolated) operating systems in a single computer. Performance is the key variable when virtualization techniques are used in real-time embedded systems. One of the most important features of a hypervisor is that it should limit overhead so that applications can be executed at almost the same speed as on the native system. The underlying architecture also plays an important role. Special design-related challenges arise from the specialization and customization of target platforms when they are used in embedded systems. The challenge is to maintain some degree of flexibility to increase the reuse of software components.

Additionally, middleware techniques have shown to be of great value when mastering integration. Traditional operating systems architectures have a large body of code running in privileged mode. Because nothing constrains the behavior of privileged code, it can violate the system integrity, particularly in relation to data isolation and controlled information flow. The multiple independent partition level architecture simplifies high assurance certifications by moving system functions from the separation kernel layer to the middleware layer. The resulting separation kernel is significantly smaller, simpler and much more conducive to certification at high level assurance.

As a result, middleware systems can facilitate the development of applications and are designed to provide common services to the partition domains and to the applications running over other network devices. To better develop applications for smart homes and health care and medical devices, middleware should include routing algorithms for wireless mesh networks and have a robust backbone to integrate and share data, voice, and video between embedded computers and mobile devices. An efficient routing algorithm is also vital if information is to be transmitted and received by many processors with their respective transmitting devices and be routed efficiently between them in a coordinated manner.

In sum, this introduction has presented some vital considerations concerning sensors and their integration into embedded systems, which when used in wireless technologies offer a future that integrates the individual, home, and outside world in ways that were unimaginable just a generation ago. The chapters we present in this book are intended to give our readers some basic concepts about embedded systems and wireless communications while encouraging them to further study this very fascinating and relatively new area. Precisely because we see such potential for future growth, it is important for us, as editors, as it is for all of the contributing authors, that our readers gain a greater vision of the future and what they might one day to which they might one day wish to contribute.

## REFERENCES

Aquino-Santos Raúl, Apolinar González-Potes, Víctor Rangel-Licea, A. García-Ruiz Miguel, L.A. Villaseñor-González and Arthur Edwards-Block. 2008. Wireless Communication Protocol Based on EDF for Wireless Body Sensor Networks. *Journal of Applied Research and Technology*, Vol. 6, No 2.

Aquino-Santos Raúl, Apolinar González-Potes, Armando Villaseñor-González Luis, Alfons Crespo, Jaime Sánchez and José R. Gallardo. 2009. Simulación de Algoritmos para Regular el Flujo Vehicular y la Comunicación entre Vehículos móviles Autónomos Utilizando Redes *Ad Hoc*. *Revista Iberoamericana de Automática e Informática Industrial*, vol. 6, núm. 1, pp. 75–83.

Butazzo Georgio. 2006. Research trends in real-time computing for embedded systems. ACM SIGBED Review. Volume 3, Issue 3 (July 2006). Special issue on major international initiatives on real-time and embedded systems pp. 1–10. Year of Publication: 2006.

Chen, S., H. Lee, Y. Chu and C. Chen. 2008. A variable control system for wireless body sensor network. IEEE International Symposium on Circuits and Systems. pp. 2034–2037.

Chung, W., C. Yau and K. Shin. 2007. A cell phone based health monitoring system with self analysis processor using wireless sensor network technology. 29th Annual International Conference of the IEEE EMBS.

Falck, T., J. Espina, J. Ebert and D. Dietterle. 2006. BASUMA—The sixth sense for chronically ill patients. Proceedings of the International Workshop on Wearable and Implantable Body Sensor Networks (BSN'06), pp. 57–60.

Ganti, R., P. Jayachandran, T. Abdelzaher, J. Stankovic. 2006. A software architecture for smart AtTIRE. Proc 4th Int. Conf. Mobile Syst. Appl. Services.

Hallinan Chistopher. 2007. Embedded Linux Primer Ed. Prentoce Hall.

Hollabaugh Craig. 2006. Embedded Linux, Hardware, Software and Interfacing. Ed. Addison-Wesley, February 2006.

Kailanto, W., Hyyärinen and J. Hyttinen. 2008. Mobile ECG measurement and analysis system using mobile phones as the base station. Second International Conference on Pervasive Computing Technologies.

Knight, J., A. Schwirtz, F. Psomadelis, C. Baber, H. Bristow and T. Arvanitis. 2005. The design of the Sens Vest. Personal Ubiquitous Computing, vol. 9, pp. 6–19.

Krco, S. 2003. Implementation solution and issues in building a personal sensor network for health care monitoring. Proceedings of the 4th Annual IEEE Conference on Information Technology Applications in Biomedicine. pp. 350–353.

Lorincz, K. and M. Motetrack Welsh. 2005. A robust, decentralized approach to rf-based location tracking. Lecture Notes in Computer Science, pp. 63–82.

Mamykina, L., E. Mynatt and D. Kaufman. 2006. Investigating Health Management Practices of individuals with Diabetes. Proceedings of the SIGCHI Conference on Human factors in Computing Systems, pp. 927–936.

Marco, A., R. Casas, J. Falco, H. García, J. Artigas and A. Roy. 2008. Location based services for elderly and disables people. Computer Communications, 31.

Ming-Hui, J., L. Ren-Guey, K. Chen-Yan, W. You-Rui, D. Frank, D. Tse-Ping and H. Kuan-Tsae. 2005. Sensor Network Design and Implementation for Health Telecare and Diagnosis Assistance Applications. Proceedings of the 2005 11th International Conference on Parallel and Distributed Systems (ICPADS'05), vol. 2, pp. 407–411.

Poon, C., Z. Yuan-Ting and B. Shu-Di. 2006. A Novel Biometrics Method to Secure Wireless Body Area Sensor Networks for Telemedicine and M-Health. IEEE Communications Magazine, vol. 44, issue 4, pp. 73–81.

Schoitsch, Erwin. 2003. Embedded Systems—Introduction, European Research Consortium for Informatics and Mathematics, ERCIM News No 52.

Schwiebert, L., S. Gupta and J. Weinmann. 2001. Research Challenges in Wireless Networks of Biomedical Sensors. Proceedings of the 7th annual International Conference on Mobile Computing and Networking, pp. 151–165.

Shnayder, V., B. Chen, K. Lorincz, T. Fulford-Jones and M. Welsh. 2005. Sensor networks for medical care. 3rd International Conference on Embedded Networked Sensor Systems.

Teller, A. and J. Stivoric. 2004. The BodyMedia Platform: Continuos Body Intelligence. Proceeding of the 1st ACM workshop on Continuous archival and retrieval of personal experiences, pp. 114–115.

Zhao, Z. and L. Cui. 2005. EasiMed: A remote health care solution. Proceedings of the 2005 IEEE Engineering in Medicine and Biology, pp. 2145–2148.

Zhou, B., C. Hu, H. Wang and R. Guo. 2007. A wireless sensor network for pervasive medical supervision. International Conference on Integration Technology, pp. 740–744.

1

# Ubiquitous Computing: Applications, Challenges and Future Trends

*Jaydip Sen*

**ABSTRACT**

We are witnessing the advent and the evolution of a revolutionary computing paradigm that promises to have a profound effect on the way we interact with the computers, devices, physical spaces, and other people. This new paradigm, called ubiquitous computing, envisions a world where embedded processors, computers, sensors, and digital communication technologies are inexpensive commodities that are available everywhere. Ubiquitous computing will surround users with a comfortable and convenient information environment and a smart space that merges physical and computational infrastructures into an integrated habitat. This habitat will feature a proliferation of hundreds or thousands of computing devices and sensors that will provide new functionality, offer specialized services, and boost productivity and interaction among the devices and the users. This chapter provides a comprehensive discussion on the central trends and the emerging impacts of ubiquitous computing considering them form technical, social and economic perspectives. It also identifies various application areas and sectors that will potentially be benefitted and will be able to provide enhanced and rich experience to the users. The challenges of ubiquitous computing that require active research and technological advancements are also discussed.

## 1. Introduction

The dissemination and use of modern *information and communication technologies* (ICT) are considered to be the preconditions today for dynamic economic growth and future viability in global competition. At the same time, the processes of change triggered, enabled and accelerated by ICT are enormous. The new technologies have an ever-expanding ripple effect on the economy, public administration, science, scholarship and private life. They exert influence on social and private life. The development of mobile telephony and Internet technology during the past decade exemplifies the transformative potential of ICT.

Digital information and services are going mobile and can be called up from any location. A trend towards ubiquitous computing is emerging that is characterized by the ubiquitous and invisible use, creation, processing, transmission and storage of information. Everyday objects are becoming *smart objects*, which are linked together into networks, react to their environment, and interact with their users.

This chapter investigates the central trends in ubiquitous computing and considers them from technical, economic and social perspectives. It emphasizes the level of analysis that is located between individual case studies and the global comprehensive picture and that can be mapped onto the application areas of ubiquitous computing. It also highlights the potential of ubiquitous computing. It clearly points out that different application areas and sectors will benefit from this potentials at different speeds and with qualitative idiosyncrasies. In addition, it also makes clear that ubiquitous computing not only poses technical problems, it also comprises serious social, economic and judicial challenges that require active solutions and management. As has often been the case with technical innovations and advances, ubiquitous computing raises the question of what sort of future we are heading to. This chapter, on one hand, contributes to the greater objectivity in this direction. On the other hand, it intends to outline important issues so that a multitude of social groups and actors will be drawn to participate actively in the discussion on ubiquitous computing.

The rest of the chapter is organized as follows: Section 2 presents a comprehensive discussion on the trends of ubiquitous computing and its economic, social and technological impact. Section 3 highlights the potential application areas of ubiquitous computing. Section 4 discusses the array of technological areas which inter-work to a build ubiquitous computing framework. Section 5 discusses one of the central issues in ubiquitous computing—security, safety and privacy of the user information. Finally, Section 6 concludes the chapter while identifying some key areas where action is needed in the coming years to ensure that the variety of groups and actors in society, politics and industry are able to profit equally from ubiquitous computing technologies.

## 2. Ubiquitous Computing: Trends and Impact

For more than thirty years, microprocessor performance has doubled approximately every eighteen months. Similar progress in performance has been made in other technology parameters, such as data transfer rates in both wired and wireless networks. This trend is shaping a future in which computers will become increasingly smaller and inexpensive, and therefore abundant.

In recent years, smaller personal devices such as PDAs, smart phones, palmtops and notebooks have begun supplementing and, to some extent, replacing traditional computer platforms such as workstations, PCs and servers. Computers are integrated increasingly into everyday devices and expand their operational capacities. These so-called embedded systems, such as driver assistance systems in cars or flexible software-driven boiler regulation can be controlled using a PC in the office or via mobile telephone.

As early as in 1991, Mark Weiser, the former Scientific Director of the Xerox Palo Alto Research Center, shaped the vision of ubiquitous computing as an omnipresent infrastructure for ICTs. His work on ubiquitous computing continues to define all technological and socio-political considerations inherent to the field. According to Weiser, we can speak of ubiquitous computing once the following four criteria are met: (i) microcomputers are integrated into physical objects of any shape and displace services heretofore performed by desktop systems; (ii) these embedded systems are characterized by their small size and by their near invisibility to the user; (iii) embedded microcomputers thus augment a physical object's original use-value with a new array of digital applications and (iv) the ubiquitous availability of services lists at the centre of communication between devices and application, not the device itself. This last point is what distinguishes ubiquitous computing from the familiar mobile networks of today. Ubiquitous computing is characterized by the omnipresent and mobile availability of services themselves, regardless of the target platform. Services will be tailored to the physical capacity of a Ubiquitous Computing: Applications Challenges and Future Trends 3 specific device, whether a mobile telephone, PDA or other value-added communications device (Weiser 1991).

Advances in microelectronics and communications technology have moved the technical vision of ubiquitous computing into the realm of the possible. Early examples of ubiquitous computing in use include processor module integration into identification documents and the integration of transponders into cargo pallets that send ID numbers to a reader automatically. In professional circles, the term Ubiquitous Computing is used commonly to describe the ubiquitous ICT infrastructure aimed at feasible short and medium-term solutions.

Ubiquitous computing is viewed less as a discrete field of technology, but rather as an emerging application of information and communications technology that is integrated into the everyday world more than ever before. The goal is to meet the claim of "everything, always, everywhere" for data processing and transmission through the ubiquity of ICT systems. The following characteristics define this application paradigm: (i) *miniaturization*: ICT components are becoming smaller and more mobile, (ii) *embedding*: as ICT components are integrated into everyday objects, they transform them into smart objects, (iii) *networking*: ICT components are linked to each other and communicate generally via radio; they are therefore not part of a fixed environment or application, but are instead designed to form networks spontaneously, (iv) *ubiquity*: while embedded ICT components are increasingly ubiquitous, they are at the same time increasingly less noticeable—or even invisible—to most people, (v) *context-awareness*: ICT components use sensors and communication to collect information about their users and environment and adjust their behavior accordingly (Wagner et al., 2006).

Ubiquitous computing is thus a complementary paradigm to virtual reality. Rather than reproduce and simulate the world with a computer, ubiquitous computing turns all objects in the real world into part of an information and communications system. Ubiquitous computing will drastically change the ways in which we use computers. As computers are integrated into everyday objects, they will no longer be perceived as such and their usage will recede largely from our conscious perception. In ubiquitous computing, a variety of processes run automatically in the background and interacts on behalf of the user. The user does not have to provide explicit instructions or make decisions. Ubiquitous computing involves smart environments that are envisioned as an individual's cooperative partner. However, the seeming disappearance of the computer together with the delegation of complex processes or tasks to a ubiquitous ICT infrastructure raises serious questions. How secure are these systems? How can one determine whether they act truly on behalf of their respective users? How will the enormous amounts of data generated by these processes be handled? Can the individual user exercise the right of privacy and prevent his or her data from being passed on, stored, compared and analyzed? Does the vision of ubiquitous computing not possibly permeate our everyday environment entail an immeasurable increase in resource and energy consumption? How can a potentially fatal dependence on technology be prevented? As the majority of commercial, administration, trade and recreation transactions become impossible without ubiquitous computing support or are provided in the form of ubiquitous computing only, will this not lead to a forced usage thereof? How will ubiquitous computing develop and what impact will it have?

No one knows in which way ubiquitous computing will actually develop. Applications and elements with both negative and positive characteristics will emerge in the coming years and have far-reaching socio-economic effects.

This chapter examines the projected technological development of ubiquitous computing, its socio-economic impact and the potential dangers to security, privacy and safety. In addition, the chapter also presents a detailed discussion on the security issues in ubiquitous computing, which is based on three scenarios.

## 2.1 Characteristics of Ubiquitous Computing

Ubiquitous computing can be characterized by a set of attributes and capabilities that describe the extent of its functionality. Mobility and *ad hoc* networking capabilities have already arrived to real-world communication. Characteristics such as autonomy, context awareness and energy autarky are expected in the near future—two to five years' time. Context awareness and embedment in everyday objects are viewed as definitive and formative characteristics of ubiquitous computing. Energy autarky and the autonomy of components and systems, in contrast, are considered secondary characteristics.

It therefore stands to reason that ubiquitous computing will establish itself gradually as its characteristics develop step-by-step. Although market-ready ubiquitous computing applications are anticipated in the next four to eight years, the characteristic of autonomy is not expected for another ten years. Also, the individual characteristics vary in relevance, depending on the field of application. For example, the autarkic power supply of ubiquitous computing components and their mobility are relatively unimportant for the smart home, whereas this is a key to communications applications.

## 2.2 Technological Foundations of Ubiquitous Computing

Communications technologies and microelectronics, in particular, are key requirements for almost all ubiquitous computing applications. Although energy autarky is certainly not an important characteristic of all ubiquitous computing applications, supplying energy is clearly a central task. Maturation and availability of ubiquitous computing-relevant technologies is expected soon, within the next one to four years; nearly all of the technological requirements needed for ubiquitous computing should be met in the foreseeable future. Unresolved problems with power supplies, an inadequate human-machine interface and a lack of well-functioning security technology pose particularly serious bottlenecks to the development of ubiquitous computing. Further studies reveal the following potential technological bottlenecks.

(i) Batteries and accumulators are important requirements for ubiquitous computing. Considerable development is needed on this point, despite the availability of current solutions. Energy harvesting—the extraction of energy from the surroundings—is an alternative technology of the future.

(ii) In human-to-machine interfaces, processing natural speech is very important. Although it is a comparatively mature technology, it represents a technological bottleneck.

(iii) In the field of security, biometric identification is considered less relevant for ubiquitous computing, especially when compared to other security technologies such as trust management and identity management.

## 2.3 Drivers and Hurdles of Ubiquitous Computing

The realization of new products and services are the primary driving forces behind the development and establishment of ubiquitous computing. Depending on the area of application, the importance of cost savings and increases in energy efficiency or comfort can vary significantly. In production and logistics, for example, reducing costs via ubiquitous computing is considered very important. In military applications, however, cost savings functions as a weak motivation. Using ubiquitous computing to improve safety is considered important for motor traffic as well as the security, military and medical sectors. However, improving safety is not a motivation in other application areas for using and developing ubiquitous computing. Enhanced comfort and improved usability are seen as significant driving forces in the automotive and household industries. Potential cost savings and improved energy efficiency play only minor roles in these two areas. In medical technology, however, there is a broad spectrum of motivations for using ubiquitous computing, which is seen as a key strategy in solving several different problems in this sector.

Analogous to the driving forces behind the development and application of ubiquitous computing, there are identifiable limiting factors that could either directly or indirectly influence its evolution. These include shortcomings in human-machine interfaces and data protection, as well as technical hurdles to availability and reliability. These limiting factors are real, despite the fact that the core technologies of ubiquitous computing are expected to be market-ready within the next one to four years.

Environmental sustainability, resource consumption and legal regulation are some of the minor limiting factors for successful deployment of ubiquitous computing. Standardization, however, is of prime importance as the functional interplay of individual components is a key requirement of ubiquitous computing. The relevance of individual data protection and privacy varies depending on the specific application. While privacy is not a primary concern in production and military applications, it is an important limiting factor in security, communications and medicine.

## 2.4 Impact of Ubiquitous Computing

Ubiquitous computing will permeate everyday life—both private and working—and is therefore expected to have far-reaching consequences that will be reflected in a variety of socio-economic contexts. Both positive and negative effects are likely in equal measure at several levels. Safety and privacy, for example, make up two ends of one key pole. The following discussion presents the impact of ubiquitous computing in terms of privacy, economics, society and the digital divide.

**Impact on privacy:** in terms of privacy, slightly positive effects are expected for the application fields of security, medicine and production; moderately negative effects are expected in other application contexts. A ubiquitous computing design for privacy that conforms to data protection standards is regarded as a requirement for ensuring privacy and is preferred to the downstream concept of context-dependent data protection filters (digital bubbles). Only a system architecture that protects privacy from the outset can prevent serious conflicts in data protection from developing. In the use and processing of data, rendering all steps in the process visible and logically comprehensible seems to be of less importance. Far more crucial is a user's explicit trust in a particular ubiquitous computing system that the service provider will handle personal data responsibly. Apart from this concern, there is the danger that frequent use of a ubiquitous computing application could potentially lead to the inattentive handling of personal data. This means that the premature availability of a groundbreaking ubiquitous computing application could result in limited public attention being given to the protection of privacy during its crucial phase of implementation.

**Economic impact:** among the economic effects associated with ubiquitous computing, work efficiency in particular is expected to improve. This will become most apparent in the key economic areas of production, logistics and commerce. This will not, however, play a role in smart homes. It is worth noting that no significant efficiency gains from ubiquitous computing are expected for housework, professionals working at home, and homecare of the elderly and/or ill. The motivation for introducing ubiquitous computing into the smart home is to increase personal comfort, which is not quantifiable in economic terms. The experts in market research and technology trends domains anticipate effects similar to those resulting from the introduction of modern household appliances during industrialization. Back then, the time saved by the use of new appliances was counteracted by increased demands in hygiene and cleanliness, which resulted in extra work. Moderately positive effects are anticipated for increasing energy and resource efficiency in all areas of application. Significant increases are expected, however, in production and especially logistics. Indeed, the potential of ubiquitous computing for commerce and production is immense because of its ability to self-organize and control industrial processes. Such self-organization depends on several things, including the availability of fully developed knowledge-based systems. Developing these systems in turn, however, poses a significant challenge. Finally, newly adapted recycling procedures will be needed to allow for the re-use of ubiquitous computing components that have been integrated and embedded into everyday objects.

**Social impact:** clear positive effects are predicted in the personal activities in medicine, the home, communications and automobiles, while moderately positive effects are expected in inner and external security, and in production, logistics and commerce. Improvements in safety are anticipated primarily in military and security-related applications, especially in medical applications. The automotive branch will also profit. Overall, ubiquitous computing is not expected to produce any negative *rebound effects*, which would offset or even negate its positive effects. This is true for work and attention efficiency, resource consumption and for the ability of human being to orient and locate themselves in their environments.

**Winners and losers:** several social groups will either win or lose in ubiquitous computing. The elderly and persons with little experience with technology stand out as groups that could benefit from ubiquitous computing and be disadvantaged by it as well. How might this paradox come about? The first generation ubiquitous computing applications are likely to demand a certain level of knowledge and requirements that will result in a temporary division between the winners and the losers. However, this divide should subside as the functional logic of later generation ubiquitous applications is automated and cost-effective mass production sets in. Once the technology matures and becomes publicly available, the less educated will begin to profit from it. It is, therefore, imperative that a ubiquitous computing infrastructure offer accordingly easy access in technical, financial or intellectual terms. Otherwise, a digital divide between those with and those without access to ubiquitous computing will emerge. Other groups that could suffer disadvantages as a result of ubiquitous computing include small businesses and retail, political minorities, critics or skeptics, marginalized groups and persons with unusual backgrounds.

**2.5 Ubiquitous Computing Security**

In ubiquitous computing, a large number of smart objects communicate with one another or with the user. Many of these interactions should be as inconspicuous and situationally dependent as possible, and thus occur semi-automatically. The user will no longer take conscious note of the actions of the smart objects. The usual means of control and correction within IT systems will not apply here. Protecting system security in ubiquitous computing is therefore very important. This includes security, privacy and safety. Security involves preventing unauthorized persons from viewing and therefore potentially manipulating confidential data. Likewise, communications must remain confidential and may not be interfered with. The meeting of digital identities must be at least as trustworthy as meeting in person. Known and tested security technologies and methods are already available. However, they need to be adapted to the peculiarities of ubiquitous computing, especially the frequently limited performance of smart object hardware and to the marked decentralization of infrastructure, services and objects. There is an inherent conflict between the goal of ubiquitous computing of accurately identifying persons, objects and messages (authenticity), and the desire for anonymity (to prevent data trails from the outset). This problem can be, to some extent, addressed through digital pseudonym technology. In addition, the competing security goals of authenticity and anonymity must be considered individually for each application.

Ubiquitous computing will see the accumulation of vast amounts of data that can provide a comprehensive overview of an individual, his or her behavior, finances and health. These huge sets of data and the spontaneous networking of smart objects will make it impossible for the users of ubiquitous computing to trace where their personal data are stored, how they are used and how they might be combined with other data. In ubiquitous computing, data protection is therefore an essential requirement for protecting privacy. The networking of ubiquitous computing is not limited to individual states because several services will not fully evolve until they are made available across borders. This internationalization requires standardized international regulations guaranteeing privacy protection. Currently, there are highly disparate approaches to preserving the private sphere in the digital world. These differences are illustrated clearly by the gap between Europe's strict legal regulations and the comparatively open, self-regulatory approach in the United States. The global networking of smart objects and services, which is anticipated in the long run, will necessitate the creation of a standardized international regulatory regime for data protection in ubiquitous computing.

The invisible nature of ubiquitous computing and the complexity of its networking could mean that system failures and malicious interference may go unnoticed, or are noticed much later. In some ubiquitous computing applications such as medicine, traffic system control or self-organized production lines, this could put human lives in danger and lead to extensive property damage. In applications where safety is crucial, the reliability of ubiquitous computing is essential. It must be guaranteed, for example, with system redundancy or a backup system.

## 3. Ubiquitous Computing Application Areas

Ubiquitous computing aims to permeate and interconnect all areas of life, thus enabling a ubiquitous flow of data, information, and, by integrating cognitive capabilities in the future, even knowledge. Mark Weiser, one of the fathers of ubiquitous computing, described this vision of a continual and ubiquitous exchange transcending the borders of applications, media, and countries as "everything, always, everywhere." This sketch offers a strongly future-oriented perspective on ubiquitous computing that is still far removed from today's reality. Although wireless Internet access, email via mobile phone, handheld computers and the like may give the impression that constant, unimpeded exchange of information is already routine, in the future the special performance characteristics of ubiquitous computing will enable an entirely new quality in the exchange and processing of data, information and knowledge. With ubiquitous computing, many of these processes will recede into the background, and most will occur partially or wholly automatically. But this new form of ubiquitous computing will not develop uniformly and synchronously in all economic and social areas. Rather, applications will be defined and implemented at different speeds in different contexts. Nine areas of application in which ubiquitous computing is already recognizable and is very likely to play a decisive role in the future are mentioned below.

**Communications:** as a cross-application, the communications area affects all forms of exchange and transmission of data, information, and knowledge. Communications thus represents a precondition for all information technology domains.

**Logistics:** tracking logistical goods along the entire transport chain of raw materials, semi-finished articles, and finished products (including their eventual disposal) closes the gap in IT control systems between the physical flow and the information flow. This offers opportunities for optimizing and automating logistics that are already apparent today.

**Motor traffic:** automobiles already contain several assistance systems that support the driver invisibly. Networking vehicles with each other and with surrounding telematics systems is anticipated for the future.

**Military:** the military sector requires the provision of information on averting and fighting external threats that is as close-meshed, multi-dimensional, and interrelated as possible. This comprises the collection and processing of information. It also includes the development of new weapons systems.

**Production:** in the smart factory, the flow and processing of components within manufacturing are controlled by the components and by the processing and transport stations themselves. Ubiquitous computing will facilitate a decentralized production system that will independently configure, control and monitor itself.

**Smart homes:** in smart homes, a large number of home technology devices such as heating, lighting, ventilation and communication equipment become smart objects that automatically adjust to the needs of the residents.

**E-commerce:** the smart objects of ubiquitous computing allow for new business models with a variety of digital services to be implemented. These include location-based services, a shift from selling products to renting them, and software agents that will instruct components in ubiquitous computing to initiate and carry out services and business transactions independently.

**Inner security:** identification systems, such as electronic passport and the already abundant smart cards, are applications of ubiquitous computing in inner security. In the future, monitoring systems will become increasingly important—for instance, in protecting the environment or surveillance of key infrastructure such as airports and the power grid.

**Medical technology:** Increasingly autarkic, multifunctional, miniaturized and networked medical applications in ubiquitous computing offer a wide range of possibilities for monitoring the health of the elderly in their own homes, as well as for intelligent implants.

Identifying each application area's potential and estimating when we can expect applications to be established is essential to a well-founded prognosis of ubiquitous computing development. Because any such assessment is based on various definitions of ubiquitous computing and depends on variable contexts, we must first describe the performance features and characteristics of ubiquitous computing and then relate them to the selected application areas.

The contemporary international discussion on ubiquitous computing remains powerfully shaped by visions of the immediate future. The current trends do not yet represent full-blown ubiquitous computing, which may well need another ten years to become established. Ubiquitous computing will develop in at least two stages: the first stage will see numerous products and applications characterized by the goals of mobility and ad-hoc networking becoming established within the next five years. This will essentially be a continuation of current trends, such as the miniaturization and integration of various electronic functions in a single chip. During this stage we can expect to see more ambitious characteristics realized, too, such as context awareness in a simplified form, e.g., in the form of user profiles. Truly ubiquitous computing, however, will emerge only in practical applications in the second stage, which will also reveal its novel characteristics. Context awareness will then actually mean intelligent, case-by-case reactions to the user's needs and to the environment.

## 4. Technologies in Ubiquitous Computing

Ubiquitous computing is not an independent technology. Instead, it describes the performance characteristics of an approach for invisible, ubiquitous, and seamless information and communication. Accordingly, a smart object cannot be described in universal terms, since so much depends on the context of its application: it might be an independent, small, mobile device. A smart object might be integrated into a house, or a household device might be transformed into a smart object by integrating the appropriate components. This variety shows that the technological basis of ubiquitous computing is not always the same; rather, different technologies will be used in each case. Still, full-blown ubiquitous computing has certain characteristics that will pertain to nearly all application areas, albeit to a varying degree. As shown in Table 1, eight technology fields seem to be the most important for realization of the six characteristics of ubiquitous computing.

The relative significance of each field cannot be determined from Table 1. For example, microelectronics is the precondition for any sort of electronic data processing, yet Table 1 lists it as essential for two characteristics, mobility and embeddedness. However, it does not reflect the fact that mobility is one of the central characteristics of ubiquitous computing, which implies more weight is assigned to microelectronics.

**Table 1: Necessary technologies for ubiquitous computing.**

| | *Mobility* | *Embeddedness* | *Ad hoc networks* | *Context awareness* | *Energy autarky* | *Autonomy* |
|---|---|---|---|---|---|---|
| Microelectronics | √ | √ | | | | |
| Power supply | √ | | √ | | √ | √ |
| Sensor technology | | | √ | √ | | √ |
| Communication technology | √ | | √ | √ | | √ |
| Localization technology | √ | | | √ | | √ |
| Security technology | | √ | √ | √ | | |
| Machine-to-machine communication | | √ | √ | √ | | √ |
| Human-machine interface | | √ | | √ | | √ |

In the realm of production technologies such as lithography and wafer technology, microelectronics is decidedly important for sensor technology too. This effect is also not captured in Table 1. In fact, communication technologies and microelectronics are the most important prerequisites for nearly all ubiquitous computing applications. It is expected that each of the technologies will mature and will be available in the near future, within a time horizon ranging from about one to four years. Among the potential bottlenecks that would have serious impact on the development of ubiquitous computing the most important are: the unsolved problems with power supply, an inadequate human-machine interface and the lack of a well-functioning security technology. The following sub-sections describe the basic principles of the individual technology fields and their significance for ubiquitous computing.

### 4.1 Microelectronics

Microelectronics—which deals with the miniaturization, development, manufacture and application of *integrated circuits* (IC)—is already an essential component in a vast number of technological devices and equipment that shape contemporary life. Consumer electronics, the automotive industry and medical technology are just a few examples of fields in which microelectronics plays a key role.

Currently, microelectronics works with structures smaller than 40 nanometers. At the present rate of development, this should drop to 22.5 nanometers by 2016 (International Technology Roadmap for

Semiconductors 2010). Significantly increased circuit density will yield considerable gains in functional capacity for a given IC size. However, the field of microelectronics is running up against physical limits that are increasingly expensive to overcome. These limits include a highly complex manufacturing process and decreased performance and durability due to undesired current leakage between the printed circuit board tracks.

Integrated circuit packaging—that is, the mounting and bonding of electronic components on a *printed circuit board* (PCB)—has undergone major changes in recent years. Increasing integration density, greater functionality and costs have been driving this development (Cosine 1999). Advanced packaging processes, such as flip chip, chip-size package and ball grid array, have assumed a key role in IC manufacturing. In addition to smaller dimensions and lower production costs, these processes offer the advantage of directly integrating packaging into the chip manufacturing process. A number of multinational technology companies manufacture advanced process packages, including Infineon, IBM and Toshiba.

The trend in microelectronics is toward greater integration density, smaller structures and falling unit costs. PCB-based system integration deals with the integration of passive and active electronic components, functional layers, electro-optical interfaces and sensor interfaces on a PCB or even on a stack of PCBs. In system integration at the semiconductor level, all digital, analogue and mixed-signal functions are integrated directly into a single IC (system-on-chip).

Current research in nanotechnology is looking at how quantum effects can be used at the sub-molecular and atomic levels to achieve further miniaturization of microelectronics.

A great many research institutions are working on microelectronics and its peripheral fields. These include various institutes such as the Fraunhofer Institute for Reliability and Micro-integration in Germany, the Inter-university Micro Electronics Center (IMEC) in Belgium, Georgia Tech and the Massachusetts Institute of Technology (MIT) in the United States, and the Industrial Technology Research Institute (ITRI) in Japan.

Overall, microelectronics is a mature and widely available technology and does not pose any bottlenecks for ubiquitous computing. Nanotechnology and polymer electronics are regarded as future technologies, but since they would be used only in niche applications, they represent significant barriers.

### 4.2 Power Supply

Supplying power to electronic systems represents an essential condition for using ubiquitous computing applications. So far, progress in chip technologies and electronics development has consistently decreased system size, reduced specific power demands and improved performance. Even so, for most applications the power supply is the largest and heaviest component, and the biggest constraint on use.

Supplying power basically poses no problems for applications plugged into the power grid and networked with cables. However, mobile and wireless systems with power self-sufficiency have become important in recent years. The older radio-supported applications consumed a lot of power, which forced the user to change or recharge the battery frequently. More recently, though, energy consumption has decreased due to progress in IC manufacturing and the deliberate, application-based reduction of active times for sensors and modules. Then, lithium-ion batteries came into use offering excellent reliability and long-term stability. The new goal is to do without batteries altogether. To this end, miniature accumulators (storage batteries) would be continuously recharged by photovoltaics, thermo-generators, miniature piezo generators or other converters, without ever needing to plug into a power supply line (Sabatier et al., 2006).

In those cases where initially equipping a device with batteries may not suffice for its entire service life, or where a constant exchange of batteries is impracticable, an ideal technology would enable autarkic, wireless energy provision from ambient sources—so-called energy harvesting. A variety of such technologies are available, some of which have been tested in practice. Each must be considered individually for each application with respect to its energy balance and efficiency. Some examples are: (i) photovoltaic generators convert energy directly based on various solar cell principles, (ii) piezoelectric generators convert mechanical energy to electrical energy by means of special piezo crystals, (iii) thermoelectric generators create electrical voltage with temperature differentials between two different metals, (iv) electromagnetic generators convert energy according to the familiar dynamo principle, (v) capacitative and electrostatic generators use capacitative or electrostatic charges in the electric field to produce energy, (vi) thermo-mechanical generators create mechanical energy which is then transformed into electrical energy, and (vii) electro-kinetic micro-channel batteries generate electricity as ion-charged fluid (e.g., salt water) passes through micro-channels or porous filters, thus separating the fluid's potentials. The fluid transport is propelled by either static or external pressure on the fluid (Yang et al., 2003).

The potential for improving battery and accumulator systems for portable applications remains great, although the composition of components has been known for years. Due to the great economic gains that could accrue from a clear technological edge, a number of international institutes are working on behalf of industry to further develop portable and alternative power sources. They include the Massachusetts Institute of Technology, the Florida Solar Energy Center and the Los Alamos National Laboratory, the Alberta Research Council, the Institute for Fuel Cell Innovation Canada, and Tokyo University of Science in Japan.

Batteries and accumulators, despite their high availability, are potential bottlenecks for ubiquitous computing applications. In light of the solutions available today, a great need is perceived by the researchers for further development to increase capacities and reduce self-discharge. While energy harvesting is considered an alternative technology for the future, micro fuel cells are not accepted as particularly relevant for ubiquitous computing. A likely reason for this is the expensive infrastructure for supplying the cells with fuel, which is more likely to become economically viable in fuel cell used as an alternative power source for cars.

### 4.3 Sensor Technology

Capturing and analyzing the real world is one of the central characteristics of ubiquitous computing. At the heart of this is the sensor as an electronic component that qualitatively or quantitatively registers characteristics of the environment and amplifies, processes and relays them as a digital signal.

The key challenges in sensor development today include reducing the size and weight of sensors and sensor systems, integration of the sensors into complex semiconductor systems, decreasing the power consumption of the sensors, increasing their performance and reliability, and developing lower-cost production technologies.

One focus of development work is the refinement of existing technologies. In system-on-chip, the sensor is no longer produced as a separate component; instead, it is directly integrated into the chip. Nanotechnology is pursuing new approaches with dramatically smaller and more sensitive sensor elements on the sub-molecular or atomic level.

The sensor technology is generally considered as a well-established technology that does not pose any serious technical barriers for ubiquitous computing. Interestingly, this assessment also applies to highly integrated, functionally sophisticated sensor networks which, as of today, are purely research topics.

### 4.4 Communication Technology

Communication technology is generally combined with information technology and referred to collectively as the ICT in order to emphasize the overlap between the two fields. In telecommunications—e.g., mobile communication, satellite communication and telephony—the important subfields are communications engineering, radio engineering, switching technology, signal transmission technology, high frequency engineering, microelectronics, technical informatics and communications networks. Since the possibility for objects to communicate is fundamental to the vision of ubiquitous computing, ICT plays an essential role.

The Institute of Electrical and Electronics Engineers (IEEE) has published a number of different standards for wireless networks. The 802.11 standard, which standardizes *wireless local area networks* (WLANs), is widely used. For short-range transmission with a reach of a few meters, Bluetooth (IEEE 802.15.1) has rapidly become popular for communication between devices in a *personal area networks* (PANs).

Beyond the existing solutions, a number of approaches will be decisive in the further development of mobile communication such as WiMAX (802.16 e/m), wireless USB, *long term evolution* (LTE) etc. In the realm of automated industrial production, there is an effort to replace the existing field bus system with an industrial Ethernet—an expanded Ethernet based on the IEEE 802.3 standard—which would meet real-time demands.

The decisive players in the ICT field—especially with regard to wireless communication—are the major mobile telephone providers such as Deutsche Telecom, Vodafone and the Japanese company NTT DoCoMo, makers of network components such as Cisco and Lucent, and suppliers of digital devices such Nokia, Motorola, Siemens, Ericsson, IBM, Intel, Microsoft, Sun and Toshiba. These companies are increasingly working with representatives of standards organizations to develop complete solutions for information and communication. The technology and market trends experts recognize communication technology as a key driver for ubiquitous computing. However, they also perceive a certain risk of technological bottlenecks in the further development of the Internet standard IPv6 for additional address space, and Mobile IP for mobile communication.

**4.5 Localization Technology**

One interesting facet of ubiquitous computing is that equipping smart objects with appropriate transmitters and receivers enables precise localization. Location-based services offer services based on the geographic locations of the users. For these services to be delivered, use of appropriate localization technology is an absolute necessity.

There are currently three types of localization systems: satellite-supported, cellular-supported and indoor localization systems. The satellite-based *global positioning system* (GPS) is the most widely used localizations and navigation system. It was commissioned by the United

States Department of Defense for the military and has been used worldwide since 1998. Mobile telephone companies can pinpoint the location of mobile telephones with the support of the cellular network, because the telephones must establish a connection to the transmitters. Cell ID—the cell-based localization technology functions in both GSM and UMTS networks. It identifies the base station to which a mobile telephone has established a connection, and thereby determines the location of the mobile user. The precision of this process depends on cell density and size and may vary between 0.1 and 35 kilometers. In some circumstances, mobile telephone companies can determine the position of a mobile telephone with a precision of few meters through techniques such as triangulation.

True standardization in the field of localization systems does not exist. So far, GPS is the only global navigation system that civilians can use. Current work focuses mostly on improving localization techniques. This primarily means increasing precision and sizing down the equipment. At the moment, precision for civil users of GPS is specified at 10 meters. Indoor systems can achieve much greater precision—usually less than a meter—depending on the type and number of sensors employed.

Apart from the American military, which is responsible for GPS, many research institutions (e.g., the Fraunhofer Gesellschaft) and companies are involved in developing localization systems. The indoor localization system RADAR from Microsoft Research, for example, is based on the IEEE 802.11 standard. RADAR is realized completely as a software extension and requires no additional hardware infrastructure.

Many technology experts feel that localization technology is the least relevant for ubiquitous computing among all the fields in Table 1. Although, localization is a mature technology, possible bottlenecks may occur with indoor systems.

### 4.6 Security Technology

A central feature of ubiquitous computing is that nearly all smart objects can exchange information. Security is thus essential in ubiquitous computing. The fact that most communication in ubiquitous computing is wireless makes the role of security all the more important because radio communication can be manipulated more easily.

Many ubiquitous computing applications intervene in central areas of life without the user always being aware of it. Thus, safety, which ensures that no damage occurs from system failures and operator error, is very important. A further security requirement in ubiquitous computing is data protection to safeguard the individual's privacy. It must guarantee that personal data can be accessed by authorized third parties only.

The three important aspects of security, e.g., security, safety and privacy in ubiquitous computing are discussed in detail below.

**Security:** it deals with protection from intentional attacks by third parties. The fundamental objectives of a security mechanism are to ensure confidentiality, integrity, non-repudiation, availability, anonymity and authenticity.

The authenticity of an entity is understood as its genuineness and credibility, which is verifiable on the basis of its unique identity and characteristic qualities. The integrity of data means that unnoticed manipulation of data is impossible. Confidentiality means that only authorized persons can access information. Non-repudiation means that entities cannot disclaim the actions they have performed. Anonymity is changing personal data in such a way that even with a reasonable degree of effort, it is impossible to match the data to the person (Eckert 2004).

Many different technologies can be used in authentication. Passwords continue to be a widespread as a way for users to authenticate themselves on a device such as a PC. The four-digit *personal identification number* (PIN), which is a simplified variant of a password, is widely used for authentication on mobile telephones and automated teller machines. Biometric methods are also used increasingly. For objects without biometric characteristics, RFID can be used in automatic identification processes for authentication.

In order to guarantee the *integrity* of data, a checksum is usually calculated from the data. Manipulation can be discovered by comparing the checksum to a reference value. Checksums are typically generated by using hash values, which transform data of any length into a unique value of fixed length. The most commonly used methods are SHA1 and MD5.

*Confidentiality* is ensured by encrypting data. Two basic types of encryption exist. In symmetric methods, the same secret key is used in both encryption and decryption. In asymmetric methods, a public key (accessible to anyone) is used in encryption, and a private key (available only to authorized persons) is used in decryption. Symmetric encryption uses the Advanced Encryption Standard (AES) as well as an older method, 3DES, which is based on the old Data Encryption Standard (DES). For asymmetric encryption, the method normally used is RSA. Newer asymmetric encryption methods are based on the discrete logarithm problem in elliptic curves, whereas RSA security relies on the integer factorization problem. An advantage of the elliptic encryption techniques is that they require less computing power for the same level of security. This method is thus well suited for devices with limited resources, which makes it especially interesting for ubiquitous computing applications.

*Non-repudiation* is achieved with digital signatures. The relevant data or a checksum corresponding to it are encrypted with a private key that is available only to the signer. The freely available public key enables verification of the sender. A *public key infrastructure* (PKI) is used in administering the public keys, which allows the ownership and validity of a key to be confirmed by a trustworthy, neutral authority. For trust management in ubiquitous computing, the *trusted platform module* (TPM) will in all probability play a leading role. The TPM is a smart card securely built into a device. However, the module is associated with a system, not a user.

The security objectives of *anonymity* and *authenticity* would appear to contradict each other. An approach to resolve this contradiction is *pseudonymization*, which alters personal data by an assignment rule so that they can no longer be linked to a natural person without knowledge of the rule. Disguising individuals' true identities with pseudonyms and administering these identities falls under the rubric of identity management. The aim of identity management is to ensure that identities and the personal data linked to them are consistent, reliable, secure and constantly available.

In contrast to the other security objectives, anonymity cannot by assured by any explicit cryptographic mechanism. However, Frank Stajano has described a protocol that enables anonymous bidding by different parties at an auction. With his *Resurrecting Duckling* security policy model, Stajano has developed a scheme for connecting devices to one another, without requiring a third, trustworthy authority. This model is ideally suited for use in ubiquitous computing applications (Stajano 2007).

The Internet Protocol (IP), the most widely used protocol for communication between various digital devices via different network technologies, has been expanded with the IPSec security standard, which supports authentication and encryption at the IP packet level. Special resource-conserving versions of the IP, such as the uIP stack developed by Adam Dunkels and the Swedish Institute of Computer Science, enable the IP protocol to be used in the world of ubiquitous computing (The uIP TCIP/IP Stack, Swedish Institute of Computer Science 2007).

**Safety:** it is also referred to as reliability. Reliability of a system refers to its ability to work correctly according to its intended purpose. The effect of lack of reliability of a ubiquitous computing system can be critical depending on the area in which it is used. In medical technology, for example, the reliability of a system can be a matter of life and death. In logistics, ubiquitous computing can improve the ability to plan and thus lead to greater reliability (Buhl et al., 2001). In real-world, every technical system is error-prone. Safety in a narrower sense refers to a system's capacity to be fail-safe - to avoid responding to system errors by spiraling out of control and thus endangering the system itself or its environment. At the same time, the system should be fault-tolerant (BSI Publication on Fingerprints 2004).

As a general rule, it is hard to develop systems with high safety levels in ubiquitous computing due to the high complexity and networking of large number of disparate units. On the other hand, ubiquitous computing can help improve the safety of other systems. For instance, use of a ubiquitous computing infrastructure can help increase the safety and efficiency of health care in a hospital (Bohn et al., 2003).

**Data privacy:** the objective of data protection is to safeguard the individual's privacy. In ubiquitous computing, data protection is more crucial than in conventional information systems. The large number of smart objects and their spontaneous networking undermine the overall system's controllability. Ubiquitous computing increases the user's dependence on a large number of background processes that are not transparent. The networking of individual objects and the distributed nature of services make it hard to discern the connections between an action in ubiquitous computing and its consequences for the transmission and processing of one's own data.

According to most of the technology experts, security technologies are relevant to most ubiquitous computing applications, though many of them believe biometrics is necessary only in selected applications. There are some ubiquitous computing applications where security technology is not relevant. These are probably either "trivial or fun applications", or else isolated applications without an extensive network, where attacks are fairly unlikely.

### 4.7 Machine-to-Machine Communication

Ubiquitous computing systems will be highly distributed systems with thousands or millions of spontaneously interacting components. The standardization of appropriate machine-machine interfaces and their development are thus extraordinarily important for ubiquitous computing.

In machine-to-machine communication, i.e., in P2P computing, Sun Microsystems' JINI (Java Intelligent Network Infrastructure) has established itself as an open industry standard for system development at the hardware level, while the JXTA standard (named for "juxtapose") is important for communication in the higher layers of the protocol stack. Similar standards have been defined for P2P communication by HP with its Chai appliance plug and play, and by Microsoft with universal plug and play (UPnP).

Service-oriented architectures (SOA) describe a design model for shared use of reusable distributed systems. The aim of service-oriented architectures is to vastly simplify the integration of applications. The basic design principles of SOA are open standards that support the use of Internet protocol-based web services, an integration platform for secure message transport (enterprise service bus), and a dedicated integration instance. Another design model for developing SOAs is the industry standard CORBA (Common Object Request Broker Architecture) for communication between object-oriented software components, which was developed by the industry consortium Object Management Group (OMG).

As the networking of smart objects becomes increasingly complex, the potential for danger also rises sharply. It is no longer feasible to explicitly program all interactions among the objects, because there are just too many possible combinations. In response to this problem, work has been done for some years on software agents that have their own semantic model of their environment. As knowledge-based systems, they are able to make autonomous decisions and adapt to their environment as a learning system. Agents can thus act independently on behalf of their human users. They translate generally formulated commands into concrete actions, choosing independently among different courses of action without consulting the user. The research field of socionics is examining how agents organize among themselves. Human interaction mechanisms are being deliberately applied as a social model for understanding the decentralized coordination of agents. A great many research groups worldwide are working on software agents, including the Software Agents Group at the MIT Media 47 Ubiquitous Computing: Trends and Impacts Laboratory and the DFG Priority Program Socionics—Investigating and Modeling Artificial Societies.

Most of the experts in the domain of ubiquitous computing believe that the relevance of machine-to-machine communication in ubiquitous computing is average compared to the other technology fields.

### 4.8 Human-Machine Interface

The smart objects of ubiquitous computing require developers who design user interfaces that move beyond the formerly dominant monitor/keyboard principle. Most objects will have a variety of interfaces to their environment, but these will not include visualization components. Moreover, there will be many implicit interactions in which the user will have little or no involvement in the computing process, to avoid flooding the user with information. Even so, the user must be given the option of controlling the activities of ubiquitous computing by means of an appropriate human-machine interface.

The human-machine interface is not a self-contained field of technology. It is instead an interdisciplinary challenge that draws on such fields as computer science, ergonomics, the cognitive sciences and microelectronics.

The design of human-machine interfaces is an important activity for most of the leading players in consumer electronics and computer systems. Companies such as Microsoft and Siemens maintain their own usability labs in order to test their products. The auto industry and its suppliers, e.g., Toyota, BMW and Mitsubishi, or their suppliers Emmerson and Siemens VDO, are also working intensively on the interfaces of their driver assistance systems.

A central challenge for the human-machine interface is to construct a semantic model of the real world, which would allow the meaning of a spoken sentence to be understood, for example. Such models have been developed for specific domains such as medicine, but a general model is yet to evolve. These developments are currently getting a strong boost from the *semantic web initiative* of the Internet standards organization- the World Wide Web Consortium. Semantic Web comprises a collection of standards for classification systems such as RDF and OWL, which model the real world in networks of concepts. Whether and how this approach might impact real applications has yet to be seen.

Technology experts believe that human-machine interfaces will play a rather average role for the evolution of ubiquitous computing. Speech technology is seen as particularly relevant, but also as a possible technological bottleneck. The visionary approaches of gestures and implants are believed to be less relevant for further evolution of ubiquitous computing.

## 5 Security in Ubiquitous Computing

The exchange of information among large numbers of smart objects is a central property of ubiquitous computing. Smart objects differ considerably from other objects in terms of their input/output capabilities, sensory interfaces and application processes. Ubiquitous computing processes are, for the most part, designed to run as inconspicuously and situationally dependent—and therefore semi-automatically—as possible. It is thus essential that information be exchanged between authorized persons and/or objects only. Data and information must be allocated clearly and protected from manipulation and espionage. The secure identification of and communication between objects and persons in ubiquitous computing is therefore of vital importance. Equally important is the protection of users in the event of smart object failure or malfunction.

As the first ubiquitous computing systems are introduced, technical issues in security, safety and privacy will quickly become decisive.

As it stands now, ubiquitous computing is for the most part a technological vision. However, the general shape of future security architectures in ubiquitous computing as well as their associated opportunities and risks can be outlined. Using three different scenarios, this chapter explores the unique security challenges posed by ubiquitous computing in terms of the objectives of security, privacy and safety.

Identifying persons and objects is both an application of ubiquitous computing (e.g., digital IDs) and a central internal operation required to make ubiquitous computing services safe and reliable. Because different technical approaches will be used to identify objects and persons, it is essential that we be able to distinguish between the two. Keeping this in mind, the first scenario considered in this chapter is concerned with the identification of objects. The second scenario examines the identification of individuals via biometric universal identification. Both scenarios rely on technology currently in use and include the Trusted Platform Module (TPM) for object identification and the ICAO Standard for biometric identification systems. The third ubiquitous computing scenario discussed in this chapter involves the interaction between vehicles and telematics systems. It has been chosen since, according to most of the technology experts in ubiquitous computing, this field is going to be one of the earliest applications of ubiquitous computing. This scenario takes a visionary look into the future and draws upon the findings of current research projects.

### 5.1 Scenario 1: Object Identification with TPM

In object identification, objects are recognized according to firmly assigned characteristics. We can distinguish three system components: (i) an *identification number* (ID) serves as a unique code for each object which is stored in a carrier, usually a chip, (ii) a reader that identifies the carrier-object via the ID, and (iii) the carrier-object and the reader exchange the ID and other information by means of a communication interface. As shown in Figure 1, the following are the potential threats (BIS Publication on Pervasive Computing 2006):

**Falsification of content:** if an attacker gains unauthorized access to an ID carrier, it can change and/or falsify content.

**ID falsification:** the attacker acquires an ID and potentially accesses an object's secured information. It then uses this information to fake its identity vis-á-vis a reader.

card containing account data and a student ID containing matriculation information (Home page of Austrian citizen card, 2005). Belgium, Finland, Italy and Switzerland have already introduced similar citizen cards (GLOBAL PLATFORM, 2006) (Hoffman et al., 2008).

One of the major properties of a UID in ubiquitous computing is that it allows for the unambiguous digital identification of its owner. With the implementation of digital personal identification, identification via knowledge can be distinguished from identification through ownership and biometric characteristics. For example, if a person is identified only via the possession of the UID carrier, such as a handheld, only the UID carrier is recognized. This form of identification has the same security problems as in the case of object identification. Furthermore, it is important that only the person assigned the UID has access to the UID carrier.

In ubiquitous computing, identification via secret passwords and PINs plays only a subordinate role, because the implicit communication of smart objects with users and among each other does not permit an opportunity to enter a PIN or similar signature. In the personal identification of the UID, human biometric signatures carry out a more secure identification than recognition via object possession. A person can be identified based on individual biometric data or a combination thereof.

In addition to its ability to represent different IDs from different contexts, a UID would also contain both changeable and unchangeable information.

This could even occur within one context itself. For example, while biometric identification data would remain unchangeable, other data on the ID could be changeable. This, for example, this could mean allowing for a change of address without having to issue an entirely new UID. Other UID functions that facilitate frequent changes will most likely also become the rule.

**Biometric Identification System of UID:** Even before the attacks on the World Trade Center on 11 September 2001, the International Civil Aviation Organization (ICAO) had already defined standards for digital travel documents with biometric functions and communication abilities via RFID.

According to ICAO standards for the security mechanisms underlying the UID, a global *public key infrastructure* (PKI) is required to sign and check digital documents. Each participating country must create a two-stage PKI consisting of exactly one *country signing certification authority* (CSCA) and at least one *document signer* (DS). The CSCA is the top certification authority in a country for travel documents. There is no global certification authority in the world. This is the only way to guarantee that each country has complete control over its own keys. Document Signers are authorized to sign digital documents and make the actual physical UIC carriers, such as the German Bundesdruckerei (the now privatized federal printer in Germany). For the UID, either public or private institutions could assume the role of the highest UID certification authority, similar to the CSCA. There can be several UID certification authorities in one country, or one certification authority for several countries. In the following UID scenario, the CSCA is replaced by the term top UID certification authority. Private or public institutions could also take on the function of a Document Signer. Figure 3 depicts the relationships between several actors involved in Universal ID infrastructure.

The key-pair created by the top UID certification authority is used exclusively for the certification of DS institutions. Each DS has at least one key-pair that can be produced by the DS itself. The private key is used exclusively for signing digital documents, while the public key must be signed by the top UID certification authority. The duration of use for the

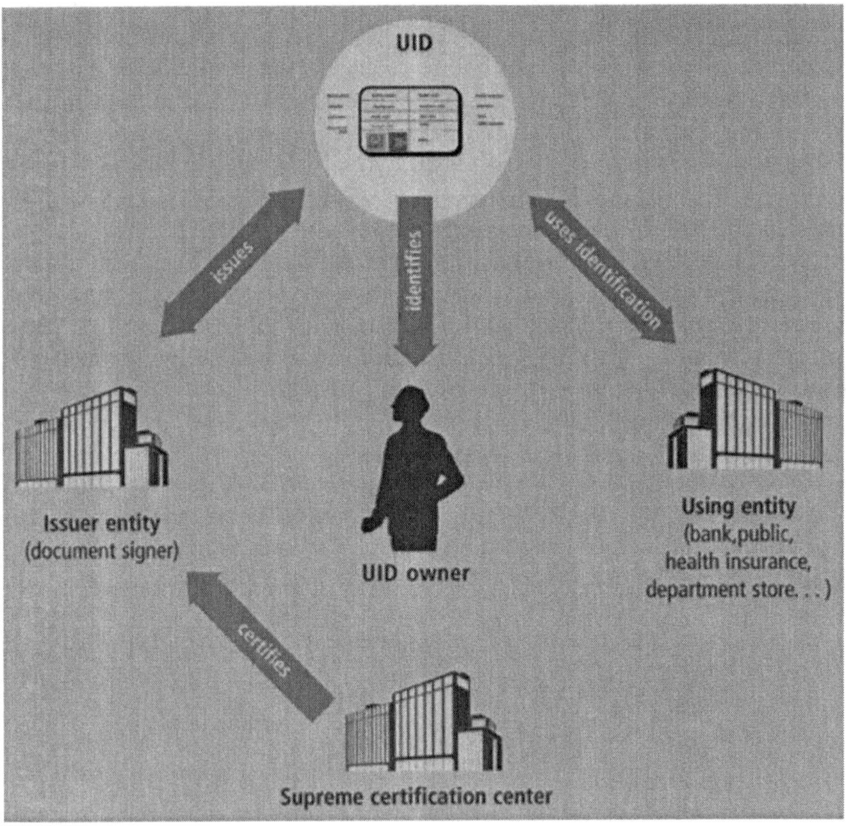


**Figure 3.** Relationships between the actors in a Universal ID system.

DS' private key is much shorter than that of the public key. This permits the fewest number of UIDs to be affected if the key is compromised (ICAO Technical Report on Biometrics 2004).

The ICAO has four planned mechanisms to achieve the security objectives of authenticity, integrity and confidentiality in personal identification. These mechanisms are: (i) passive authentication, (ii) active authentication, (iii) basic access control, and (iv) extended access control (ICAO Technical Report on PKI 2004).

**Passive authentication:** it allows for the verification of the authenticity of the data on the chip. It also determines whether the data have been changed. However, there is no check whether or not the UID carrier is the authentic carrier-object for the chip. During the authentication, the document security object (DSO) is read, by which the document signer can be verified. The DSO is a data structure signed by the DS according to RFC 3369 (ICAO Technical Report on PKI, 2004). The signature of the DSO is examined with the help of the public key of the DS. The validity of the Document Signer's certificate, signed by the top UID certification authority, can be verified with the public key of the top UID certification authority. The relevant data are read and the associated checksums (hash values) are examined. If the signatures of the hash values are correct, one can assume that the data has not been changed.

The ICAO also specifies two optional schemes designed to increase the security of the authentication: Basic Access Control and active authentication.

**Active authentication:** it serves to protect against UID cloning. It does not, however, provide protection from unauthorized users reading the content of the UID card. The UID features a key-pair of one private and one public key. The public key is signed by the document signer. The public key for active authentication must be linked to the corresponding UID carrier and to the corresponding biometric data. Otherwise, an attack involving the use of another UID carrier than the one presented as an oracle designed to answer active authentication requests is possible. For this reason, active authentication is accompanied by optically scanning the person with the UID carrier.

**Basic access control:** it guarantees the confidentiality of data contained in the UID chip. The ICAO adopted Basic Access Control because communication between identification documents and a reader can be intercepted within a range of up to several meters, as was demonstrated by the German Federal Office for Information Security in experiments involving standard RFID systems (BSI Publication on RFID 2004). There are two secret keys stored on the UID that are designed to ensure that the data contained in the UID chip is read by authorized readers only. Following ICAO specifications, the keys can be derived from optically readable data on the UID carrier. A wristband provides relatively little space in comparison to a card. The ability to store optically readable data on a small surface is therefore essential. Data could, for example, be stored using a 9 × 9 point matrix (dotcode) on a one square centimeter surface with high information density and error correction capabilities. This data would then be read using a compliant reader. To access the data stored in the UID, a reader must have a sightline connection to the optically readable data. This prevents data from being read secretly.

**Extended access control:** the top UID certification authority in each country determines who has access to the biometric data of UID carriers. UID issuers associated with other top certification authorities must agree to the respective access rules of the original top certification authorities. This authorization can be refused or revoked at any time. The UID carrier verifies the reader's authorization and thus guarantees its validity. The UID chip then checks the reader certificate as part of a challenge-response protocol. However, one significant risk remains: as long as the certificate is valid, a reader could continue to accept the top UID certification authority of a UID, even if the authorization for that UID has been recently revoked.

**Opportunities and risks:** a universal identification would spell for an initial increase in user comfort. The user would be able to identify him/herself securely for a number of digital communication transactions, e.g., in public transport, at the doctor's office, in a hotel, at the bank and at public offices etc, without having to remember several PINs and passwords. In addition to this high degree of user friendliness, a UID would allow strong biometric identification mechanisms to be used for almost every process in which the user must be identified as a person. Although this type of strong identification exists in some cases today, e.g., personal IDs must be shown to purchase a prepaid mobile phone, open a bank account or when signing a purchase or rental contract, the use of an omnipresent UID could make this a norm. The ease and availability of the UID could prompt the creation of a new set of procedures even for secondary processes and small financial transactions. For example, the date of birth contained on the UID would make it possible to determine the age of a person attempting to purchase cigarettes. The UID could also be used as the only ticket needed for access to public transport. People who do not own a UID would be excluded from these processes, which could lead to a splitting of society into UID owners and non-owners. It is therefore imperative that UID issuing be an inexpensive process. It should also be guaranteed that all processes can be used without a UID and that this alternative access is not associated with any specific consequences—such as high fees or less comfort.

In a UID scenario, several data can be stored on a UID, some of which are very sensitive. Moreover, all institutions wishing to use the authentication for their processes must have access to the data on the UID. These institutions thus have access to very private data and are able to create user profiles over time without the knowledge of the UID owner. It would also be possible for the UID owner to manipulate his/her own data. An adolescent, for example, might attempt to change his age so that he could gain access to prohibited products such as tobacco and alcohol. Specific data must be protected from possible manipulation by other issuers or the owner of the card himself, and data from other issuers must be protected against an unintentional overwrite. The data should be stored on a medium so that only authorized users can access it.

For the user, the loss of a UID would be much more serious than the loss of individual cards. If the UID carrier, either in the form of a card or a wristband, featured a sufficiently strong level of access protection that prevented its use by anyone other than the user himself, then it would be possible to issue a copy of the ID, or even several identical IDs, so that if one is lost, its integrated functions could be replaced quickly. However, considering the many different issuers involved, this would prove a complex process. One potential solution to this problem would be to establish a top certification authority similar to the CSCA associated with the electronic passport. Such a UID trust centre would manage all the necessary data and would therefore be able to issue a new medium quickly. However, given the large amount of data contained in a UID, this would involve a significantly more complex process. Furthermore, as the manager of the entire data of an individual, such an issuing authority would be able to establish unwanted cross-linkages at any time. The potential for misuse and the high sensitivity of data must be thoroughly considered in deciding whether such a UID trust centre should be state-run, private or a combination thereof. Given the crucial importance of such a centre for the entire UID system, its operator would have to meet very high security demands.

### 5.3 Security Requirements for Future Ubiquitous Computing

Ongoing miniaturization allows for the ubiquitous presence of computers and smart objects. This phenomenon, along with the interlinking of digital networks and attendant ubiquitous computing, is leading progressively toward the context-aware and adaptive exchange of information. The concept of cyberspace is thus being turned inside-out. No longer depicted as a digital simulation, the real world will become part of a ubiquitous digital network. In a ubiquitous computing world, a digital terrain will exist in parallel to the physical landscape. This terrain will not limit itself to humans and their communication. It will potentially involve and depict the environment in its entirety.

While today's users are usually conscious of the ICT applications they are using, several exchange and application processes will run independently in ubiquitous computing. In ubiquitous computing paradigm, smart objects will interact seemingly imperceptibly with one another and the services will emerge as they are available without explicit human intervention. Information processing will be an endless background activity.

Ubiquitous computing systems thereby exhibit a unique set of characteristics as compared to other information systems. Ubiquitous computing systems are largely decentralized and complex systems involving a multitude of service providers and operators. This also means that ubiquitous computing systems will be vulnerable to several different types of disturbances, attacks and technical malfunctions. Human beings will not take conscious note of several interactions in ubiquitous computing.

One can, therefore, assume neither an explicit consent of the user to an activity nor any knowledge of the possible consequences of his/her actions. In addition, vast amounts of data will be involved with ubiquitous computing, which could easily result in personal data or data attributed to a person being collected and evaluated automatically.

It is still far too early for a final word on future security architectures in ubiquitous computing. Yet, it is abundantly clear that data security must be a central technological component, if user acceptance is to be ensured. As with any information system, the data protection principles of data minimization, transparency and consent must also apply in ubiquitous computing. The data security technologies in a ubiquitous computing of the future must be easy-to- use, understandable and free of discrimination. User must have the freedom to choose at all times for themselves which data they want to transfer to which service providers, and how these data are used. It is especially important that users be able to switch off any of the ubiquitous computing services that interact with them at any time.

## 6. Conclusion and Future Trends

Today, ubiquitous computing is still a vision of technology. Extensive development work will be necessary to realize all of its characteristics, such as autarkic power supply, machine-machine communication, human-machine interface and security technologies. Apart from RFID-based logistics and security systems, there are very few ubiquitous computing applications currently in existence. Yet the dissemination and use of the Internet and mobile telephones over the past decade suggests how quickly ICT can develop, affect and even transform large segments of society.

In the short run, ubiquitous computing is the continuation of the Internet. The development of ubiquitous computing is typified by two characteristics that may appear contradictory at first glance. On one hand, only a few ubiquitous computing applications exist at present. On the other hand, many experts believe that numerous applications will be realized within the next one to five years. Most likely, these early smart objects will offer integration of different functions, which will include, in particular, certain sensory capabilities and data exchange via mobile broadband, enabling connection to the Internet. As a logical consequence, the first ubiquitous computing applications will probably draw heavily on what is already realized in the Internet. Audiovisual and data communication will merge, existing media ruptures will be overcome, and the possibility of digital communication will become ubiquitous. Ubiquitous computing offerings will be called up via a multitude of everyday electronic devices, while the services themselves will be provided by a central, Internet-based IT infrastructure. The close correspondence between web services and the aspirations of ubiquitous computing is also reflected in what are expected to be its early uses. Since mobility is a central characteristic in the early stage of ubiquitous computing, one can assume that enabling Internet connectivity from any device will be the main focus. Smart objects will thus represent a materialization of available online services. In the coming years, ubiquitous computing will be typified not by the refrigerator or range automatically connecting to recipes, but by ubiquitous access to information and services available over the Internet.

In the long run, ubiquitous computing will noticeably change processes in the personal, economic and public realms. In the long term, as ubiquitous computing is integrated into everyday objects and they become networked, information and control systems that are presently centralized in industry, transport, the service industry and the public sector can be decentralized and expedited. In the personal realm, ubiquitous ICT will bring new functions to everyday objects and more opportunities to communicate anytime and anywhere. To gain a sense of the potential and the limits of emergent ubiquitous computing, its central forerunner, the Internet, can serve as a model. With the dissemination of the Internet in society and the economy, a number of social and economic processes have indeed changed significantly or even fundamentally. In economic terms, though, one cannot speak of a purely Internet-based society and economy. Still, there are individual industries that are undergoing profound change and that find themselves in crisis due to digitalization and the Internet. The music and film industries furnish the most vivid example, since their products, which are already available in digital form anyway (CD or DVD) can easily be copied to other digital media.

Based on experiences with the Internet and its constant development and expansion, as well as on the characteristics of ubiquitous computing, one can identify central fields that will shape ubiquitous computing in the future. The often-implicit actions and the high degree of networking in ubiquitous computing make it hard to explicitly perceive and control system activities. This makes it all the more important for users to be able to totally trust the services and content of ubiquitous computing. This trust involves:

- The non-repudiability and accurate content of services
- The security of information in ubiquitous computing, and
- The careful treatment of personal data

Implementation of these standards will need to rely equally on technological, organizational and regulatory instruments.

On the technological side, there is still a great need for research and development in many areas. A broad spectrum of challenges awaits industry, science and research support, especially in regard to autarkic power supply, the human-machine interface and security technologies. The formulation and implementation of technical standards will exert a decisive influence on the introduction and impact of ubiquitous computing. Open standards—like PC architecture in the 1980s and the Internet in the 1990s—seem particularly well-suited to separating the development of the infrastructure from that of the products based on it. This will make it possible to create a broad range of freely-combinable ubiquitous computing products that compete unrestrainedly with one another. Technology suppliers for the ubiquitous computing infrastructure, service providers, user associations and standards organizations are thus called upon to agree on appropriate open standards, in order to counter tendencies toward monopolization.

Like the Internet, ubiquitous computing is not beyond the law. User acceptance of a system will depend heavily on rules that will guarantee that they can have the necessary trust in its services and content. These include ensuring that business and legal actions are legally binding, defining liability for services and faulty content that cause damages, and protecting the privacy of data. Even though ubiquitous computing does not, in principle, create any new requirements beyond those of other distributed information and communication systems, its invisibility and ubiquity will generally require that existing rules be adapted. This is particularly clear regarding data protection. It remains to be seen whether the existing instruments for data protection will prove adequate over the long run in light of the complexity of information processing and networking in ubiquitous computing. Here, service providers must develop transparency mechanisms, for instance, so that the individual can foresee the long-term consequences of accepting or rejecting an exchange of information.

Due to its functional logic, ubiquitous computing cannot be confined to individual countries. Internationally uniform rules are thus an essential requirement for the widespread introduction of technology. At present, though, very disparate regulatory approaches stand in the way of this in such realms as consumer law, data protection, freedom of speech and the press, and copyright law. On top of this come highly varied approaches to how such rules should be implemented—for example, in the form of voluntary commitments, certificates or laws. Internationally uniform rules are urgently needed here, with effective mechanisms for verifying and enforcing compliance. Otherwise, certain ubiquitous computing services with specific access barriers may be offered or usable only in certain countries, while they are banned entirely in other countries, as is currently the case in China with restrictions on Internet access and search machines.

An essential question in establishing a ubiquitous computing infrastructure is how to ensure simple, cost-effective and unimpeded access to services. This is especially important when certain public and commercial services are offered solely—or at least preferentially—via ubiquitous computing systems. This could affect purchasing age-restricted products like alcohol and tobacco, for instance, or accessing health care services. Particularly in regard to public services, care must be taken that society is not split into participants and non-participants in ubiquitous computing, with noticeable disadvantages for non-participants.

Ubiquitous computing has the potential to dramatically change production and business processes. This can mean not only that processes will become increasingly decentralized and flexible, but also that newly available information will increase market transparency. This would tend to favor small and flexible companies. On the other hand, large companies are better positioned to make the major investments initially required for ubiquitous computing and to set standards for cross-company ubiquitous computing infrastructures that benefit themselves most of all. Companies, their associations and economic policymakers are called upon to ensure that the opportunities of ubiquitous computing are equally open to all companies.

Considering that ubiquitous computing is in its infancy and the necessary technologies are at best partially available or not at all, it is impossible to foresee all of its social implications. However, the technological vision of ubiquitous computing is clearly recognizable in its contours and its technical implementation and application within the next ten years seems generally possible; this will entail fundamental socio-economic challenges, as discussed above. Clearly, ubiquitous computing is a highly dynamic paradigm that harbors a tremendous potential. The challenge lies in following its development attentively, and shaping it actively to exploit its positive effects while avoiding its potential negative effects as much as possible.

## REFERENCES


Austrian Citizen Card. 2005. URL: http://www.buergerkarte.at/index.en.

Bohn, J., F. Gärtner and H. Vogt. 2003. Dependability issues of pervasive computing in a healthcare environment. In: Proceedings of the 1st International Conference on Security in Pervasive Computing, LNCS, Vol. 2082, pp. 53–73, Springer-Verlag, Heidelberg, Germany.

BSI Publication on Fingerprints. 2004. Evaluation of fingerprint recognition technologies—BioFinger, August 2004. URL: https://www.bsi.bund.de/EN/Publications/publications_node.html.

BSI Publication on Pervasive Computing. 2006. Pervasive computing: trends and impacts (2006). Technical Report, 2006. URL: https://www.bsi.bund.de/SharedDocs/Downloads/DE/BSI/Publikationen/Studien/Percenta/Percenta_eacc_pdf.pdf?

BSI Publication on RFID. 2004. Security aspects and prospective applications of RFID systems (2004). Technical Report, October 2004. URL: https://www.bsi.bund.de/ContentBSI/EN/Publications/RFID/RIKCHA_en_htm.html

Buhl, H.U., A. Huther and B. Reitiesner. 2001. Information age Economy. 5, Internationale Tagung Wirstschaftsinformatik 2001, Physica-Verlag, Heidelberg, Germany.

Cosine, P. 1999. Wafer bumping for wafer-level CSP's and flip chips using stencil printing technology. In: Proceedings of 12th European Microelectronics and Packaging Conference and Exhibition- IMAPS EUROPE '99, Harrogate.

Eckert, C. 2004. IT security: Concept, procedure, minutes. 3rd Auflage, Oldenbourg-Verlag.

Global Platform. 2006. Concise guide to worldwide implementations of global platform technology. Internal Report, February 2006. URL: http://www.globalplatform.org

Hoffman, M., S. Heikkinen, G. Hornung, H. Thuvesson and C. Schnabel. 2008. Privacy-enhanced personalization in ambient environments. In: Proceedings of the 19th IEEE International Symposium on Personal, Indoor and Mobile Radio Communications (PIMRC 2008), pp. 1–6.

International Civil Aviation Organization. Biometrics deployment of machine readable travel documents. 2004. ICAO Technical Report, Version 2.0, 2004. URL: http://www.policylaundering.org/archives/ICAO/Biometrics_Deployment_Version_2.0.pdf

International Civil Aviation Organization. 2004. PKI for machine readable travel documents offering ICC read-only access (2004). ICAO Technical Report Version 1.1, 2004. URL: http://www.csca-si.gov.si/TR-PKI_mrtds_ICC_read-only_access_v1_1.pdf

International Technology Roadmap for Semiconductors, Update Overview. 2010. URL: http://www.itrs.net/Links/2010ITRS/2010Update/ToPost/2010_Update_Overview.pdf.

Sabatier, J., M. Aoun, A. Oustaloup, G. Gregoire, F. Ragot and P. Roy. 2006. Fractional system identification for lead acid battery state of charge estimation. *Journal of Signal Processing*, Vol. 86, No. 10, pp. 2645–2657, October 2006.

Stajano, F. 2007. Security for Ubiquitous Computing, John Wiley & Sons.

The uIP TCIP/IP Stack, Swedish Institute of Computer Science. 2007. URL: http://www.sics.se/~adam/contiki/contiki-2.0-doc/a00364.html.

Trusted Computing Group, TPM Main Design Principles. 2011. Specification Version 1.2, March 1, 2011. URL: http://www.trustedcomputinggroup.org/resources/tpm_main_specification.

Wagner, P.A., L.M. Hilty, P. Amfalk, L. Erdmann and J. Goodman. 2006. Experience with a system dynamics model in a prospective study on the future impact of ICT on environmental sustainability. In: Proceedings of the IEMSs 3rd Biennial Meeting Summit on Environmental Modeling and Software', International Environmental Modeling and Software Society, Burlington, USA.

Weiser, M. 1991. The computer for the 21st century. *Scientific American*. **265(3):**S.94–104.

Yang, J., F. Lu, L.W. Kostiuk and D.Y. Kwok. 2003. Electrokinetic micro-channel battery by means of electro-kinetic and micro-fluidic phenomena. *Journal of Micro-mechanical and Microengineering*. Vol. 13, 2003, pp. 963–970.

# 2

# Orchestrating Mobile Applications: A Software Engineering View

*Reyes Juárez-Ramírez, Guillermo Licea, Itzel Barriba, Victor Izquierdo and Alfonso Angeles*

**ABSTRACT**

Mobile applications have proliferated greatly; their use varies from personal applications to business systems. Still, although there are many mobile applications in the market, there is not much formal research about the engineering processes used to develop them. These applications possess characteristics similar to desktop applications. As a result, the fundamental principles of software engineering, as well as agile techniques, could be applied to construct mobile applications and systems. However, due to the limitations of mobile devices and communication aspects, they have additional constraints and face specific challenges. Some of the most significant constraints are the following: portability, efficiency, and usability. To develop robust mobile applications that satisfy fast-moving markets, in this chapter we propose the use of a well-structured development process, integrating effective principles and practices of the agile approach. Moreover, to achieve more user-oriented applications, we propose the integration of practices from usability engineering and human-computer interaction. We present a case study exemplifying the use of our proposal.

## 1. Introduction

Today, mobile devices have proliferated greatly and have a vast array of uses; for example, since the last decade mobile phones have been used as a medium for personal interaction using services such as voice (Nilsson, 2005), and SMS. Nowadays, these devices are becoming more powerful and are being built with functionality, which was, until recently, only available for laptops and desktops computers (Serhani et al., 2010; Wirth 2008, The 2020 Science Group 2006). Taking advantage of this capacity, other uses have emerged for mobile phones (Delic and Vukasinovic 2006; Petrova 2007; Wirth 2008); they serve as utility devices incorporating personal information management, such as contacts, calendars, etc. Additionally, a more recent use is as a medium of entertainment, playing music, videos, and videogames. Also, mobile phones have incorporated utilities such as e-mail managers and Internet browsing. Furthermore, there is a growing desire in organizations to perform more of their business functions using mobile phones (Delic and Vukasinovic 2006; Gruhn 2007; Microsoft Research 2008). Mobile applications are also used in the context of health care, remote sensing, and remote monitoring and control.

By today, there are 300,000+ mobile applications in the market (Wasserman 2010), and their production is growing faster than ever. Since the end of the last decade, businesses are completely, or at least partially mobilizing their information systems for internal and external operations (Gruhn 2007). For example, mobile technology is used for sales and service channels and connecting field workers to the company through mobile devices.

All the mobile devices have the following common nominator (Abrahamsson 2005): *software plays a crucial role*. This software is like any other; that is, it has requirements and is implemented in a programming language. In spite of great proliferation, there is not much formal research that explores the engineering processes used to develop mobile applications. Wasserman (2010) states "*the existing body of knowledge is highly pragmatic, with lots of guidelines and many pieces of sample code as examples.*" There is much work to do to understand the characteristics of mobile applications besides defining structured development processes.

Commonly, mobile applications are considered quite small. It is also believed that small software teams developed most of them (Abrahamsson et al., 2005). However, mobile applications are becoming more complex; they are moving from small recreational applications to more business oriented and critical uses (Wasserman 2010).

Due to increasing numbers of users worldwide who utilize mobile devices for demanding applications, such as videos and social networking, special attention must be focused on *efficiency* and *interoperability* for mobile computing (GLG 2010). Efficiency applies to many aspects of mobile computing, such as algorithms, power consumption, network access, data storage, distributed computing, visualization, *user interface, ergonomics*, etc. Interoperability focuses on making various mobile devices collaborate effectively, efficiently and reliably with each other as well as with other computers and devices (e.g., automobiles, refrigerators, ATMs, etc.) by means of open standards.

On the other hand, in the case of emerging market areas, such as the case of mobile Web applications, the following challenges are faced (Spriestersbach and Springer 2004): (1) increasing the usability and the suitability of applications and (2) increasing the efficiency of the development process. Improving the quality of the development process can help to achieve quality characteristics for the products and programs, especially such as portability and maintainability, which can benefit the use of applications through different target platforms.

Wasserman (2010) argues that the most classic software engineering (SE) techniques can be transferred easily to the mobile applications domain. On the other hand, Gruhn (2007) argues that a process-based approach is actually useful to develop mobile applications, indicating that the software process deserves particular support for: (1) mobile business processes analysis, (2) mobile architectures specification, (3) mobile interfaces design, (4) defining mobility supporting infrastructure.

Regardless of these possibilities, taking into account the characteristics of emerging mobile applications, the need to apply well-structured software engineering processes in order to assure the development of secure, usable, high-quality applications is clear (Wasserman 2010). Processes and people involved in the construction of mobile applications need to consider the challenges to overcome. A software engineer designing and developing mobile applications and systems must still have the rigorous training and hands-on experience in fundamentals of computer science, software engineering principles, programming languages, and architectures and tools for creating quality software (GLG 2010).

Also, there is the need to specialize existing software engineering approaches, using basic principles, but adapting some aspects to the restrictions that the mobile market imposes. An area emerging recently is mobile software engineering (MSE). As Gruhn (2007) stated, MSE is essentially software engineering, but it must be seen as a specialization area that demands knowledge of additional computing technologies, tools and techniques to design *efficient* and *reliable software* that *meets the specific requirements* of the user and the mobile computing environment (GLG 2010).

The market for mobile applications presents a significant dynamism, as stated in Abrahamsson (2007): "*Mobile information systems are subject to frequent requirements changes in terms of changing business needs and technology, and their market is highly competitive.*" Given the circumstances, organizational processes need to support rapid information sharing and short feedback cycles (Abrahamsson et al., 2005) to obtain products that meet requirements and restrictions, in addition to gaining knowledge to prepare new developments. This agility is not only necessary at the level of the development process to make adjustments during the product implementation, but agility is also crucial for interaction with the client and end-users.

In this chapter we present a process-oriented proposal for developing mobile applications. This proposal has a software engineering focus; but we integrate some practices from usability engineering (UE) and human-computer interaction (HCI) into the development process. This consideration of human factors helps achieve quality attributes, such as usability and other characteristics of mobile applications. The suggested process consists of a well-structured set of phases. Each phase generates specific artifacts, which contribute to the quality of the final product. Additionally, we present a set of important aspects of the agile approach in order to integrate them into the development process for mobile applications.

This chapter is organized as follows. Section 2 emphasizes mobile applications characteristics, justifying a requirements documentation model. Section 3 presents the suggested development process, emphasizing the phases, activities, and work products. Section 4 describes aspects for agile implementation. Section 5 presents a case study for developing a mobile application for iOS devices, emphasizing the phases implemented for a project. Section 6 exposes some aspects of agility practiced during this project implementation. Section 7 presents some of the lessons learned implementing both the structured process and the agile approach. Finally, section 8 outlines conclusions and future work.

## 2. Characteristics of Mobile Applications

Mobile devices have special characteristics, which restrict the implementation of the software to operate them; therefore, mobile applications are also characterized. Some of the most representative characteristics (Allen 2010, GLG 2010) are listed in Table 1.

Developers have to think about ways to implement applications that use these limited characteristics as efficiently as possible (Abrahamsson 2005).

In spite of the proliferation of mobile technologies and systems, the development of mobile applications confronts some limitations and faces particular challenges. Some specific constraints (Gruhn 2007; Ming-Chun and Shyan-Ming 2005; Paspallis and Papadopoulos 2006) are listed in Table 2.

**Table 1: Characteristics of mobile applications.**

| *Characteristic* | *Description* |
|---|---|
| *CH_01. Small display size.* | Mobile devices come in many sizes and shapes, setting limits on their input and output media. |
| *CH_02. Small memory size.* | Limited memory restricts storage of instructions and data. |
| *CH_03. Small battery capacity.* | Battery capacity limits information and instructions processing. |
| *CH_04. Network bandwidth limitations.* | Bandwidth limitations restrict performance of mobile applications. |
| *CH_05. Multiple accessories.* | Aspects such as real and/or virtual keyboard, global position system, microphone, and camera(s) must coexist. |
| *CH_06. Multiple networking protocols.* | Aspects of protocols, such as routing overhead and other characteristics can limit the performance of mobile applications. |
| *CH_07. Real World interaction.* | Almost all new devices include sense handling -accelerometer responding to device movement (motion sensor), and touch screen responding to gestures, etc. |
| *CH_08. Always on.* | Vibrate option and connectedness of the phone is a key part of the interface with the human user. |
| *CH_09. Individual and personal use.* | Privacy of information stored on personal devices is very important. For instance, the cell phone is typically tied to an individual who has a phone number, contacts and other very personal data. |

**Table 2: Constraints faced in developing mobile applications.**

| |
|---|
| *CS_01.* Each major handset manufacturer has its own Operating Systems, SDK and implementation, making it difficult to build applications for use across all mobile handsets. |
| *CS_02.* Mobile functionality provided by packaged applications is generally not easy to extend and not open to incorporate functionalities from other systems. |
| *CS_03.* Security considerations are closely related to corporate security policies. |
| *CS_04.* The combination of software aspects with telecommunications yields complex emergent behavior. |
| *CS_05.* There is a necessity to support all kinds of mobility -physical mobility (user/device mobility), logical mobility (mobile code, stateless or stateful). |
| *CS_06.* A clear distinction between logical and physical components and connectors to allow realistic modeling of mobility influence is necessary. |
| *CS_07.* The inclusion of the communication network model implies considerations for programming those aspects. |

Also, mobile applications involve important contextual aspects to be considered, such as those shown in Table 3.

**Table 3: Contextual aspects to consider in developing mobile applications.**

| |
|---|
| *CX_01.* People use mobile devices for different objectives. This implies differences in the pattern of usage. |
| *CX_02.* The variation in the context in which they are used complicates the establishment of product line patterns. |
| *CX_03.* The technological environment that enables mobile applications also implies restrictions of communication, hardware and software. |

Mobile applications imply interactive software, which require significant elements of usability. Most software engineering approaches applied to user interface (UI) design have been conceived based on the experience from developing traditional PC-based systems. However, technological advances, over the last decade, have brought considerable changes (Molina et al., 2003) to electronic devices such as computers, mobile devices and displays. These new devices require attention to design their interfaces (Shyan-Ming 2009; Paspallis and Papadopoulos 2006) to improve the user experience.

Some of the more important quality models, such as McCall (McCall et al., 1977), Boehm (Boehm et al., 1978), and ISO 9126 (ISO 2001) suggest usability as one of the primary quality attributes. Usability, as a quality characteristic of a system or product, can be approached from the point of view of different disciplines, such as SE, HCI, and UE. In each discipline, most efforts have been made as stated in previous works (Lethbridge 2006; Knouf 2009; Blevis 2007). However, the problem of usability coverage still remains (Kushniruk and Patel 2004).

Nowadays, the *"Lack of usability is the most critical problem facing software engineering"* (Lethbridge 2006). It is useful to combine other disciplines, such as UE and HCI. Usability engineering is a generic practice, traditionally used to develop any kind of engineering product, and it can be applied to design software systems (Nielsen 1992). In practice, UE is associated with HCI (Knouf 2009; Blevis 2007) since it considers user and tasks analysis practices, which determine the aspects that the user interface should have. This merging of interdisciplinary practices suggests the inclusion of processes for system evaluation with the user interaction, existing since the early phases.

Emphasizing the necessities to improve development of mobile applications, Table 4 indicates important areas for consideration in the context of mobile software engineering (GLG 2010). We will consider them opportunities for improvement.

**Table 4: Areas to consider in the context of mobile applications development.**

| |
|---|
| *OA_01*. Highly efficient systems and algorithms for low power consumption. |
| *OA_02*. Innovative techniques to manage the access to wireless network. |
| *OA_03*. Smart ways of distributing and reducing the computation workload. |
| *OA_04*. Intuitive, easy-to-use and appealing user interfaces to access, manipulate and visualize information. |
| *OA_05*. Interoperable solutions based on open standards that support collaboration across different mobile devices as well as with traditional desktop environments. |
| *OA_06*. Open source tools that promote collaborative, and low cost development of mobile software. |
| *OA_07*. Reliable security and privacy measures related to access and sharing of data. |

Taking into account all of these characteristics and opportunity areas, the need for a streamlined approach to the requirements and restrictions for mobile applications is clear. Based on this, it is necessary to consider the quality characteristics defined by quality models to analyze requirements and restrictions of mobile applications.

The ISO 9126 (ISO 2001) defines a set of characteristics that specify what the software (product) does to fulfill the user's requirements, such as: functionality, usability, reliability, efficiency, maintainability, and portability. Also, it defines sub-characteristics, which extend the quality attributes of a product.

Taking into account this set of attributes and considering the characteristics of mobile applications, it is possible to emphasize the utility of using a quality model to guide the construction of mobile applications. Spriestersbach and Springer (2004) suggest that in the particular case of mobile Web applications, the impacted product characteristics from ISO 9126 are the following: *functionality* (suitability, security), *usability* (understandability, learnability, operability), *efficiency* (time based, resource based), *maintainability* (changeability, stability), *portability* (adaptability). Georgiadis and Stiakakis (2009) indicate that *reliability* (information accuracy) is also impacted.

The quality models, such as McCall (McCall et al., 1977), Boehm (Boehm et al., 1978), and ISO 9126 (ISO 2001) can seem complex to be implemented, especially for mobile applications. We can suggest a single model for the requirements specification: FURPS+. This is a model developed by Robert Grady (Grady 1992) in Hewlett Packard. FURPS considers the following characteristics: *functionality, usability, reliability, performance,* and *support*. This model later was extended by Rational Software—now IBM Rational Software, adding the "+" (Jacobson et al., 1999; Kruchten 2000), which considers the following constraints and requirements: *design restrictions, implementation requirements, physical constraints*, and *interface requirements*. Table 5 shows some sub-characteristics of FURPS, and the description of the constraints.

Most of the characteristics and constraints of mobile applications can be expressed in terms of the FURPS+ model. For example, characteristics of mobile devices, such as *CH_01, CH_02,* and *CH_03* are referred in the physical constraints of FURPS+. Characteristics *CH_05, CH_07,* and *CH_08* are referred in the interface requirements. Characteristic *CH_09* is referred in functionality (security). The constraints *CS_01* and *CS_02* can be referred in supportability (portability). Contextual aspects also can be referred in the FURPS+ model. For example, *CX_01* can be referred in functionality (generality). Likewise, the opportunity areas for improvement in mobile applications can be expressed in FURPS+. For example, *OA_03* can be referred in performance (efficiency).

Based on this, we suggest that the FURPS+ model can be used to specify requirements to implement mobile applications.

**Table 5: The FURPS+ model description.**

| *Characteristic* | *Description* |
|---|---|
| Functionality | Feature set; capabilities; generality; security. |
| Usability | Human factors; aesthetics; consistency; documentation. |
| Reliability | Frequency/severity of failure; recoverability; predictability; accuracy; mean time to failure. |
| Performance | Speed; efficiency; resource consumption; throughput; response time. |
| Supportability | Testability; extensibility; adaptability; maintainability; compatibility; configurability; serviceability; installability; localizability; portability. |
| Design Restrictions | Specify or restrict the system design, limiting the possibilities for design. This concerns special restrictions, for example, the need to use a relational database. |
| Implementation Requirements | Refer to considerations for implementation. For instance, the use of a specific programming language, rules for programming, or compliance with standards. |
| Physical Constraints | Refer to considerations for hardware, such as form, weight, size, etc. |
| Interface Requirements | Consider external elements with which to interact. These elements can be subsystems, specific peripheral devices, or external systems. |

## 3. The Process-oriented Approach

To construct robust mobile applications, we propose a process-oriented approach (POA) with specific phases, activities and work products. In this approach, we merge aspects from SE, HCI, and UE. This approach is supported by the proposals exposed in most related works (Pimenta and Barthet 1996; Kim and Corrington 2002; Molina et al., 2003; Ferre 2003; Walenstein 2003; Seffah and Metzker 2004; Seffah et al., 2005), which clearly suggest merging software engineering with other disciplines to have more user-oriented software applications. Here we present a set of phases adjusted from the proposal presented by Zhang et al. (2005), indicating specific activities and work products. Also we considered the recommendations exposed in Harper et al. (2008). Fig. 1 shows the main phases of the process.

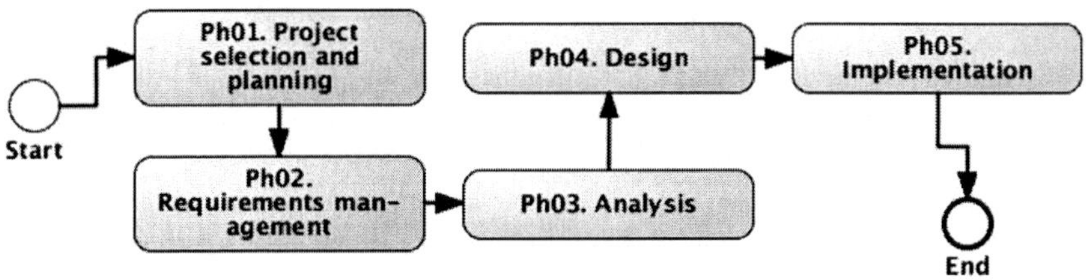


**Figure 1.** Main phases of the development process.

### Ph01. Project selection and planning

Any software development needs to start with this phase. Project selection and planning is executed by generating a project plan, which contains activities and work products, such as those shown as output of an activity in Fig. 2. Activities *Ph01.A1*, *Ph01.A2*, and *Ph01.A3* represent basic activities performed for any engineering project. *Conducting risk analysis* (*Ph01.A4*) is performed to ensure the successful completion of each deliverable. The main result of this phase is a project plan containing the work products, which are indicated as output of activities.

### Ph02. Requirements management

After defining a project plan, requirements management is one of the most important phases for a software development project. Fig. 3 shows the main activities for this phase.

*Requirements elicitation* (*Ph02.A1*) involves the most interaction with end-users. Different techniques can be used to elicit requirements, such as

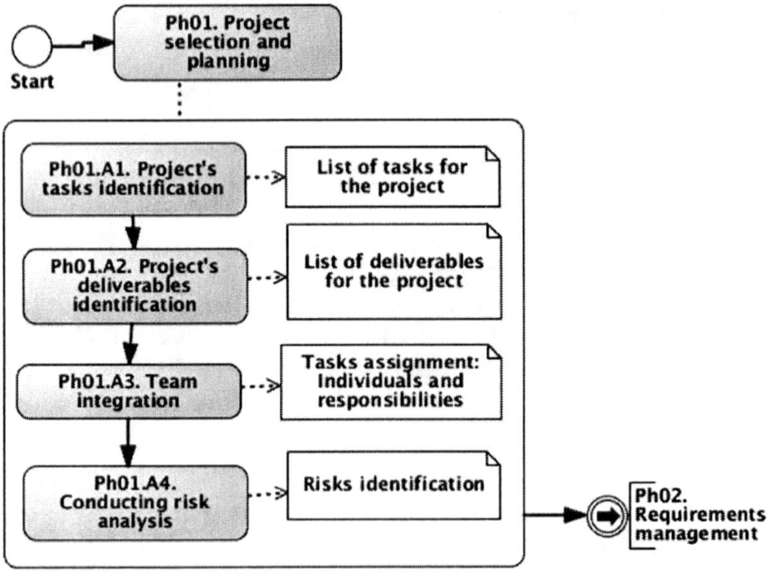


**Figure 2.** Project selection and planning.

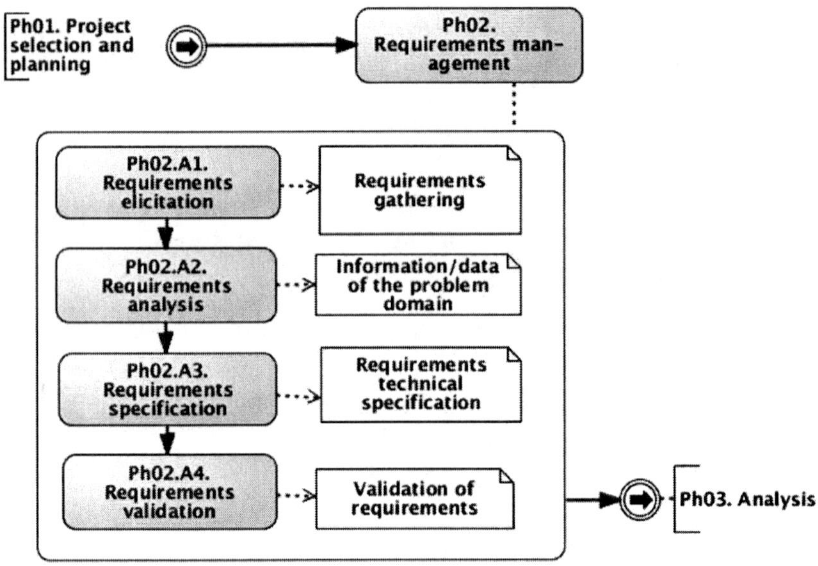


**Figure 3.** Requirements management.

observation, interviews, document reviews, etc. Requirements gathering may be documented in an informal style. *Requirements analysis* (*Ph02.A2*) allows us to specify the information and data domain of the system to build. Technical language can be used to represent the information and data. Some of the most used techniques are the following: business process modeling, activity diagrams, and data flow diagrams.

As a result of the analysis, we can have the *requirements specification* (*Ph02.A3*). A single model can be used to specify the software requirements, such as FURPS+. Requirements specification is like a contract between end-users and developers. This helps them to have the same perception of the system (Zhang et al., 2005).

*Requirements validation* (*Ph02.A4*) is a process that certifies that requirements are consistent with the intentions of the client and end-users, occurring simultaneously with elicitation and specification, to assure that ideas and concepts described are clearly expressed and well identified.

**Ph03. Analysis—context analysis focus**

With the requirements specification for the system to build, we need to perform an exhaustive analysis of the context in which the system will act. This phase represents the combination of the usability engineering and human-computer interaction with software engineering. Most of the activities listed here come from UE and HCI and should be inserted in the software development process. Fig. 4 shows the main activities to perform in this phase, containing sub-activities, which are explained here.

The *context analysis* (*Ph03.A1*) includes understanding the *technical, environmental* and *social* settings where the information system will be used (Zhang et al., 2005). It examines whether and how the interaction between physical and social environment with the physiological and psychological characteristics of the user would impact the human-computer interaction. *Technical context analysis* (*Ph03.A1.SA1*) considers aspects such as technology infrastructure, platforms, hardware and system software, and network/wireless connection. *Physical context analysis* (*Ph03.A1.SA2*) considers aspects such as the following: Where are the tasks carried out? What entities and resources are implicated in task operation? What physical structures and entities are necessary to understand observed task actions? *Organizational context analysis* (*Ph03.A1.SA3*) considers the larger system where the new information system is embedded and the interactions with other entities in the organization. *Social and cultural context analysis* (*Ph03.A1.SA4*) considers the social or cultural factors that may affect a user's attitude and eventual use of the information system.

Context analysis is highly valuable when creating mobile applications due its great changeability (Zhang et al., 2005). The analysis should be done in an agile way, taking into account the volatility of the market.

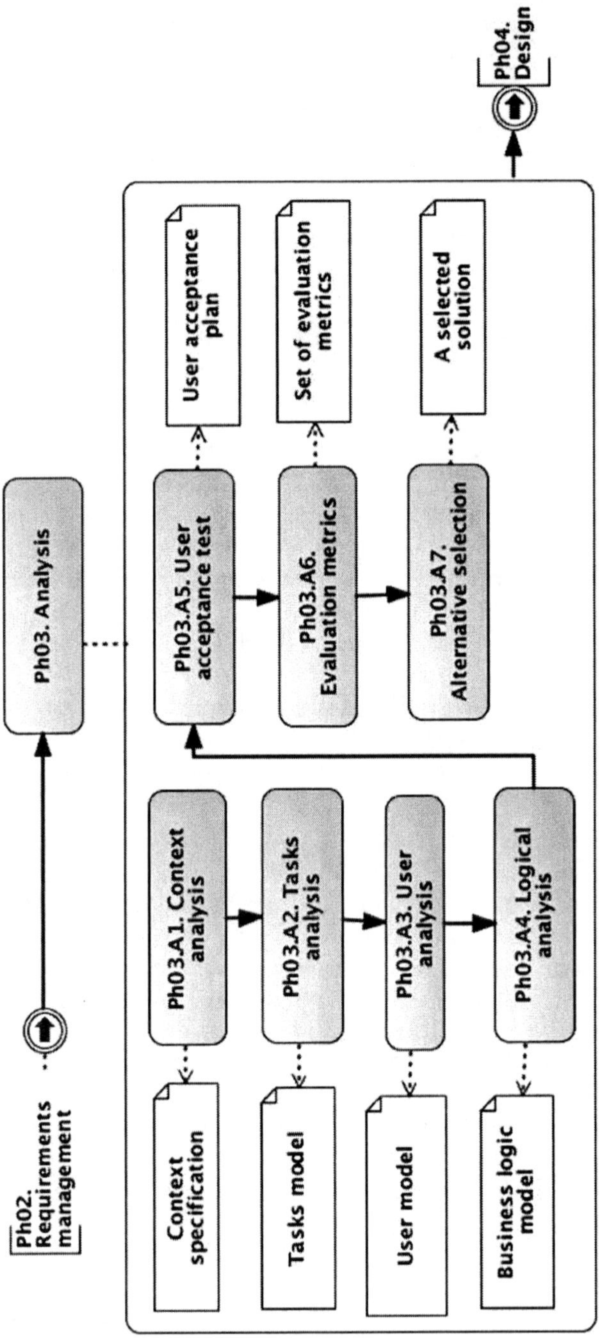


**Figure 4.** Analysis phase.

The *tasks analysis* (*Ph03.A2*) is concerned with understanding what users' goals are and what they do to achieve their goals (Zhang et al., 2005). It also includes scenarios and conditions under which humans perform the tasks. The most used techniques to perform the task analysis are *use cases* and *scenarios*. For each user task, the cognitive, affective, and behavioral analysis should be performed. *Cognitive analysis* (*Ph03.A2.SA1*) identifies the cognitive needs of processes involved to perform a task. *Affective analysis* (*Ph03.A2.SA2*) identifies and predicts the user reactions to the system stimulus. *Behavioral analysis* (*Ph03.A2.SA3*) identifies and predicts the user behavior after using the system. What's more, the task analysis also involves the workflow analysis and the work distribution. The *workflow analysis* (*Ph03.A2.SA4*) identifies and establishes task flows, branches, back sequences and intentions, using activity diagrams and hierarchical task analysis. *Work distribution* (*Ph03.A2.SA5*) identifies and establishes general work to be done by the users and the system. Actions are classified as (Zhang et al., 2005) *user actions* (inputs, selections, etc.) and *system actions* (outputs, selection options), and all relevant information is specified for each action.

The *user analysis* (*Ph03.A3*) identifies the target users and their characteristics. It includes demographic data, traits and intelligence, job or task related factors (Zhang et al., 2005). *Demographic data* includes aspects such as age, gender, education, occupation, cultural background, special needs, computer training and knowledge, experience with similar systems and products, etc. *Traits and intelligence* includes aspects such as cognitive styles, affective traits, and skill sets or capability. Job or task related factors include aspects such as job characteristics, knowledge of application domain and job familiarity, frequency of computer use for the job, and usage constraints.

The *logical analysis* (*Ph03.A4*) identifies all the business logic, which represents the operability of the system being built.

The *user acceptance test* (*Ph03.A5*) determines the likelihood of target users' accepting the usefulness of the system (Zhang et al., 2005). A *user acceptance test* is proposed right after the requirements are determined. *Evaluation metrics* (*Ph03.A6*) specify the goals of human-computer interaction. Evaluation metrics include user task-related aspects, such as learnability, effectiveness, efficiency, and satisfaction. This defines the usability dimensions, and user experiences. Evaluation metrics are derived from the context analysis, task analysis, and user analysis. These metrics guide the tasks in the design phase and provide the benchmarks for the summative evaluation in the implementation phase.

In the case of the *alternative selection* (*Ph03.A7*), before transforming all gathered and structured information into design ideas, there is a need to *select the final alternative* for the proposed information system. This selection can be done after reviewing some prototypes, which can be implemented during the requirements specification or after the task and data analysis. The selected strategy is judged in terms of the most likely to lead to the most desirable system, from functionality, usability and user experience perspectives (Zhang et al., 2005).

**Ph04. Design**

The design phase is also a key stage in the development process. In this phase all the characteristics of the system are stated. Fig. 5 shows the activities for this phase, containing sub-activities, which are explained here. In the case of the *database design* (*Ph04.A1*), a formal structure for the database is proposed. This activity is realized exclusively when the application involves interaction with a database.

The *interface specification* (*Ph04.A2*) includes metaphor, media, dialogue, and presentation designs (Zhang et al., 2005). The *metaphor and visualization design* (*Ph04.A2.SA1*) helps the end-user to develop a mental model of the system. Metaphors concern something like scenarios, possibly graphical, where possible user interactions are represented. *Media design* (*Ph04.*

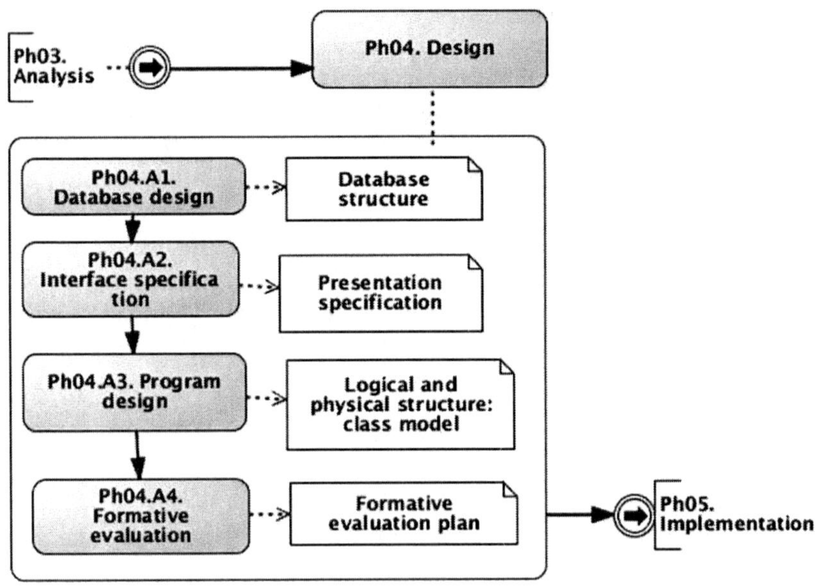


**Figure 5.** Design phase.

*A2.SA2*) is concerned with selecting appropriate media to meet specific needs. Media concerns graphical objects used to present concepts to the user, such as graphic objects, photos, etc. *Dialogue design* (*Ph04.A2.SA3*) focuses on how information is provided to and captured from users during a specific task. To specify the dialogue design many existing interaction styles can be used, such as menus, form-fill-ins, natural language, dialog boxes, and direct manipulation. *Presentation design* (*Ph04.A2.SA4*) concerns the decisions on display layout and incorporation of metaphors, media, and dialog designs with the rest of the displays.

The *program design* (*Ph04.A3*) concerns logical and physical structure for the program (Zhang et al., 2005). In the case of the logical structure, when the system involves object oriented programming, a set of classes can be implemented. A class model can be implemented for the logical analysis, dialogue model, and user model. The physical structure can be expressed using a well know technique, such as the component deployment diagram.

*Formative evaluations* (*Ph04.A4*) identify defects in usability and user experience designs thus inform design iterations and refinements (Zhang et al., 2005). A variety of different formative evaluations can occur several times during the design stage of an information system.

**Ph05. Implementation**

In this phase the user interface is specified, sketched, materialized, and tested, and business logic is implemented. Fig. 6 shows the activities and work products involved in this phase.

*Prototyping support* (*Ph05.A1*) is considered in the analysis and design phases, but it is also considered as part of the implementation phase, just because prototypes actually can involve the use of tools for user interface implementation or programming languages. *Programming* (*Ph05.A2*) is the concluding part of the development process. This phase generates software components, libraries, and modules. *System testing* (*Ph05.A3*) is performed at the unit test and integration test levels. Unit testing verifies software components' functionality separately. Integration testing verifies proper integration of different but related components or modules. When *installation* (*Ph05.A4*) is performed, some of the product quality characteristics are tested, such as interoperability and portability.

*Documenting* (*Ph05.A5*) is also a concluding activity, consisting of the creation of technical reports and user and installation manuals. Moreover, programming, system testing and installation, and *maintenance* (*Ph05.A6*) plans and guidelines are generated.

*Summative evaluation* (*Ph05.A7*) takes place after the system is developed to confirm whether the evaluation metrics or other industry

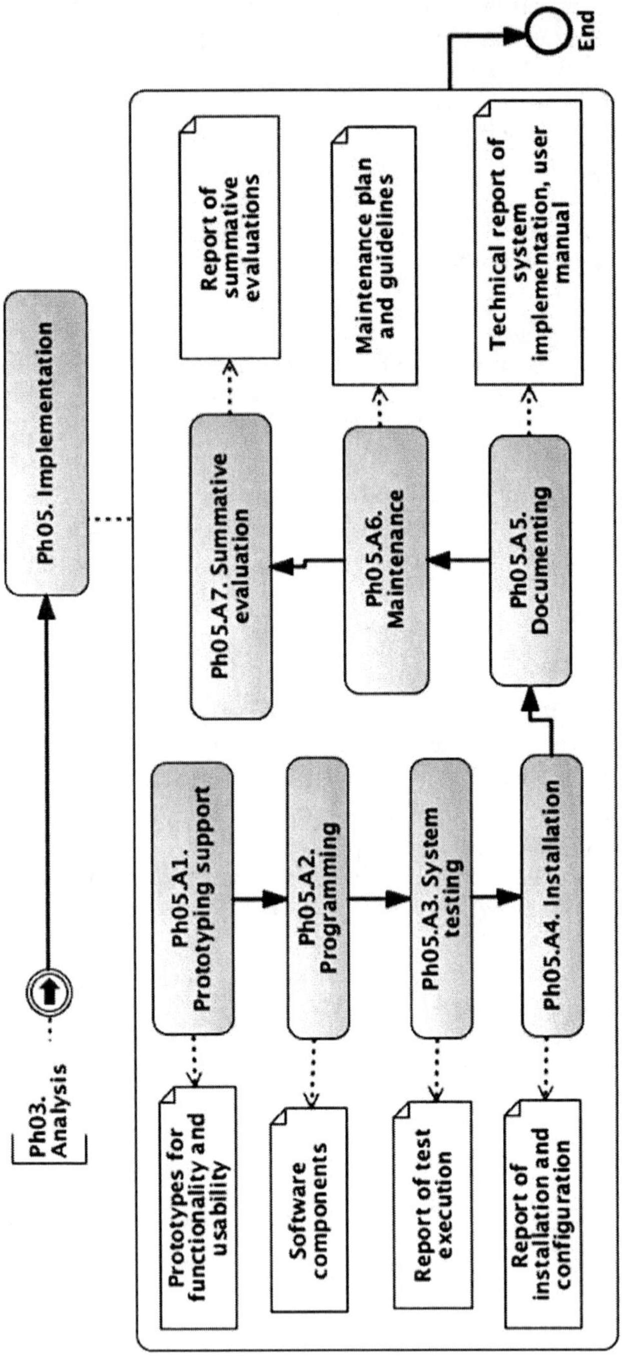


**Figure 6.** Implementation phase.

standards are met (Zhang et al., 2005). In addition, more evaluation is performed after the system is installed and being used by targeted users for a period of time. In this case, this evaluation gathers feedback to understand the actual behavior toward system use.

## 4. Considerations in the Agile Approach for Mobile Applications

The market for mobile applications is constantly moving forward. The mobile telecommunications industry has shown itself to be comprised of a highly competitive, uncertain, and dynamic environment (Lal et al., 2001). Organizations operating in this type of business environment need to react rapidly to changing market needs (Abrahamsson et al., 2005). This situation represents more characteristics and constraints for mobile applications, incrementing those mentioned in section 2. Agile software development solutions can be seen as providing a good fit for the mobile environment.

Recently, agile development methods for information systems have received considerable attention from practitioners (Aydin 2004; Conboy 2009; Vidgen and Wang 2009). One of the reasons is that it seems that agile methods, to some degree, can be adaptable to different project situations. However, there is no evidence about the employment of specific agile practices to construct mobile applications (Abrahamsson 2007). On the other hand, concerns have been raised whether agile methods are suitable for any given information systems development domain (Vidgen and Waang 2009). Indeed, quite little is known empirically about the validity of agile methods in most of the industrial domains (Abrahamsson 2007). In general, little empirical research has been conducted on this subject (Aydin 2004), not only in the industry context, but also in the academic environment (Abrahamsson et al., 2002).

Still, when there are companies trying agile practices, most problems need to be solved. For instance, for certain mobile development scenarios the stakeholder's identification is not well realized, especially in terms of end-user identification due to the multitude of software distribution channels (Spataru 2010). In these conditions, the agile principles are not applied, which involve a close contact with customers. It is important to remark that the continuous contact with end-users is almost a rule in the agile approach. End-users could be consulted during two stages of product development (Spataru 2010): (1) During requirements gathering in order to identify all their needs and desires, and (2) after releasing the product, to measure acceptance and obtain feedback.

Our current work is focused in implementing an agile approach in combination with a structured development process just like one described in section 3. In order to achieve this objective, we first consider the basic characteristics of the agile approach.

The Agile Software Development Manifesto includes the following value statements (Agile Alliance 2001): Individuals and interactions over processes and tools (*VS_01*), working software over comprehensive documentation (*VS_02*), customer collaboration over contract negotiation (*VS_03*), and responding to change over following a plan (*VS_04*). Agility denotes the quality of being agile; readiness for motion; nimbleness, activity, dexterity in motion (Abrahamsson et al. 2002). Other fundamental aspects of agile development are the following (Boehm and Turner, 2003): simple design principles (*FC_01*), a large number of releases in a short time frame (*FC_02*), extensive use of refactoring (*FC_03*), and seeing change as an advantage (*FC_04*).

Agile development methodologies also emphasize the following important aspects (Abrahamsson et al., 2004; Spataru 2010): individuals on the project team represent one aspect to assure the project's success (*IA_01*), producing deliverables and work products (*IA_02*), collaboration with business owners and clients (*IA_03*), responding quickly to necessary change (*IA_04*). Specific studies on agile implementations indicate more aspects. In terms of the process focus, Spataru (2010) cited nine main elements involved in the different practices throughout the development cycle: phasing and placing (*DP_01*), architecture line (*DP_02*), mobile test-driven development (*DP_03*), continuous integration (*DP_04*), pair programming (*DP_05*), metrics usage (*DP_06*), agile software process improvement (*DP_07*), off-site customer (*DP_08*), and user-centered focus (*DP_09*).

Some of the most significant advantages of agile methodologies (Abrahamsson et al., 2004; Conboy 2009; Vidgen and Wang 2009) are shown in Table 6.

**Table 6: Advantages of agile methodologies.**

| |
|---|
| *AD_01*. Face to face communication and continuous inputs from customer representative leaves no space for guesswork. |
| *AD_02*. Team is focused on producing a deliverable product to satisfy customer requirements. |
| *AD_03*. An adaptive team which is able to respond to the changing requirements. |
| *AD_04*. Faster speed in responding to market. |
| *AD_05*. A reduced budget. |
| *AD_06*. Less defects in the final product—the end result is the high quality software in least possible time duration and satisfied customer. |
| *AD_07*. Fewer "surprises" and scope changes. |

On the other hand, agile methodologies also have drawbacks (Cockburn 2001). Some of the most significant disadvantages are shown in Table 7.

**Table 7: Drawbacks in agile methodologies.**

| |
|---|
| *DS_01.* Because agile methods are not process-oriented and require quick response to change, a lack of documentation is often a primary characteristic. |
| *DS_02.* There is a lack of emphasis on necessary designing and documentation. |
| *DS_03.* In the case of some software deliverables, especially large ones, it is difficult to assess the effort required at the beginning of the software development life cycle. |
| *DS_04.* When changes come quickly, it is difficult to avoid resistance from stakeholders and complications to end-users training. |
| *DS_05.* The project can easily get taken off track if the customer representative has not clearly defined the final outcome. |
| *DS_06.* Only senior programmers are capable of taking the kind of decisions required during the development process; therefore, there is no place for novice programmers, unless teamed with experienced people. |

Comparing the characteristics of the agile approach with the structured process-oriented approach presented in section 3, it seems that there are some contradictions. For instance, POA promotes documentation while the agile approach does not; in practice, agile approach practitioners consider the lack of documentation as a disadvantage. Another example is that POA promotes a clear definition of the desired system, using structured techniques for requirements specification, which is something like a contract. The agile approach does not promote negotiation, and if the final outcome is not clear, then the project can easily be disbanded, which is a disadvantage. Likewise, there are other contradictory aspects. The challenge is to combine aspects from both approaches, depending on project characteristics.

In the case of mobile applications, taking into account characteristics such those mentioned in section 2, we consider it necessary to combine aspects of the POA and agile methodologies. POA provides robust techniques to define requirements and perform the system analysis, which generate a clear idea of the desired product or system. The agile approach provides simplicity and continuous end-user contact, which helps to achieve the desired product.

In order to try the agile approach, we also consider some lessons learned from practitioners and researchers, such as those mentioned in Abrahamsson et al. (2004). In this case, we consider the following enablers of agility (Vidgen and Wang 2009): *time pacing* (*EN_01*), *self-management with discipline and routine* (*EN_02*), *co-evaluation of business value* (*EN_03*), *sustainable working with rhythm* (*EN_04*), *sharing and team learning* (*EN_05*), and *collective mindfulness* (*EN_06*). On the other hand, it is necessary to consider inhibitors such as the following (Vidgen and Wang 2009): *event pacing* (*IN_01*), *centralized management* (*IN_02*), and *lack of resources allocated to exploration* (*IN_03*).

In our current work we are focused on overcoming the disadvantages presented by the agile approach and the inhibitors. These are some of the motivations for trying to combine a process-oriented approach with agility.

## 5. Case Study: A Mobile Application for Remote Monitoring

In this section we discuss a case study consisting of a security system using video cameras (four cameras) controlled remotely. This system was called *iSysCam*, which is operated by iOS devices. We developed this application for both iPhone and iPad devices. We implemented this system using Objective-C, in the Xcode 3.2.3 framework, which integrates Interface Builder, a graphic tool to create user interfaces.

This case study represents collaboration between the academy and industry, in this case Universidad Autónoma de Baja California (UABC) and GPPI Telecomunicaciones.

People from UABC and GPPI participated in this project. We had the following roles: project manager, project supervisor, technical advisor, team leader, business analyst, UI designer, architect, programmer, tester, and documenter. Table 8 shows the amount of people playing each role.

**Table 8: Roles and people participating in the project.**

| *Role* | *UABC* | *GPPI* |
|---|---|---|
| Project manager | 1 | 1 |
| Project supervisor | 1 | 1 |
| Technical advisor | 1 | |
| Business analyst | 2 | 1 |
| UI Designer | 1 | |
| Architect | 1 | |
| Programmer | 3 | |
| Documenter | 1 | |
| Team leader | 1 | 1 |

We planned seven months for the project. In this case, the project not only involves development of a commercial software application, but also research and academic issues, taking time prearranged for the project. Additionally, this includes administrative activities and infrastructure gathering, which also take time. As a result, time available for software development activities was reduced.

In this case, we will explain how we implemented the phases of the structured process-oriented approach presented in section 3. Later we will explain some aspects of the agile implementation.

### Characteristics of the Application

The application implemented can be considered as a small system; it includes 22 use cases, which were extracted from the functional requirements. Table 9 shows some of the use cases implemented. *CU_01:* system log in; *CU_02:* password recovery; *CU_04:* visualize streaming of all cameras; *CU_05:* visualize streaming of one camera; *CU_07:* rotation of a camera.

For these functionalities we identified two end-users: a client (guard) and an administrator. The client is the main user of the system at the level of streaming consulting. The administrator is a kind of user with permissions to perform operations, such as manipulating the cameras and adjusting images.

Since new mobile devices combine GUI and NUI aspects, for the user interface we identified both elements. Also, the dialogue elements were identified for each use case. Table 9 shows these elements.

**Table 9: Use cases for *iSysCam*—identification of GUI and NUI elements.**

| *Use case* | *GUI elements* | | *NUI elements* |
|---|---|---|---|
| | *Graphic objects* | *Dialogue elements* | |
| **CU_01:** | | | |
| Basic flow | Window 1: TextField, Button<br>Window 2: Message | Message (Success) | Tap, swipe |
| Alternative flow | Window 3: Message, Button | Message (Error) | Tap |
| **CU_02:** | | | |
| Basic flow | Window 4: TextField, Button | Message (Success) | Tap, swipe |
| Alternative flow 1 | Window 5: Label, Button | Message (Error) | |
| **CU_04:** | | | |
| Basic flow | Window 6: UITable<br>Window 7: UITable<br>Window 8: UIWebView, Button | | Tap, swipe<br>Tap, swipe<br>Tap |
| **CU_05:** | | | |
| Basic flow | Window 9: UIWebView, Button | | Tap |
| Alternative flow 1 | Window 12: UIWebView | | Swipe |
| **CU_07:** | | | |
| Basic flow | Window 12: UIWebView | | Pinch out |
| Alternative flow | Window 11: UIWebView, Button | | Tap |

In order to define the user's profile, we considered aspects such as user experience, physical, and cognitive skills. To define this profile we applied a survey including 54 potential users. Specific details of the user's profile and activities and work products can be consulted in Juárez-Ramírez et al. (2011), which describes in depth the requirements and analysis phases, emphasizing the user and tasks analysis.

### The Implemented Phases

In this case we have implemented most of the phases discussed in section 2, as we can see in Tables 10–12. Table 10 shows that the first two phases (project selection and planning and requirements management) were completely implemented. Only the result of the requirements validation was not explicitly documented.

Almost all of the activities from the analysis phase were implemented (see Table 11). In some cases, the results of the activities were not documented in a formal way (using a template or a standard). That is, they were performed but not documented in a requirements specification format or a template for the use cases description. In the case of activities not performed, this is because they were not considered in the project plan. It is important to mention that, for us, this project had opened the possibility to integrate practices from UE and HCI to the software development process, applied to the mobile applications. Based on this, we are considering forms to document new activities such as *Ph03.A6* and the others that were not explicitly documented in this project.

**Table 10: Project selection and planning and requirements management phases.**

| *Activity* | *Completion* | *Observations* |
|---|---|---|
| *Ph01.A1* | √ | Documented |
| *Ph01.A2* | √ | Documented |
| *Ph01.A3* | √ | Documented |
| *Ph01.A4* | √ | Not explicitly documented |
| *Ph02. A1* | √ | Documented |
| *Ph02. A2* | √ | Documented |
| *Ph02. A3* | √ | Documented |
| *Ph02.A4* | √ | Less documented |

The activities and sub-activities from the design phase were completely performed, as we can see in Table 12. In the case of *Ph04.A2.SA1* and *Ph04.A2.SA3* there are partial releases of documentation. Again, like

**Table 11: Analysis phase.**

| *Activity* | *Sub-activity* | *Completion* | *Observations* |
|---|---|---|---|
| *Ph03.A1* | | √ | |
| | *Ph03.A1.SA1* | √ | Documented |
| | *Ph03.A1.SA2* | √ | Documented |
| | *Ph03.A1.SA3* | √ | Documented |
| | *Ph03.A1.SA4* | √ | Not explicitly documented |
| *Ph03.A2* | | √ | |
| | *Ph03.A2.SA1* | √ | Partially documented |
| | *Ph03.A2.SA2* | √ | Not explicitly documented |
| | *Ph03.A2.SA3* | √ | Not explicitly documented |
| | *Ph03.A2.SA4* | √ | Documented |
| | *Ph03.A2.SA5* | √ | Partially documented |
| *Ph03.A3* | | √ | Documented |
| *Ph03.A4* | | √ | Documented |
| *Ph03.A5* | | √ | Not explicitly documented |
| *Ph03.A6* | | X | (Not considered before de project) |
| *Ph03.A7* | | √ | Documented |

in the analysis phase, we didn't use defined templates of standards for documentation of some activities and sub-activities.

In the case of the implementation phase, almost all the activities were performed. Yet again, specific activities from UE and HCI, such as *Ph05.A7*, were not considered explicitly at the beginning of the project. However, in this case, the application was installed in the mobile devices and tested by a period of time. This evaluation collected feedback in understanding the actual behavior toward system use.

## 6. Implementing the Agile Perspective

Before describing how the agile approach was implemented, it is important to comment on some aspects that enabled agility:

i) The nature of the system implemented: mobile application, small size.

ii) The period of time to implement the system: reduced to three months.

iii) The amount of individuals in the development team: not a big team.

iv) The interaction with the business owners and the communication plan defined.

**Table 12: Design and implementation phases.**

| *Activity* | *Sub-activity* | *Completion* | *Observations* |
|---|---|---|---|
| *Ph04.A1* | | √ | Not explicitly documented |
| *Ph04.A2* | | √ | |
| | *Ph04.A2.SA1* | √ | Not explicitly documented |
| | *Ph04.A2.SA2* | √ | Documented |
| | *Ph04.A2.SA3* | √ | Not explicitly documented |
| | *Ph04.A2.SA4* | √ | Documented |
| *Ph04.A3* | | √ | Documented |
| *Ph04.A4* | | √ | Not explicitly documented |
| *Ph05.A1* | | √ | Documented |
| *Ph05.A2* | | √ | Documented |
| *Ph05.A3* | | √ | Documented |
| *Ph05.A4* | | √ | Partially documented |
| *Ph05.A5* | | √ | Documented |
| *Ph05.A6* | | √ | Partially documented |
| *Ph05.A7* | | X | (Not considered before de project) |

We tried each process phase as a sprint, with specific deliverables. Inside each phase we also have sub-sprints. We had continuous meetings to reach consensus on implementation topics, to review deliverables, and to plan specific activities.

In this section we show what phases and activities were implemented applying agile principles and practices. First, we have to mention that we respect the value statements of the agile manifesto, however, we tried to make a balance between the two extremes. For instance, referring to *VS_01*, we used a well-defined process for development, but we also paid special attention to individuals of the development team (as a key element to the project's success) and interactions with business owners and end-users. Additionally, referring to *VS_02*, we focused on artifacts and working deliverables, but we also implemented a comprehensive documentation. Referring to *VS_03*, the requirements specification, performed at the beginning of the project, actually served as a guide to achieve the desired system. Also, referring to *VS_04*, we defined a project plan, which allowed us to obtain partial deliverables and the final product; but in this case, the few changes presented were well supported and implemented. This way of implementation is supported by the recommendations exposed in Boehm and Turner (2003), which suggests making a balance between agility and discipline.

In the next paragraphs we indicate for each phase which aspects were applied in general. Also, for some activities we indicate specific agile aspects applied.

**Project selection and planning:** *DP_01, EN_03*

*Ph01.A1: DP_01, IA_03.* Task identification was done in collaboration between business owners and development team.

*Ph01.A2: IA_02, IA_03.* Deliverables identification was done in collaboration between business owners and development team - *IA_03.*

*Ph01.*A3: IA_01, EN_06.

*Ph01.A4: IA_03, EN_03.* Risk analysis was done in collaboration between business owners and development team - *IA_03.*

**Requirements management phase:** *IA_03, AD_01*

*Ph02. A1: IA_02, IA_03, DP_09, AD_01.* The gathering of requirements was performed in collaboration between the business owners and the development team - *IA_03.*

*Ph02.* A2: FC_04, IA_03, DP_09.

*Ph02.* A3: IA_02, AD_02, AD_07.

*Ph02.A4: IA_03, DP_09, AD_06.* The requirements validation was performed in collaboration between business owners and development team, which allowed ensuring the correct product - *IA_03.*

**Analysis phase**: *IA_02, IA_03, DP_09, AD_01, AD_07.*

*Ph03.*A1: IA_02, AD_01, AD_07.

*Ph03.A1.SA1: IA_03, AD_04.*

*Ph03.A1.SA2: IA_03, AD_04*

*Ph03.A1.SA3: IA_03, AD_04, DS_05.*

*Ph03.A1.SA4: DP_09.*

*Ph03.*A2: IA_02, DP_09.

*Ph03.A2.SA1: DP_09.*

*Ph03.A2.SA2: DP_09.*

*Ph03.A2.SA3: DP_09.*

*Ph03.A2.SA4: DP_09.*

*Ph03.A2.SA5: DP_09.*

*Ph03.A3: IA_03, DP_09.* The user analysis was performed in constant collaboration with a business representative - *IA_03.* The business representative provided end-users information.

*Ph03.A4*: IA_03, DP_09.

*Ph03.A5: IA_03.* The user acceptance test was performed in constant collaboration with a business representative. This provided a kind of end-user perspective for acceptance. This was performed using prototypes.

*Ph03.A7: FC_01, IA_03, DP_09, AD_01*. The alternative selection was performed in collaboration with a business representative. This was performed using prototypes.

**Design phase**: *AD_01, AD_02, AD_03, AD_06, AD_07*

*Ph04.A2*: *FC_01, FC_04, IA_02, DP_09*

*Ph04.A2.SA1: FC_01, IA_03, DP_09*. Visualization design was performed in constant collaboration with a business representative - *IA_03*.

*Ph04.A2.SA2: FC_01, IA_03, DP_09*. Media design was performed in constant collaboration with a business representative - *IA_03*.

*Ph04.A2.SA3: FC_01, IA_03, DP_09*. Dialogue design was performed in constant collaboration with a business representative - *IA_03*.

*Ph04.A2.SA4: FC_01, IA_03, DP_09*. Presentation design was performed in constant collaboration with a business representative - *IA_03*.

*Ph04.A3*: FC_01, DP_02.

*Ph04.A4*: *FC_01, DP_02*.

**Implementation phase**: *DP_07, DP_09, AD_01, AD_02, AD_03, AD_06, AD_07*

*Ph05.A1: FC_01, IA_02, IA_03, AD_01*. Prototyping was performed in constant collaboration with a business representative - *IA_03*.

*Ph05.A2*: *FC_02, DP_03, DP_04, DP_05*.

*Ph05.A3*: *DP_04, DP_09, AD_06*.

*Ph05.A5*: *DP_04, DP_09*.

In the case of the disadvantages for the agile approach, we overcame them as indicated below; however, we first need to emphasize that due to the fact that the mobile application implemented is a small system, the documentation was easy to implement and manage.

*DS_01*: We used a process-oriented approach as basis for the software development. This allowed us to have a clear set of activities to perform. Also, deliverables were well identified (artifacts for each phase) and implemented. We have documented almost all the phases and activities. There were few changes in the system requirements, so it was easy to manage them.

*DS_02*: As mentioned before, we implemented documentation for almost all the phases and artifacts. We used templates and recommendations to document each artifact. Also, we paid sufficient attention to the design phase, which was also documented.

*DS_03*: Due to the small size of the system, deliverables were well identified at the beginning of the project. They were dimensioned and assigned to the team elements, considering the effort required to implement these artifacts and the capacities of developers.

*DS_06*: Still when we had novice programmers on the development team, we combined efforts with experienced people (e.g., team leader and technical advisor) and we made decisions for various phases of the development process, especially for design and programming. This helped us avoid errors and unrecoverable aspects in design and programming.

## 7. Lesson Learned Using a Combination of the Process-oriented Approach and the Agile Implementation

It is important to remark that, by using a project plan, the following enablers took place: *EN_02, EN_03, EN_04, EN_05* and *EN_06*. In the case of "self-management with discipline and routine," team members followed the guide of activities assigned, keeping in mind the work products being generated. We equipped them with resources allocated to exploration, and they reached for information and knowledge about the required implementation topics. On the other hand, in the case of "sustainable working with rhythm," team members always worked on pace, generating the artifacts specified and required for each partial release. Moreover, one of the most valuable experiences concerns "sharing and team learning"; in this case, due to the constant work meetings, team members shared their ideas, information and knowledge, supporting the development work on each stage.

Due to the small size of the mobile application implemented, it was possible to manage partial releases of documentation for some activities and sub-activities. However, there are some aspects that can be improved in this context. It is necessary to use a configuration technique to manage the documentation in order to organize partial releases so they can be localized and consulted during the development cycle. Also, it is necessary to integrate partial releases of documentation for some activities into the technical documentation of the project.

## 8. Conclusions and Future Work

In this chapter we have presented a proposal for a structured software development process. This proposal contains a software engineering focus, emphasizing how some practices can cover the main characteristics and nature of mobile applications. Also, considering that usability is one of the most important quality attributes for mobile applications, we have integrated aspects from the usability engineering and human-computer interaction. Additionally, we presented a set of important aspects of the agile approach, emphasizing useful principles and practices for developing mobile applications.

Still, when we didn't use a specific agile methodology, such as SCRUMP, in the case study we experimented with a combination of the process-oriented approach and agile principles and practices. The project was a success.

Currently, we are trying to integrate an agile methodology, defining specific stages to be considered as sprints, indicating specific deliverables. Furthermore, we are defining a minimum level of documentation to be generated, to have the possibility to implement rapid changes in the system requirements, without affecting the initial documentation. Additionally, we are selecting a set of tools for project management and configuration management for documentation.

### Acknowledgements

This work was done under the support of the Mexican Government through the CONACYT and the Secretary of Economy, granting proposal No. 140022 in the program "Convocatoria de Proyectos de Investigación, de Desarrollo o de Innovación Tecnológica 2010." We are grateful for the hard work performed by all undergraduate students who participated in this project. Also, thanks to the graduate students who contributed to the drafting of this chapter. Likewise, thanks to GPPI employees who participated in this project.

## REFERENCES

Abrahamsson, P. 2005. Keynote: Mobile software development—the business opportunity of today, pp. 20–23. In: Benediktsson et al. [eds.] The Proceedings of the 3rd SWDC-REK International Conference on Software Development, Reykjavik, Iceland, May 27-June 1, 2005. University of Iceland.

Abrahamsson, P. 2007. Agile software development of mobile information systems. pp. 1–4. In: J. Krogstie, A. Opdahl and G. Sindre. [eds.]. Advanced information systems engineering. Lecture Notes in Computer Science 4495/2007. Springer-Berlag, Berlin Heilderberg.

Abrahamsson, P., A. Hanhineva, H. Hulkko, T. Ihme, J. Jäälinoja, M. Korkala, J. Koskela, P. Kyllönen and O. Salo. 2004. Mobile-D: An agile approach for mobile application development, 174–175. In: J.M. Vlissides and D.C. Schmidt [eds.] The Proceedings of the 19th Annual ACM Conference on Object-Oriented Programming, Systems, Languages, and Applications (OOPSLA'04), October 24–28, 2004, Vancouver, British Columbia, Canada. Vancouver. ACM.

Abrahamsson, P., A. Hanhineva and J. Jäälinoja. 2005. Improving business agility through technical solutions: a case study on test-driven development in mobile software development. pp. 227–243. In: R.L. Baskerville, L. Mathiassen, J. Pries-Heje and J.I. DeGross [eds.]. Business Agility and Information Technology Diffusion. IFIP International Federation for Information Processing 180/2005. Springer Boston.

Abrahamsson, P., O. Salo, J. Ronkainen and J. Warsta. 2002. Agile software development methods. *Vtt Publications.* **478:**167–168.

Agile Alliance. 2001. Agile Software Development Manifesto. Retrieved from Manifesto for Agile Software Development: http://agilemanifesto.org/

Allen, S. 2010. The future of mobile software engineering. Retrieved from: http://www.ultrasaurus.com/sarahblog/2010/10/the-future-of-mobile-software-engineering/

Aydin, M. 2004. An agile information systems development method in use. *Turkish Journal of Electrical Engineering Computer Sciences.* **12:**127–138.

Blevis, E. 2007. Sustainable interaction design: Invention & disposal, renewal & reuse, 503–512. In: M.B. Rosson and D.J. Gilmore [eds.] The Proceedings of the ACM Conference on Human Factors in Computing Systems. San Jose, California, USA, 2007. ACM Press, New York, NY, USA.

Boehm, B.W., J.R. Brown, H. Kaspar, M. Lipow, G. McLeod and M. Merritt. 1978. *Characteristics of Software Quality.* TRW series of software technology. Amsterdam: North Holland.

Boehm, B. and R. Turner. 2003. Balancing Agility and Discipline: A Guide for the Perplexed. Addison-Wesley, Boston, MA. USA.

Cockburn, A. 2001. Agile Software Development. 1st edition. Addison-Wesley Professional, Reading, Massachusetts. USA.

Conboy, K. 2009. Agility from first principles: reconstructing the concept of agility in information systems development. Information Systems Research. **20:**329–354.

Delic, N. and A. Vukasinovic. 2006. Mobile payment solution—symbiosis between banks, application service providers and mobile network operators, pp. 346–350. In: The Proceedings of the Third International Conference on Information Technology: New Generations. IEEE Computer Society, Washington, DC. USA.

Ferre, X. 2003. Integration of Usability Techniques into the Software Development Process, pp. 28–35. The Proceedings of the International Conference on Software Engineering (ICSE'03), IFIP, Portland, Oregon, May 03–10, 2003. IEEE Computer Society, Washington, DC. USA.

Georgiadis, C.K. and E. Stiakakis. 2009. Key issues for the quality assessment of mobile commerce services, pp. 148–153. In: G. Eleftherakis, C.K. Georgiadis, P. Ketikidis and D. Stamatis [eds.] The Proceedings of the Fourth Balkan Conference in Informatics (BCI'09), 17–19 Sep. 2009, Thessaloniki, Greece.

Gerson Lerhman Group (GLG). 2010. Software engineering for mobile applications and systems. Retrieved from: http://www.glgroup.com/NewsWatchPrefs/Print.aspx?pid=51393&cb=1.

Grady, R.B. 1992. Practical software metrics for project management and process improvement. Prentice Hall, Englewood Cliffs, New Jersey. USA.

Gruhn, V. 2007. Mobile software engineering. University of Leipzig. Retrieved from: http://www.iasted.org/conferences/2007/innsbruck/se/pdfs/GruhnSE2007.pdf.

Harper, R., T. Rodden, Y. Rogers and S. Sellen. 2008. Being Human: Human-Computer Interaction in the year 2020. Technical Report, Microsoft Research Cambridge (2008).

International Standard Organization (ISO). 2001. ISO/IEC 9126-1 Information Technology: Product Quality—Part 1: Quality model. June 2001.

Jacobson, I., G. Booch and J. Rumbaugh. 1999. The Unified Software Development Process, Addison Wesley Longman Publishing Co. Inc., Boston, MA. USA.

Juárez-Ramírez, R., G. Licea, I. Barriba, V. Izquierdo and A. Ángeles. 2011. Engineering the Development Process for User Interfaces: Toward improving Usability of Mobile Applications. pp. 65–79. In: H. Cherifi, J.M. Zain, E. El-Qawasmeh [eds.]. Communications in Computer and Information Science 167, Part II. Springer-Berlag, Berlin Heilderberg.

Knouf, N.A. 2009. HCI for the real world, pp. 2255–2564. In: D.R. Jr. Olsen, R.B. Arthur, K. Hinckley, M.R. Morris, S.E. Hudson and S. Greenberg [eds.] The Proceedings of the 27th international conference on human factors in computing systems CHI 09—extended abstracts, Boston, Massachusetts, USA, 2009. ACM Press, New York, NY. USA.

Kim, S-K. and D. Carrington. 2002. Integrating Use-Case Analysis and Task Analysis for Interactive Systems, 12–21. In: P. Stooper and P. Muenchaisri [eds.] The Proceedings of the Asia Pacific Software Engineering Conference, APSEC'02, 4–6 December 2002, Gold Coast, Queensland, Australia. EEE Computer Society Washington, DC. USA.

Kushniruk, A.W. and V.L. Patel. 2004. Cognitive and usability engineering methods for the evaluation of clinical information systems. *Journal of Biomedical Informatics*. **37**:56–76.

Kruchten, P. 2000. The Rational Unified Process An Introduction—Second Edition, Addison Wesley Longman Publishing Co. Inc., Boston, MA. USA.

Lal, D., D.C. Pitt and A. Beloucif. 2001. Restructuring in European telecommunications: modeling the evolving market. *European Business Review*. **13**:152–156.

Lethbridge, T.C. 2006. Integrating HCI and Usability into Software Engineering: The Imperative and the Resistance—CapCHI Talk—Sept. 27, 2006. Retrieved from: http://www.capchi.org/documents/capchi_lethbridge_060927.pdf.

McCall, J.A., P. K. Richards and G. F. Walters. 1977. Factors in Software Quality. *National Technology Information Service*. 1:2–3.

Microsoft Research. 2008. Being human: Human-computer interaction in the year 2020. R. Harper, T. Rodden, I. Rogers and A. Sellen [eds.]. Microsoft Research Ltd., Cambridge, England. Retrieved from: http://research.microsoft.com/en-us/um/cambridge/projects/hci2020/downloads/BeingHuman_A3.pdf.

Ming-Chun, C. and Y. Shyan-Ming. 2005. An adaptive mobile application development framework. pp. 765–774. In: L.T. Yang, M. Amamiya, Z. Liu, M. Guo and F.J. Rammig. [eds.]. Embedded and Ubiquitous Computing—EUC 2005, LNCS 3824. Springer Berlin Heilderberg, New York. USA.

Molina, J.P., P. González, M.D. Lozano, F. Montero and V. López-Jaquero. 2003. Bridging the gap: Developing 2D and 3D user interfaces with the IDEAS methodology. pp. 379–390. In: Interactive Systems. Design, Specification, and Verification, Lecture Notes in Computer Science, 2003, Volume 2844/2003. Springer, Berlin/Heidelberg.

Nielsen, J. 1992. The Usability Engineering Life Cycle. *Computer*. **25:**12–22.

Nilsson, E.G. 2005. Design guidelines for mobile applications. SINTEF Report STF90 A06003, ISBN 82-14-03820-0.

Paspallis, N. and G.A. Papadopoulos. 2006. An approach for developing adaptive, mobile applications with separation of concerns, pp. 299–306. In: The Proceedings of the 30th Annual International Computer Software and Applications Conference (COMPSAC 2006), I, 17–21 September 2006, Chicago, Illinois, USA. IEEE Computer Society.

Petrova, K. 2007. Mobile learning as a mobile business application. *International Journal of Innovation and Learning*. **4:**1–13.

Pimenta, M.S. and M.F. Barthet. 1996. Context Modeling for an Usability Oriented Approach to Interactive Systems Requirements Engineering, 315–321. Proceedings of the IEEE International Symposium and Workshop on Engineering of Computer Based Systems (ECBS 96). Friedrichshafen, Germany, 1996. IEEE Computer Society Washington, DC, USA.

Seffah, A. and E. Metzker. 2004. The obstacles and myths of usability and software engineering. *Communications of the ACM—The Blogosphere*. **47:**71–76.

Seffah, A., M. Desmarais and E. Metzker. 2005. HCI, Usability and Software Engineering Integration: Present and Future. pp. 37–57. In: A. Seffah, J. Gulliksen, M.C. Desmarais [eds.]. Human-Centered Software Engineering—Integrating Usability in the Software Development Lifecycle, Human-Computer Interaction Series 8, II. Springer Netherlands.

Serhani, M.A., A. Benharref, R. Dssouli and R. Mizouni. 2010. Toward an efficient framework for designing, developing, and using secure mobile applications. *International Journal of Human and Social Sciences*. **5:**272–278.

Spataru, A.C. 2010. Agile development methods for mobile applications. M.S. Thesis, University of Edinburgh.

Spriestersbach, A. and T. Springer. 2004. Quality attributes in mobile web application development. pp. 120–130. In: F. Bomarius and H. Iida [eds.] Product Focused Software Process Improvement, LNCS 3009: The Proceedings of the 5th International Conference on Product Focused Software Process Improvement, PROFES 2004, Kausai Science City, Japan, April 5–8, 2004. Springer-Berlag, Heilderberg, Germany.

The 2020 Science Group. 2006. Toward 2020 Science. Microsoft Corporation. Retrieved from: http://research.microsoft.com/en-us/um/cambridge/projects/towards2020science/downloads/t2020s_report.pdf

Vidgen, R. and X. Wang. 2009. Coevolving systems and the organization of agile software development. *Information Systems Research*. **20:**355–376.

Walenstein, A. 2003. Finding Boundary Objects in SE and HCI: An Approach Through Engineering-oriented Design Theories, pp. 92–99. In: R. Kazman, L. Bass and J. Bosch [eds.] The Proceedings of the International Conference on Software Engineering (ICSE'03), Portland, Oregon, May 03–10, 2003. IFIP 2003.

Wasserman, A.I. 2010. Software Engineering Issues for Mobile Application Development, pp. 397–400. In: R. Gruia-Catalin and K.J. Sullivan [eds.] The Proceedings of the FSE/SDP Workshop on the Future of Software Engineering Research, FoSER 2010, November 7–8, at the 18th ACM SIGSOFT International Symposium on Foundations of Software Engineering, 2010, Santa Fe, NM, USA, November 7–11, 2010. ACM New York, NY.

Wirth, N. 2008. A brief history of software engineering. *IEEE Annals of History of Computers*. **30**:32–39.

Zhang, P., J. Carey, D. Te'eni and M. Tremaine. 2005. Integrating human-computer interaction development into SDLC: A Methodology. *Communications of the AIS*. **15**:512–543.

3

# Secured Professional Use of Mobile ICT Devices

*Jeanne Schreurs, Ahmad Al-Huneidi and Tom Princen*

**ABSTRACT**

Nowadays employees are often on the road wasting time when travelling and when waiting at airports or other transportation facilities. Companies and employees try to change this lost time into productive time. Laptops, smart phones and other mobile appliances can be used to do professional work at a distance. But using mobile appliances to access the company network to retrieve information can be risky. Authentication and privacy are two crucial points. Browser attacks and mail client attacks are the biggest threats to mobile applications.

Mobile security management can be structured as a 3-steps procedure; first, securing the user mobile devices and the local data; second, securing user communication via interne, or the transition of the data and; third, securing user access to systems on the company server.

Mobile learning is one of the possible options on distance professional applications, where learner authentication is a key issue. Also, mobile learning applications require the integration of security design in the development process.

The effective professional use of mobile applications can be guaranteed if an information security management policy is formulated and implemented in the organization.

## 1. Professional Use of Mobile Devices and the Security Problem

Companies and institutions are operating on a global scale and their staff members are using mobile devices and experiencing security problems. Each organization needs to manage security and implement more stringent security measures.

## 1.1. Global Business and Mobility of Professional Staff

As long as computers were wired and the communication was limited to internal communication, computer security meant nothing more than closing the door at the office so that nobody could access the device or steal it. But, in the early eighties, things changed forever. PC's entered into the professional computing arena. Researcher Fred Cohen introduced a new term: "computer virus". He developed some software which took care of hazardous software and got rid of it Cohen (1984). Ever since then, a lot of things changed.

Nowadays information is quickly becoming one of the most crucial assets in business and it is stored and accessed in business networks that are increasingly vulnerable to attacks. However, these systems are increasingly becoming more open as world commerce expands Doherty and Fulford (2005).

Today's employees are often on the road. This results in a loss of time when travelling and when waiting at transportation centers. Companies and employees try to change this lost time into productive time. Laptops, smart phones and other mobile appliances can be used to do professional work on distance. Mobile learning is a distance professional application.

## 1.2. Security Management and Implementation of Security Measures Needed in the Organization

A lack of security management results in business losses. Without its information system, an organization cannot function effectively. Just think about the image of customers switching to another supplier, losing income, etc. Graux and Dumortier (2009). The longer these systems are down, the worse things get for the business. So one could conclude it is important to protect them for any threat that might materialize. A company has to share data with its employees and because these employees can be virtually everywhere they must have access at a distance. This brings the question "Where is safe"?

As an example, we can address is the problem of an employee wanting to access a mobile learning platform of his company datacenter from his hotel room or on the airport.

Several threats linked with using mobile computing can attack the business, but the organization is not always aware of the risks. On a mobile platform, a lot of data can be published. However, it seems now that companies just cannot publish their data because of the risk of security threats. The question is now how this problem should be tackled. How can companies publish crucial insider data without having to worry about this data getting stolen and sold to the competitors?

Users can face several substantial problems when using smart phones to access a company network to retrieve information. According to Tsiantis et al. (2007) authentication and privacy are two crucial points.

A company has to share data with its employees and because those employees can be virtually everywhere, they must have access at a distance. This begs the question, "What is safe"? As an example, we can address the problem of an employee wanting to access a mobile learning platform, which is located within the company datacenter, from his hotel room or in at an airport with a mobile phone.

What are the possible threats? There are device oriented threats, application threats and the threats related to transferring data via wired or wireless networks. According to the NIST (2008) the most important threats for mobile devices are theft, loss, unauthorized access, malware, spam, electronic eavesdropping, electronic tracking, cloning and server resident data. But there are also many threats to mobile applications. Cross (2007) stated that the biggest threats are browser attacks and mail client attacks. The transmission problem is the biggest with smart phones. There is a large variety of networks to secure. Smart phones can be part of PAN, LAN or a WAN. All these situations have to be considered. And the goal is to find a universal solution to all these possible ways of transferring data so that administration and management is relatively simple.

## 2. Security Management of Mobile Business: A 3-step Procedure

### 2.1. Security in a Mobile Professional Activity

To secure the mobile computing applications and the infrastructure, the company needs to follow a 3-step procedure. First, it must secure the mobile device; second, it must secure the connection between it and the company network and third, it needs to secure the company network and the company servers.

### 2.2. Securing the User Mobile Devices and the Local Data

It is very important to secure the access to the device. Because it is a small device, it can get lost easily. For example, it can be forgotten in the plane, train, etc. The access to the device can be protected by using a PIN code or an access password. A rule of the security policy can be that after a number of wrong attempts, the system becomes blocked and the user has to wait some time before trying to access the device again, Research In Motion (2010). The application data and the static data of the PDA need to be protected.

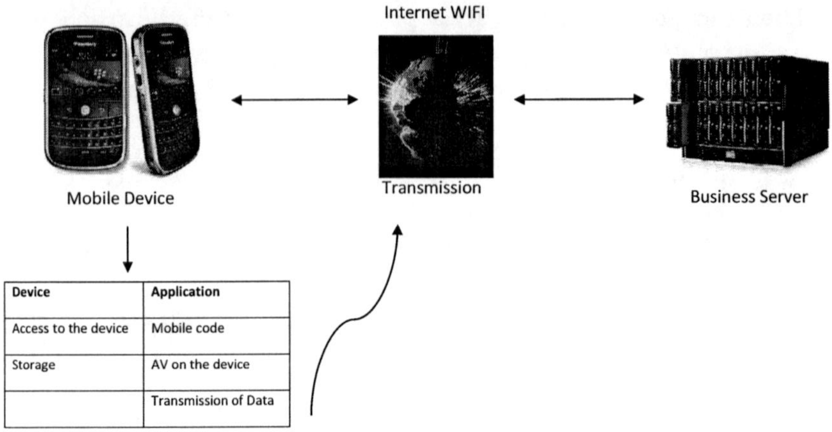


| Device | Application |
|---|---|
| Access to the device | Mobile code |
| Storage | AV on the device |
| | Transmission of Data |

**Figure 1.** The three stages security problem.

*(Color image of this figure appears in the color plate section at the end of the book.)*

When using a mobile application like mobile learning, there is always some local data. This data has to be downloaded from the server and it is being used locally to make the system work. For example, a spreadsheet can be downloaded and stored locally. When the device is stolen, this data is still there and it will be lost. This kind of problem can be easily solved by implementing some security measures. The vulnerable kinds of memory on mobile devices are the data on the object's permanent memory and on the removable memory.

These can be protected by implementing encryption technology to local memory and to removable memory cards.

### 2.3. Securing Transition of Data and Access to Systems on the Company Server

Mobile wireless access to organizations' information is very critical. Due to the sensitivity of the transmitted documents and emails across public wireless networks, the information should be protected with a high level of security.

When an employee accesses the company network, he always leaves traces. These traces can be used by unauthorized people. For example, an employee downloads the latest confidential product specifications that he has just received via e-mail. In case he loses his device and if access to it is not blocked, this confidential data will become public and can be used by a competitor.

RIM, the manufacturer of the Blackberry handheld, has so far done a good job of building security into its Blackberry handhelds and the

Blackberry Enterprise Server (BES). It is much better than the competing mobile wireless email solutions such as Pocket PC, McNurlin et al. (2009).

A variety of data transmission techniques exist and each of them has its own strengths and weaknesses. Each transmission technique offers a solution for a different type of attack. It delivers its specific built-in security features and solutions.

### *Virtual Private Networks (VPNs)*

The Virtual Private Network (VPNs) data transmission solution uses protocols such as Internet Protocol Security (IPSec), which is the most widely used in VPNs.

By using VPNs, organizations are challenged to secure the remote access point. If a hacker manages to compromise a home user's machine, an open door is provided into the corporate network. In this case, the organizations should assure the security of a machine that is not under their direct control. Nevertheless, security solutions providers have reacted to this problem with a flurry of devices that can be remotely managed and have security features built-in. According to Frost and Sullivan (2002), the remote access capacity of VPNs is the most attractive incentive for corporate deployment. When wireless transmissions are transported through a VPN and it is guaranteed to be more secure than with a WEP only.

### *Wireless Transport Layer Security*

The standard Wireless Transport Layer Security (WTLS) is derived from transport layer security and it enables security on mobile appliances running WAP. WTLS has been classified into three classes according to the WAP 1.1 specification, Class 1, Class 2 and Class 3. Class 3 encompasses the capabilities of the other two classes and also authenticates clients. Mutual authentication is also possible between the server and consumer through certificate exchange. The WTLS Class 3 session l only commence after both parties check the certificates. This level of security is not sufficient as the WAP protocol encrypts data at the user level, deciphers it at the operator level and then encrypts it again. The gateway decrypts the data to establish where it should be sent to reach the destination. This could present a hacker with an opportunity to access the message. A perfect solution to this would involve positioning the WAP server behind a trusted network, Lopez (2009).

### *Wireless Public Key Infrastructure (WPKI)*

Public Key Infrastructure (PKI) allows the distribution of encrypted digital certificates as authentication means during a transaction. The certificate authority validating the digital certificate Wireless PKI for the wireless environment is based on WAP standards. The main components of the WAP/WPKI system are as follows:

- Mobile phones containing the WAP identity module (WIM).
- WAP gateway with certificate-based identity authentication.
- Registration authority.
- Back-end PKI infrastructure with access to certification authority infrastructure.

*The WIM* within the handset stores the security keys and certificates and is a tamper-proof hardware module. SWIM and dual chip are the two current techniques applied to WIM. The SWIM method involves the WIM security features being stored on a subscriber identity module (SIM) card. The dual chip approach involves a smartcard been placed temporarily in the phone. This allows segregation between the network subscription and other applications, Lopez (2009).

***User Authentication***

The user authentication market refers to user identity equipment, which encompasses tokens that produce passwords, universal serial bus (USB) keys, smartcards, smartcard readers, server seats and connected software, software tokens and biometrics. According to Frost and Sullivan 2002, the user authentication market is expanding and will continue to do so. A number of issues have restricted user authentication from becoming the main approach for authenticating users onto networks. These include:

- Standards and support issues.
- High equipment price.
- The market failing to consider network security to be a prime issue in general.
- Most of the user authentication market, in fact 80%, is composed of the hardware token market.

Market expansion in this area will be accelerated by the capability of smartcards and USB tokens to store digital certificates.

## 2.4. The Security Management Solution for a Blackberry

Blackberry Security Features become the most effective security solution available for a mobile wireless handheld, NIST (2008). Some of these features use Industry-standard AES or Triple-DES encryption to encrypt all data in transit between the BES gateway and Blackberry handhelds, with unique encryption keys generated for each handheld. This prevents eavesdropping or tampering with data as it travels to and from the Blackberry handheld over public wireless networks. The per-handheld encryption key is obtained using a one-time cradle-based or wireless "activation" communication with the BES gateway, where a systems administrator provides the Blackberry user an activation password over the telephone. A combination of the per-handheld encryption key and a per-handheld Personal Identification Number (PIN) authenticates the handheld to the BES enterprise gateway, ensuring that one Blackberry handheld cannot obtain information destined for another Blackberry.

Moreover, Blackberry users can use a password to lock the handheld while not in use, with password complexity enforcement and a limit on the number of times the password can be entered before the handheld memory is wiped clean. In addition, data stored on the Blackberry handheld can be encrypted using a content protection key derived from the handheld password, which prevents saved emails, attachments, contacts, etc. from being transferred from the handheld to other devices (which do not have the content protection key) and also prevents unauthorized third-party software applications installed on the Blackberry handheld from accessing protected data. Only third party applications digitally signed by RIM can decrypt and access protected handheld data.

On the other hand, system administrators can centrally monitor and control the security of remote Blackberry handhelds from the BES management console, with a fine-grained policy model governing all aspects of the handheld security model, such as enforcing use of the handheld password and content protection (encryption). This means that security is not left to the end user, but can be centrally enforced in accordance with enterprise policy. The BES periodically transmits policy settings to each handheld wirelessly, ensuring that all handhelds are using the latest security policy. System administrators can also respond to a lost or stolen Blackberry by wirelessly sending command messages to the handheld that either wipe clean the device, or set/reset the handheld password and lock the device (in case it is found). As soon as the lost or stolen Blackberry handheld comes into wireless range, the command message takes effect (unless the device is off or the wireless component disabled).

Moreover, system administrators can ensure that all Blackberry handheld communications flow via the BES gateway by disabling direct device-to-device communications (for example SMS messages), again using centralized policy settings. Most Blackberry handheld models have limited direct device-to-device communications capabilities, compared to other mobile wireless devices (such as IrDA ports, USB file drivers) and those they do have, such as Bluetooth, can be deactivated by centralized policy settings. System administrators can also control which third party software applications can be downloaded and installed onto Blackberry handhelds, and create application policies governing which data (for example, saved emails and attachments) can be accessed by these applications.

All these effective security features have made Blackberry very popular for corporate use because they assure sufficient protection of the local data, McNurlin et al. (2009).

## 3. Mobile Learning and Security Problem

A recent and significant change in learning environments is the demand for mobility. An important question arises about the security of mobile learning. According to Cross (2007), two main types of attacks exist, the browser attacks and the e-mail client attack. Because the browser is executing code locally, it can be insecure. This dangerous code (VBScript, Javascript) can be part of html pages. In mobile learning, the user opens html pages during learning and as a result these html attacks can appear in mobile learning.

The level of security will be different for different types of applications, because sometimes data is stored in the local device and in other cases, application users only read the data without downloading it on the local device. Security is also linked with information ownership, determining access and manipulating rights, Tsiantis et al. (2007).

With respect to security of a mobile learning system, we see two main points: authentication and privacy protection, Tsiantis et al. (2007). The user has to prove his identity, but on the other hand the privacy of the learner has to be secured. The mobile learning application has to be secured by an authentication mechanism, as we show in Figure 2.

This authentication model illustrates how authentication is more than simply accessing the system. In fact, it is an integral part of the learning system and of the application itself. This means that security is already built into the development process of a mobile learning solution. According to Aljawarneh (2011), in the "web engineering security methodology", security components in all phases of the SDLC development model are integrated.

Examples of mobile learning systems are the Blackberry mobile learning platform. It is a joint application with Desire2learn, Judy Brown (2009). This system is already implemented in universities and in professional learning practice. Other examples are Apple iPhone and Apple iPod Touch. Apple developed a complete mobile learning application based on ITunes store and functions as podcasts, being audio and video files that are primarily downloaded by the users on their iPods or iPhone.

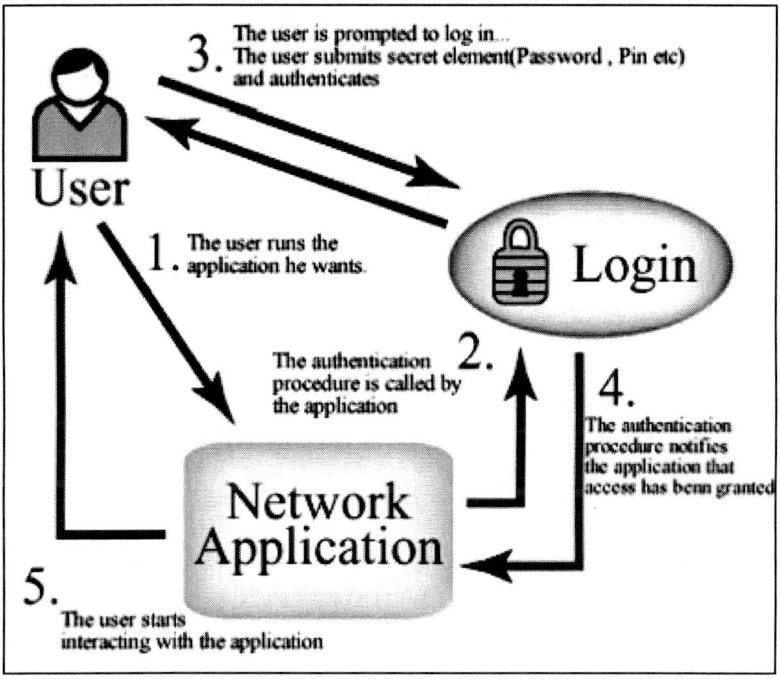


**Figure 2.** The process of authentication, Tsiantis et al. (2007).

## 4. Organizational Policy to Manage Professional Use of Mobile Appliances

### 4.1. The Need for a Security Policy

Companies know that security threats exist and how to prevent and respond to them. The implementation of security controls requires careful planning and inclusion of all employees in the organization. A security policy can be stated as follows: "Policies act as clear statements of management intent and demonstrate that employees should pay attention to information security", Wood (1995). Depending on the company's size, financial resources, and the degree of threat, a security policy has to be defined that finds the right balance between overreacting and the vulnerability of exposing the computer systems to any and every hack. Security management policies are the foundation and the bottom line of information security in an organization. A well-written and implemented policy contains sufficient information on what must be done to protect information and people in the organization. It is about computer usage guidelines for staff in the course of their job duties. Its goal is to improve information availability, integrity and confidentiality, from both inside and outside the organisation.

There are three kinds of security policies which are being used: (i) first, the enterprise information security policy on high management level (ii) second, an issue- specific security policy addressing specific issues and (iii) third, a system-specific security policy, Whitman and Mattord (2009).

To set up a security management policy, the following stepwise approach can be followed:

- Identification of the assets to be secured
- Identification of the threats and the probability of occurrence of it
- Decision on which security technologies to implement
- Monitoring and reviewing the process continuously for improvement

ISO 17799 can be used as a basis to develop such a security policy.

### 4.2. The Security Policy Applied in Practice: Case KBC Bank

KBC is a bank and has a lot of data that is being modified on a very frequently basis. It is impossible to encrypt all data. One can imagine that the real-time ATM transactions of the customers have to be executed immediately. The challenge for the bank is to keep the strategic data in the company and to prevent data leakage. In the organization, data can be lost in a lot of places including print screens, e-mails, USB sticks, mobile hard disks;

The policy of KBC is to protect the perimeter. The perimeter in KBC, having implemented blackberries as professional handhelds for management can be seen presented in the figure 3 below:

For KBC, data is under control if it is used as follows:

- For the right purpose: doing transactions for a customer.
- By the right person: a staff member or employee.
- With the right technology: via a secure desktop or other appliance.

KBC is aware of the risk and is assessing the risks following a stepwise procedure. KBC defined a firewall policy. The main idea is that the employee (the user of the data function as a firewall). It involves the use of a whitelist. Everything on the whitelist is OK, the other things simply cannot be done! Of course employees can take shortcuts, but this is not be tolerated. KBC installed special control teams to address these problems.

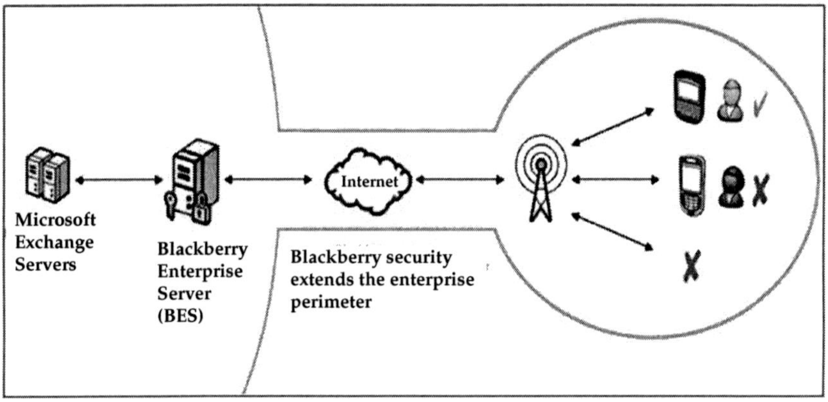


**Figure 3.** Adapted from Perimeter Security, Lambert (2005).

### 4.3. A Security Policy for a Company Implementation of Blackberry Handhelds

The situation of a handheld professional solution is more complex than in case of a desktop, due to the different operating systems; ex: Apple iPhone OS, Windows Mobile, …etc. KBC decided to limit the choice of handhelds to the blackberry, because of the strong security it possesses via its Blackberry Enterprise Server. Based on this company policy, the probability of problems is minimized.

To develop a mobile security policy, the implementation of a server solution is set forward. A development can be based on a large set of facilities made available by Blackberry. The RIM environment is being advised, Lopez (2009). In this architecture, a 3 layer security system is implemented: (i) first, the data on the appliance and the application (ii) second, the protection against malware and hacking and (iii) third, the data transmission.

## 5. Conclusion

The paper is the result of a literature study, of our experiments with PDA systems and is based on communication with experts of KBC Bank regarding technology and company policy. We believe that a company can experience attacks while working at a distance using mobile appliances connected to company servers.

The greatest threats to mobile applications are browser and e-mail attacks. Earlier, we explained the 3-step procedure for securing mobile applications.

The transmission problem is the greatest with smart phones. There is a large variety of networks to secure. Smart phones can be part of PAN, LAN or a WAN. All these situations have to be considered. And our goal was to find a universal solution to all these possible ways of transferring data so that administration and management is relatively simple.

In mobile learning, the main point is the authentication. Also, the security management of m-learning solutions has to be integrated in the learning system.

To guarantee effective mobile security management, a company policy must be developed.

## REFERENCES

Cohen, Frederick. 1984.Computer Viruses—Theory and Experiments, Retrieved October 7, 2011 from: http://all.net/books/virus/index.html

Cross, M. 2007. Web Application Security. United States and Canada: O'Reilly Media, Inc.

Doherty, N.F. and H. Fulford. 2005. Do Information Security Policies Reduce the Incidence of Security Breaches: An Exploratory Analysis. *Information Resources Management Journal*, Vol. 18, No. 2, pp. 21–39.

Frost and Sullivan. 2002. Wireless Security—what is out there? *Network Security*, Volume 2002, Issue 5, 31 May 2002, pp. 5–6.

Graux, H. and J. Dumortier. 2009. Privacywetgeving in de praktijk. Recht & ICT, 1. Brussel: UGA.

Judy, Brown. 2009. Exploring Mobile Learning: Part One of the mLearning Series, Research In Motion & Desire2Learn Incorporated.

Lopez, M. 2009. Succesful Mobile Deployments Require Robust Security. Lopez Research LLC.

Martin, Lambert. 2005. *Blackberry Security Network Security*, Volume 2005, Issue 6, June 2005, pp. 18–20.

McNurlin, C.B., R.H. Sprague and T. Bui. 2009. Information Systems Management in Practice. Upper Sadle River, New Jersey, 07458: Pearson Education Inc.

NIST. 2008. Guidelines on Cell Phone and PDA Security. NIST.

Research in Motion. 2010. Security Features Blackberry Enterprise Server. Retrieved April 7, 2010, from uk.blackberry.com: http://uk.blackberry.com/ataglance/security/features.jsp

Shadi, Aljawarneh. 2011. A webengineering security methodology for e-learning systems. Isra University, Jordan, *Network Security*, Volume 2011, Issue 3, March 2011, pp. 12–15.

Tsiantis, L.E., E. Stergiou and S. Maragariti. 2007. Security Issues in e-learning systems. Patra, Greece: Hellenic Open University.

Whitman, E.M. and J.H. Mattord. 2009. Principles of information security. Canada: Thomson Course Technology.

Wood, C.C. 1995. Writing infosec policies. *Computers & Security*. 1995. **14(8)**:667–74.

# 4

# Intrusion and Detection Systems in Wireless Networks

*Mohamed Elboukhari, Mostafa Azizi and Abdelmalek Azizi*

**ABSTRACT**

The rapidly expanding area of mobile and wireless computing applications in recent years has definitely redefined the concept of network security. Even though wireless computing has opened a new world with its advanced technology, there is no doubt that its popularity is still rising. However, the biggest concern with either wireless or mobile computing applications in security. While an attacker needs physical access to a wired network in order to trigger an attack, a wireless network allows anyone within its range to passively monitor traffic or even start an attack. One of the countermeasures can be the use of Wireless Intrusion Detection Systems WIDSs. There is presently intensive research being carried out in this area. Some researchers have been focused on the development of theories to allow reasoning about detection, the development of wireless intrusion, anomaly detection systems and preventive mechanisms and tolerance functions. This chapter will discuss the major models, theories, and mechanisms developed to protect wireless networks/systems against intrusions, threats, and anomalous behaviors. The objectives of this chapter are to: (1) describe the basics of intrusion detection systems; (2) discuss security problems in wireless environments; (3) present some WIDSs approaches presented by current research activities; (4) study important architectures already developed by researchers; and (5) discuss the aspects of tolerance and prevention.

## Introduction

Recent years have seen an explosive growth in the number of internet working systems due to the evolution of computer systems and the Internet. Such systems must remain not only "up-and-running" but also be "secure"—against any form of unwanted harmful actions such as attack, misuse, and abuse. It is very important that security mechanisms of systems be designed to prevent unauthorized access to system resources and data. Actually, completely preventing breaches of security appears to be unrealistic. However, we can try to detect these intrusion attempts so that actions may be taken to repair the damage. This domain of research is called Intrusion Detection.

An Intrusion Detection System (IDS) is the equivalent of a burglar alarm, one that is configured to monitor information gateways, hostile activities, and known intruders. An IDS is a tool that knows how to parse and interpret network traffic and/or host activities. This data can range from network packet analysis to the contents of log files from firewalls, routers, and servers, local system logs and access calls, network flow data, and more. An IDS often stores a database of known attack signatures and can compare patterns of activity, traffic, or behavior it sees in the data it is monitoring against those signatures to recognize when a close match between a signature and current or recent behavior occurs. By analogy, an IDS does for a network what an antivirus software package does for files that enter a system: it inspects the contents of network traffic to look for and deflect possible attacks, just as an antivirus software package inspects the contents of incoming files, active Web content, e-mail attachments, and so forth to look for virus signatures (patterns that match known malware) or for possible malicious actions (patterns of behavior that are at least suspicious, if not downright unacceptable).

An IDS applied to the wireless networks is known as a Wireless Intrusion Detection System (WIDS). Today, wireless technology can actually be seen almost everywhere. Unfortunately, this technology is always under fire when it comes to security, which is a topic that has now become more imperative and indispensable when dealing with transferring and storing sensitive computer information. Wireless technology is still new, and as it becomes more common among many types of computer networks, hackers or intruders will find more ways to intrude a wireless system. While the number of risks and vulnerabilities continues to rise, security engineers have attempted to slow and hope to eventually halt the many types of wireless network intrusions. A WIDS is a system for detecting such intrusions.

WIDS is a network component that protects networks by detecting wireless attacks, which target wireless networks with specific characteristics and features. Wireless intrusions can belong to two types of attacks. The first type targets the fixed part of the wireless network, such as IP spoofing, MAC spoofing, and denial of service (DoS), and the second type of these attacks targets the radio part of the wireless network, such as noise flooding, the access point (AP) rogue, and wireless network sniffing.

In order to detect such attacks, the WIDS deploys approaches and techniques provided by intrusion detection systems (IDSs) protecting wired networks. Among these approaches, one can find the anomaly-based approaches and signature-based. The first approach aims at detecting any deviation of the "normal" behavior of the network entities. The second approach consists in matching user patterns with stored attack patterns (or signatures).

## Basics of Intrusion Detection Systems

An intrusion occurs when anyone, usually a hacker, attempts to break into or misuse a computer system and an Intrusion Detection System is a system for detecting such intrusions. An IDS continuously monitors packets on a network and attempts to discover whether a break into the system has been attempted. The IDS can also try to determine other intrusions such as an attempt to cause a 'denial of service' attack.

### Different Type of IDS

An IDS may run on the target machine monitoring its own traffic (host-based) and only packets from the device are monitored and an IDS will provide alerts when suspicious activity is detected. Alternatively, an IDS can run on an independent machine watching all network traffic (net-based). The IDSs are placed at strategic points in the network to provide maximum monitoring of all inbound and outbound traffic. There are IDSs that detect intrusions based on specific signatures of known malicious threats. On the other hand, there are IDSs that detect intrusions based on comparing traffic patterns against a baseline and looking for anomalies. The baseline will identify what may be considered "normal" for that network including services, protocols, ports and IPs. This kind of IDS will alert against traffic that is anomalous, or significantly different from the established baseline. Additionally, there are passive IDSs and reactive IDSs. Passive IDS simply detect intrusion and alert the appropriate personnel, who decide which actions need to be taken. Reactive IDSs not only detect suspicious network traffic, but also take pre-defined proactive actions in response to detected intrusions.

### Example of Baseline Algorithm for Processing

There are many algorithms for intrusion detection system processing. These algorithms comprise the present state-of-the-practice and state-of-the-art in intrusion detection system processing. A given intrusion detection can implement several algorithmic approaches to intrusion processing. We will describe one such approach to intrusion detection processing as shown below (Edward 1999). The notation used is standard pseudo-code and we will illustrate the high-level idea and ignore the low-level details.

```
repeat /*iterative loop forever*/
        target_system_feed (info)
        intrusion_processing (info, result)
   if (result = intrusion) then initiate_response (result)
forever
```

The *repeat-forever* loop is used to indicate that these systems function continuously and constantly monitor traffic. The function *target_system_feed (info)* is to show the means by which information is obtained from the target system. The function *intrusion_processing (info, result)* illustrates the manner in which processing accepts an information stream, processes it, and returns a result. The comparison of the result to some intrusion activity is done in the following step. The function *initiate_response (result)* represents whatever activity is required for the result. The response depends on this result.

## Case Study: Finding Intrusions in a Hypothetical Trail

We would take an example case study in which an IDS finds intrusions in a hypothetical audit trail (Edward 1999). Suppose that the audit records the initiation of all TCP sessions in the trail log. Let us consider a simple figure as shown below wherein audit records are coming from a gateway on the perimeter network of Intranet A which is connected to the internet. IP addresses from A are the in addresses while those from the Internet are out addresses. We assume gateway IP address is gw. We also consider that inbound directed packets come into A, whereas outbound packets leave A for the internet.

Let us assume that the audit trail format is as below: <src IP add, dest IP add, src port, dest port, protocol, time, direction, success/failure of session>. Let us examine a collection of sample audit records and see if we can deduct any intrusion related information.

*<in, in, 4050, 80, TCP, 07:36:04, inbound, success>*
*<out (X), gw, 6025, 23, TCP, 07:51:12, inbound, failure>*
*<out (X), gw, 6025, 23, TCP, 07:51:55, inbound, failure>*
*<out (X), gw, 6025, 23, TCP, 07:52:17, inbound, failure>*
*<out (X), gw, 6025, 23, TCP, 07:52:58, inbound, failure>*
*<out (X), in , 3000, 23, TCP, 13:04:22, inbound, success>*
*<out (Y), gw, 5000, 23, TCP, 23:54:22, inbound, success>*

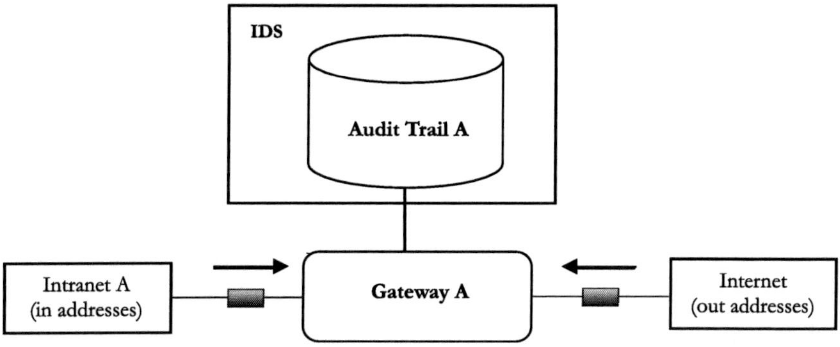


**Figure 1.** An IDS Architectural Schema.

The IDS monitors the above audit records and by looking at these records, it can determine that something suspicious is occurring. The first record presents an in source and in destination IP address for an inbound session. This is not right as inbound sessions should have outsource IP addresses and something is changing the source IP address. This type of intrusion is referred to as an IP gateway spoof. The second, third, fourth and fifth records illustrate repeated attempts by out address X to telnet A's intranet gateway. This repetitive attempt warrants some attention by the IDS and the administrator. The sixth record shows that the user at address X managed to telnet an inside address of A. This might increase some suspicion because the source IP address correlates to previous unsuccessful activity that was considered suspicious. The last record shows that at midnight some out IP address Y tried to telnet inbound and was successful to telnet the gateway. This means that the gateway was compromised. We can say from the above discussion that the attacker had been trying different ways to penetrate the network and 17 hours after the first try, he was successful.

## Vulnerabilities and Attacks in Wireless Networks

Wireless connectivity is related to specific backdoors and vulnerabilities for potential attackers that are not available in wired networks. The access to a wireless network is easier because of the nature of radio communication. A wired network can only be accessed and attacked through a physical connection, usually via Internet. The software and hardware tools needed to penetrate wireless networks are known and publicly offered ("Net-Stumbler" software, for example (NetStumbler 2011) and various hardware tools for so-called "war driving"—antennas, amplifiers, etc.).

## Wireless Vulnerabilities

There are numerous wireless threats and vulnerabilities that are studied in the literature. In the following, we distinguish two categories of vulnerabilities: those existing in 802.11 wireless networks (AirMagnet 2011) and those existing in cellular-like wireless networks (Hutchison 2004).

### *802.11 wireless vulnerabilities*

The following are typical vulnerabilities existing in the 802.11 wireless vulnerabilities.

*Easy access to 802.11 networks*

The access point (AP) uses a so-called Service Set Identifier (SSID) to differentiate networks from one another. The SSID is broadcast every few seconds in so-called "beacon frames" in order to let authorized users to find the correct network. The SSID, by default, is set to a fixed value known by everybody and this often enables easy unauthorized access to such networks.

*Unauthorized ("rogue") access points*

This vulnerability is a sort of man-in-the middle attack, where an attacker can place an unauthorized AP on the network and configure it to look legitimate to gain access to the wireless user's sensitive data.

*Unauthorized use of service*

If an authorized AP is misconfigured, this may result in enabling an attack against the whole organization's network. The unauthorized use of service opens the door to spamming and similar activities.

*Denial-of-service vulnerability*

The capacity of 802.11 wireless networks is shared by all the clients associated to a single AP. An obviously unauthorized client could start a massive transfer of data and occupy the entire available bandwidth, which would result in a denial-of-service for the rest of the authorized clients.

*MAC spoofing and session hijacking*

No frame authentication was originally defined in the 802.11 standard. An attacker can use spoofed frames to redirect traffic and corrupt Address

Resolution Protocol (ARP) tables. So, attackers can easily get the MAC addresses of the stations currently in use on the network and adopt those addresses for malicious transmissions.

*Relatively easy traffic analysis and eavesdropping*

Frame headers are not encrypted and the security against eavesdropping was initially assumed to be provided by a weak encryption algorithm (Borisov et al., 2001). The WPA (Wi-Fi Protected Access) standard extension has been introduced as a transitional solution. As a final solution to the encryption related problems within the 802.11 standard, the WPA-2 standard extension has been recently introduced (IEEE Standards, 2004) and the encryption algorithm has been changed to AES (The Advanced Encryption Standard 2011). However, the problem is in the requirement that hardware must be changed in order to implement the WPA-2 framework. The need for more resources in order to implement the AES cipher is the reason for this incompatibility.

*Possibility of higher level attacks*

An attack against a wireless network can serve as a launching point for attacks on other systems. Thus, placing a wireless LAN inside the security perimeter is therefore considered weakening the security within the perimeter.

***Cellular System Vulnerabilities and Threats***

Some cellular system vulnerabilities and threats are categorized as follows (Nichols and Lekkas 2002):

*Service interruption*

The increased capacity gained by high-speed technologies has resulted in fewer cable routes necessary to meet capacity requirements. This has decreased the number of switches. Thus, the lack of overall diversity in cabling and switching has increased the vulnerability of telecommunication infrastructures.

*Natural threats*

Natural threats comprise the category of repeated threats caused by geological, climatic, or seismic events. Severe damage resulting from natural disasters may cause long-term damage to the telecommunication infrastructures.

*Handset vulnerabilities*

Wireless messages travel through the air by passing conventional wired network for transmission to the receiver, messages may need to be changed to another protocol. This operation presents a threat because anyone can access the network at that moment.

## Wireless Attacks

A large set of attacks by a WIDS requires studying the attacker's methods and strategies. The typical attacks and malicious events that can be detected by a WIDS (Vladimirov et al., 2004) include:

### *Illicit Use*

Illicit use is a passive attack which does not cause damage to the physical network. It includes the following attacks (Mateli 2006):

*Wireless network sniffing*

Because wireless packets traverse the air, attackers equipped with appropriate devices and software can capture them. Sniffing attack methods include:

*Passive scanning:* This attack listens to each channel, which can be done without sending information.

*Service set identifier (SSID) detection*: This attack consists of retrieving SSID by scanning frames of the following types: probe requests, probe responses, beacon, association requests, and re-association requests.

*MAC addresses collecting:* To construct spoofed frames, the attacker has to collect legitimate MAC addresses that can be used for accessing AP filtering out frames with non-registered MAC addresses.

*Probing and network discovery*

This attack uses two forms of probing: active and passive. When an attacker uses passive probing, he is listening on all channels for all wireless packets. Active probing involves the attacker actively sending probe requests with no identification using the SSID configured to solicit a probe response with SSID information and other information from any active AP.

*Inspection*

The attacker can inspect network information using different tools like Airodump and Kismet (Low 2005). With these tools persons can identify IP address ranges, MAC addresses, and gateways.

### *Wireless Spoofing*

Wireless Spoofing aims to modify identification parameters in data packets. New values of chosen parameters can be collected by sniffing. Typical spoofing attacks include the following:

*MAC address spoofing*

The purpose of MAC spoofing is to change the legitimate MAC address with the attacker's MAC address.

*IP spoofing*

IP spoofing tries to change source or destination IP addresses by talking directly with the network device.

*Frame spoofing*

The attacker injects frames having the 802.11 specification with spoofed containing and due to the lack of authentication, spoofed frames cannot be detected.

### *Man in the Middle Attacks*

To realize an attack of man in the middle, two steps have to be accomplished. First, the legitimate AP serving the client must be brought down to create a state of "difficult to connect" scenario. Second, the attacker must setup a rogue AP with the same credentials as the original for purposes of allowing the client to connect to it. There are two main forms of this attack: eavesdropping and manipulation. Eavesdropping can be done by receiving radio waves on the wireless network. Manipulation requires not only having the ability to receive the victim's data but then be able to retransmit the data after compromising it.

### *Denial of Service Attacks*

Denial of Service attacks (DoS attacks) can target different network layers as discussed in the following:

*Application layer*

DoS occurs when a large amount of legitimate requests are sent with the goal of preventing other users from accessing the service by forcing the server to respond to a large number of transaction requests.

*Transport layer*

DoS occurs when many connection requests are sent. It aims at targeting the operating system of the victim's computer.

*Network layer*

DoS succeeds if the network permits to associate clients. In such a case, an attacker can flood the network with traffic to deny access to other devices.

*Data link layer*

DoS targeting the link layer can be performed as follows: Since there is a single channel that is reused, keeping the channel nodes busy leads to a DoS attack at that node. By using a special node to continually relay spurious data, the battery life of that node can be drained.

*Physical layer*

This kind of DoS can be done by emitting very strong Radio Frequency (RF) interference on the operating channel. This causes interference affecting all wireless networks operating at or near that channel.

## Intrusion Detection in a Wireless Network Environment

Due to the nature of radio propagation, the exact border between the internal and the external network is not known. Thus, exact classification of attackers into insiders and outsiders is impossible. Classification of attacks into outsider and insider attacks is not possible either. So, the security policies that use host based IDS to protect against the insider attacks and network based IDS to protect against from outsider attacks make no sense in the wireless environment. Intrusions are detected not only within the wireless network protected area, but also outside of it, keeping in mind the possibility of attacks against other parts of the network from a wireless network as well as pure interference with other wireless networks.

WIDS's can be divided into misuse-based and anomaly-based systems in the same way as the IDS for wired networks. In addition to the classical misuse and anomalies detectable in any network, wireless IDS must also detect wireless specific misuse and anomalies. The main problem in wireless misuse detection systems is the problem of distribution of the elements of the IDS. We can mention three approaches (Slobodan 2005):

*Wireless IDS Processors and Sensors are Integrated into the APs*

Over other approaches, the advantage of this approach is in total network coverage at a relatively low cost an easier network management. For the

IDS no separate hardware is needed, and this lowers the total cost of ownership. The impossibility of integrating it into existing networks for hardware incompatibility is the obviously disadvantage of this approach.

*Overlay IDS with Centralized Processing*

This approach employs dedicated radio frequency sensors deployed throughout the wireless network to be protected. The sensors transmit the data to the dedicated IDS server that performs all the processing and manages eventual responsive actions. The clear advantage of this approach is in integrating it into the existing network framework. The IDS of this type can realize total network coverage, but at a higher cost than integrated solutions. Centralized processing of all the data from the sensors demands an expensive hardware for the IDS server for the overall IDS to be efficient enough.

*Overlay IDS with Decentralized Processing*

The IDS structure used in this approach delegates some data processing to the very sensors decreasing in such a way the processing power needed at the IDS server. Obviously, the advantage of this multilevel approach over the centralized processing approach is in the cost of hardware needed for the IDS server.

Misuse-based wireless IDS may not detect new attacks. Besides, the signatures of the attacks have to be updated very often because new attacks are detected every day and many of the attacks remain undetected for quite a long time. Thus, anomaly based systems could also be a solution, but the problems of false alarms and real time operation persist in this case too.

A multilevel wireless IDS/IPS is proposed in (Pleskonjic 2003). The simplified scheme of the system is illustrated in Figure 2. This scheme uses agents on sensors, hosts, an IDS server and a reporting tool to combine host based and network based detection in a wireless network environment. The IDS cooperates with tan antivirus program, a firewall, and other security tools used to coordinate activities among them. The overall system aims at achieving the following goals:

1. To realize an efficient system to defend the wireless network
2. To give definition to attack and intrusion "axioms scope" (misuse detection)
3. To make definitions to the conclusions mechanisms ("theorems")
4. To learn in order to anticipate (anomaly detection)—there is a trade-off between the level of the system intelligence and its efficiency

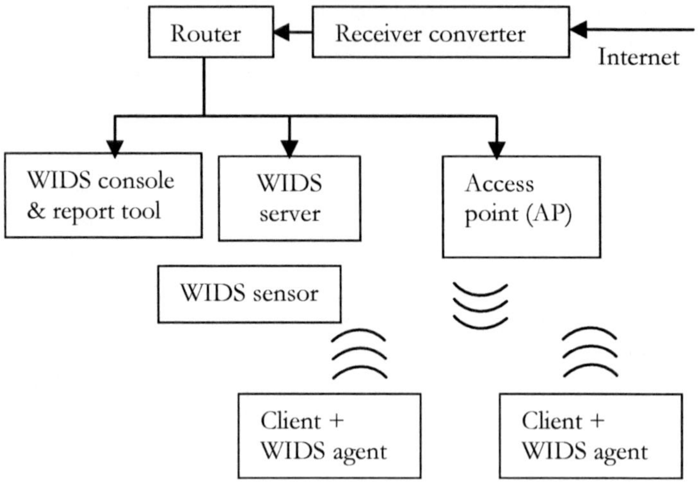


**Figure 2.** A multilevel WIDS (Pleskonjic 2003).

5. To know the wireless specific attacks
6. To take responsive actions in order to defend the system and/or the network

Neural networks and fuzzy logic have been combined in this system to achieve self learning and recognition of previously unknown attacks (anomalies), especially for those who are wireless specific. The responsive actions are triggered at both the local and global level (multilevel detection and response), which increase the efficiency of the system as the whole.

In (Gorodetsky et al., 2004), another multilevel approach to intrusion detection is mixed with situation assessment, a classification procedure that maps a label to the current state of a system based on data received from different sources. This is a general approach which is applicable to many processes, including the prognosis and handling of emergencies, monitoring, securing and recovering of critical systems like nuclear power plants and electrical power grids, command and control, prediction of terrorist intents, etc. Classification of input data IDS of this type is performed at two levels: source-based classifiers label security status of user activity either as Normal or Alert. Decisions are forwarded to the upper level of decision as asynchronous data streams. The upper level classifier combines the decisions of the lower level ones and derives the final decision on the situation status.

The method used in (Gorodetsky et al., 2004) may be especially applicable in wireless IDS, because these systems need to detect events at two completely different levels—the network level and the physical level. Completely different sensors and processing are needed to detect these two types of events. It is quite probable at the same time that the events of different types are asynchronous. An unauthorized access point, for example, may be installed by an insider without any malicious intent and this is detectable at the physical level from the very moment of installation. However, the attack may arrive later, after the vulnerability has been detected by an attacker and only then the attack may be detected at the network level.

## Architectures of IDS For Mobile Wireless Networks

According to the authors (Brutch and Ko 2003), the optimal IDS architecture for a mobile network will totally depend on the network infrastructure itself. All the nodes in a flat network infrastructure have the same level of priorities and suit applications such as conferences whereas on multi-layered network infrastructure some nodes may be separated into different clusters, each having a cluster head to allow the communication process.

### Stand-alone IDS

The architecture of stand-alone IDS is normally based on running each node separately in order to locate perpetrated intrusions. Therefore, every decision is based and focused on all the information that is collected at each and every node, as all the nodes are independent and work individually. Being totally isolated, network nodes do not know anything about the other nodes or the network itself as no data is exchanged, so no alert information is passed on. Even though restricted by its limitations, it is much more preferable for a flat network architecture which will, unfortunately, not suitable for wireless mobile network.

### Cooperative and Distributed IDS

The authors (Zhang and Lee 2003) indicate that wireless mobile networks have to adapt a cooperative and distributed intrusion detection system architecture. This is realized by the IDS agent running on top of the nodes. The IDS agent can however be complex but when analyzed closely, the IDS agent can be broken into six different modules. Figure 3 below illustrates a clear illustration of the 6 different components of the IDS agent.

In the distributed and cooperative network mentioned by (Zhang and Lee 2003), every single node has an important role to play. Each node has the responsibility for detecting any signs of intrusion and is responsible for contributing individually to the network. Thus, this can be achieved through the different parts of the IDS agent illustrated in Figure 3, where the "local data collection" would be collecting real-time that would definitely include both user and system activities within radio transmission range. The IDS also lunches a response if intrusion is detected. If an anomaly in the local data is detected on the boarder search, then the neighboring IDS agents will collectively associate themselves into the global intrusion detections actions. Figure 4 provides a detailed outline of the IDS architecture for wireless mobile networks with each node bundled with the IDS Agent which is responsible for the intrusion state and the response action.

**Local Intrusion Detection Systems**

In Albers et al. (2002), the authors proposed a collaborative and distributed IDS architecture by making use of mobile agents. Local Intrusion Detection Systems (LIDS) are mounted on every single node for either local or

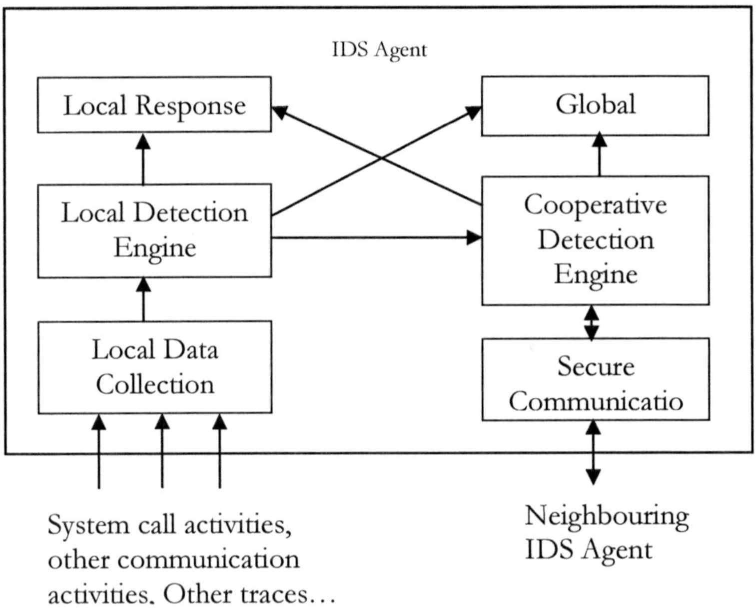


**Figure 3.** An IDS agent conceptual model.

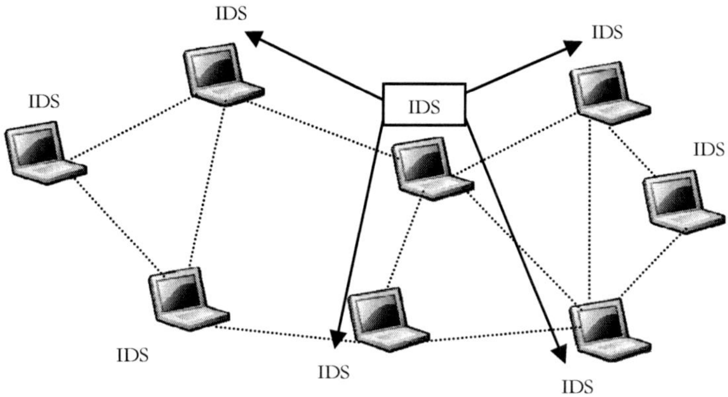


**Figure 4.** An IDS architecture of the wireless mobile network.

global concern, depending on the kind of intrusion detected. Using the LIDS architecture, there are two types of data that are exchanged and communicated among the nodes:

*Intrusion alerts data*—this would be to communicate transmit all the other nodes of any locally detected intrusion;

*Security data*—would be capturing highly favorable information for all the combined nodes.

The way the LIDS operate is that all the data that has been collected from what has been deleted from the combined or individual nodes on the network. Because each node might be running a different OS or may be using data from various sources, including application, system, or network activities might make the analysis process of LIDS harder. On making use of simple network management protocol (SNMP) all the data located in the management information base (MIB) as a source of data auditing.

LIDS can either use anomaly or misuse detection but a preferred combination of both techniques will definitely offer a better model when the local intrusion is recorded. The LIDS will initiates a response and thus inform all the other nodes belonging to the mobile network about the possible threat.

### Distributed Intrusion Detection System Using Multiple Sensors

In Kachirstu et al. (2003) a multi-sensor intrusion detection system combining a mobile agent technology is proposed. It can be further categorized into 3 main areas, each representing mobile agents with some intelligence capable of decision making, initiating a response and monitoring.

*Decision-making Agent*

In (Kachirski et al., 2003), authors describe the decision-making agent on an agent that is posted only on certain network nodes and preferably on the same nodes that the performance monitoring agent are run. Therefore, the decision-making agent gathers all the packets within its radio range subject to analysis to determine whether the network is being attacked. If for any particular reason the detection agent cannot come to a decision because of a lack of evidence, the local detection agent reports to the decision agent so that an investigation can be carried out using packet monitoring outcomes that have been collected from network monitoring sensors running locally.

*Performance Monitoring Agent*

Two distinct functions are performed in this category involving network monitoring and host monitoring. In the first instance, the host-based monitoring agent hosting the system level sensor and user-activity is run on every node. In network monitoring, sensors run on selected nodes to monitor the captured packets going through the network.

*Initiation Action Agent*

This agent is omnipresent on every node in the network because every network node hosts a host based monitoring agent. When corporative

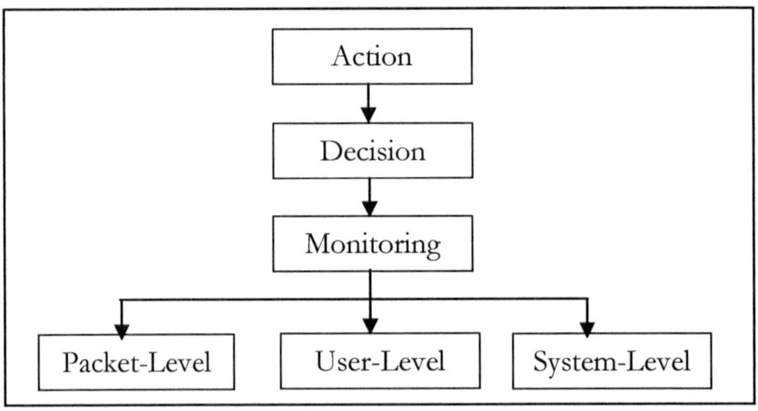


**Figure 5.** A layered architecture for mobile agent.

evidence is gathered supporting any inconsistencies that have occurred on the network, this initiating action agent can launch a termination of processes or even block any particular user from the network.

### Dynamic Intrusion Detection Hierarchical Architecture (DIDHA)

In Sterne et al. (2005) describe a dynamic intrusion detection hierarchy which is quite capable of being used by huge networks by making use of clustering. According to this proposed architecture, all nodes have the responsibility of monitoring by accumulating counts and statistics, logging, analyzing packet headers and payloads, identifying intrusions if there is enough supporting evidence, and finally alerting or reporting to cluster heads.

Besides reporting intrusions, cluster heads must be able to carry out the following functions:

*Data reduction and Integration* – thus avoiding conflicting data, fictitious and entered reports.

*Intrusion detection estimations* – many attacks depend on different sets of detected data. As a result, the data held on a single node might not be able to identify the attacks.

*Security Administration* – According to this DIDHA the layers of the hierarchy are responsible for controlling any detection and responding actions of both the clusters and the cluster heads under their supervision.

To construct the hierarchy structure, every node uses clustering, which is normally found in Mobile *Ad hoc* Networks to build up the trace routes, to then self-implement them into local precincts and identify the cluster heads which then employ clustering to form second-level clusters to suggest that the following criteria are respected:

a) Proximity
b) Connectivity
c) Storage Capacity
d) Hardening
e) Processing power
f) Energy remaining and so on

### ZONE-based Intrusion Detection System (ZBIDS)

In Sun et al. (2003), the authors propose an anomaly-based two-level non-overlapping Zone-based Intrusion Detection System (ZBIDS) can be used by separating the network into non-overlapping zones. As illustrated in Figure 6, the nodes can be classified into 2 different groups:

*Intrazone node* are independent nodes as shown in Figure 6 with nodes E, F, I, and J.

*Interzone node* are the nodes that have a physical connection to a certain node in a different zone area. An example would be nodes B, H, C and K, as illustrated in Figure 6.

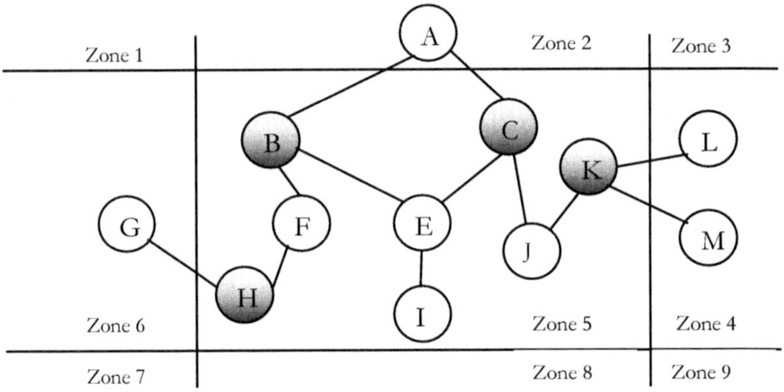


**Figure 6.** Depicting the ZBIDS architecture.

Hence each node has an IDS agent that sits upon it and run as illustrated in Figure 7. As the IDS agent supported by (Zhang and Lee 2003), the data gathering and detection agent are the major agents for collecting local data in the instance of system log files and system call activities and then putting the collected data under analysis for any form of intrusion.

Having had a look at the different architectures and the way in which they operate, the best IDS techniques and approach for WIDS would be to adopt the cooperative and distributed IDS referred to Figure 3 mentioned earlier. We will now be looking at how WIDS really adapt to the wireless network and how.

*Data Collection:* According to Figure 3, local data collection is the area responsible for gathering real-time data from different sources and on the intrusion detection algorithms. These crucial data include user activities and the system's activities within the node, as well as all the communication activities within the range of that node.

*Local Detection:* The local detection process analyzes the collected data from the local data unit to identify or trace any irregularities on the network

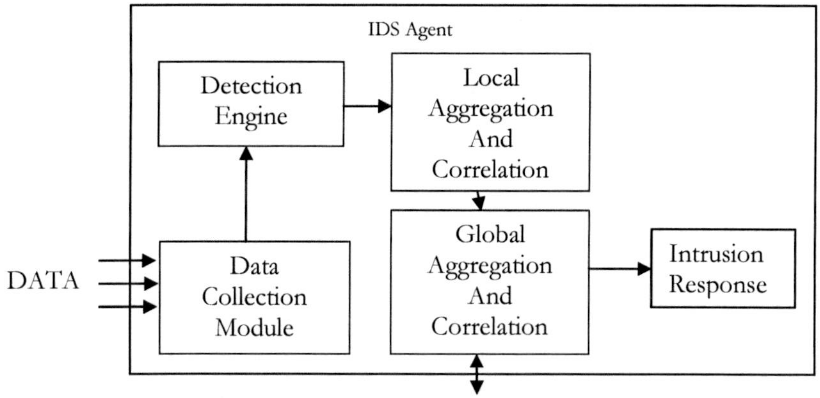


**Figure 7.** Illustration of an IDS agent in ZBIDS.

knowing that attacks will definitely increase at an alarming rate as more and more network applications tend to adopt mobility.

*Cooperative Detection:* With cooperative detection, any node on the mobile wireless network that identifies an intrusion or an irregularity can individually identify that the mobile network is under intrusion and launch a response to that intrusion. If a node indentifies an anomaly on the mobile network with weak evidence, it can start a cooperative global intrusion detection guide which normally operates by spreading the intrusion detection status information to the other nodes in the network. An example of distributed intrusion detection includes the following actions:

Step 1: The node broadcasts an irregularity state request to the closest nodes on the network.

Step 2: Each then transmits the state information manifesting an inconsistency to its closest neighbors.

Step 3: Afterwards, all nodes derive a conclusion whether the greater number of received reports identify this irregularity and, if positive, then determine that the network is subject to attack.

Step 4: And finally, any node that identifies any inconsistency or irregularity in the network can launch a response.

Careful attention should be focused as audit data from different nodes cannot be relied upon because the corrupted nodes can be sending erroneous data.

*Intrusion Response*: All intrusions are treated differently on mobile networks depending, obviously, on the type of intrusion and the different network protocols and applications used in the wireless network. An example of this response would be like re-connecting the communication channel between the nodes and then forcing a re-keying to take place or to locate the corrupted nodes and re-calculate how to anticipate for the other nodes that have not been compromised.

## Wireless Tolerance and Prevention

Intrusion prevention is considered an extension of intrusion detection technology, but it is actually another form of access control, similar to an application layer firewall. Intrusion prevention systems (IPSs) were developed to resolve problems related to passive network monitoring by placing detection systems online. Demonstrating a considerable improvement upon firewall technologies, IPSs make access control decisions based on application content, rather than IP address or ports by denying potentially malicious activity.

The same as those caused by a buffer overflow, some IPSs can also prevent yet to be discovered attacks. Aiming to strengthen wireless security, wireless IPSs monitors radio frequencies in order to detect malicious traffic. The support and development of intrusion aware survivable applications in wireless networks are key problems in the provision of wireless services.

Special aspects of intrusion tolerance include: (1) the ability to change in environmental and operational conditions to survive intrusions; (2) the management and coordination of adaptation of changes in service provision; (3) the awareness of resource statuses to respond to attack symptoms effectively; and (4) the management of resource redundancy. The following are two approaches that implement intrusion tolerance to prevent wireless attacks.

### Intrusion Tolerance Based on Multiple Base Stations Redundancy

This research discusses a redundancy to provide fault tolerance in the form of multiple base stations (BSs). Because an adversary can disallow delivery of sensor data that is routed over only one path to a given BS, a multi-path routing redundancy to improve intrusion tolerance of wireless nodes is introduced (Deng et al., 2004).

The easiest way to set up multiple paths for each node to multiple BSs is to use a flooding message: each BS broadcasts a unique request message, called REQ. Upon the reception of REQ from a BS, it records the packet sender as its parent node for that BS, and re-transmits REQ to its neighbor and child nodes. Next, the node then ignores all copies of the same REQ that it receives later. In such a way, the REQ generated by a BS will be able to flood the entire network, even though the network nodes forward that message only once. If one BS transmits its own REQ, every sensor node will have one path for it. However, this scheme cannot prevent a compromised node from BS spoofing by sending forged REQ. Every node will suppose that the forged message is generated by the legitimate BS and will forward the forged REQ. In order to defend against such an attack, each BS can authenticate the sent REQ (Deng et al., 2004).

## INSENS: Intrusion-Tolerant Routing in Wireless Sensor Networks

INSENS (Deng et al., 2003; 2005) can be employed to prevent DoS attacks, where individual nodes are not allowed to broadcast routing data. Only the BS is allowed to broadcast (Deng et al., 2003). It suggests a BS authentication using a hash function. To prevent DoS/distributed denial of service (DDoS) broadcast attacks, unicast packets must first traverse through the BS. Next, the control routing information has to be authenticated and encrypted by using symmetric cryptography. Redundant multipath routing is built into INSENS to achieve secure routing to address compromised nodes.

INSENS is performed through two phases, route discovery and data forwarding. The first phase discovers the sensor network topology. The second deals with forwarding data from sensor nodes to the BS, and vice versa. Route discovery is elaborated in three rounds:

- During the first round, the BS floods a request message to all the reachable sensor nodes in the network. The BS broadcasts a request message which is received by all its neighbors. A sensor, receiving a request message for the first time, records the identity of the sender in its neighbor set and broadcasts a request message. To counter attacks, two mechanisms are used. The first one identifies the request message initiated by the BS using a hash function while the second mechanism configures sensors with separate pre-shared keys by applying a keyed MAC algorithm to the complete path (Deng et al., 2005).
- The sensor nodes during the second round send their local information using a feedback message to the BS. After a node has forwarded its request message, it for waits a period of time before generating a feedback message.
- In the third round, based on the information received in the second round, forwarding tables are computed by the BS for each sensor node. Then, it sends them to the respective nodes using a routing update message and waits for a certain period to collect the connectivity information received via feedback messages to compute possible paths to each other node. Next, the BS then updates the

forwarding tables using entries of the form: *(destination, source, and immediate sender).*

Destination is the node ID of the destination node, source is the node ID of the node that generated this data packet, and immediate sender is the ID of the node that just forwarded this packet. Once the data packet is received, a node looks for a matching entry in its forwarding table and forwards the data packet if it finds a match (Deng et al., 2005).

## Conclusion

While wireless networks become more common in today's Information Technology (IT) infrastructure, WIDS is slowly becoming a critical component for both active and passive security monitoring. WIDS not only provides a security monitoring function, but also allows protection against threats outside the organization's network perimeter, and internally with the enforcement of the organizations wireless policy.

Thus, WIDSs have an important role in securing the network by protecting its entities against intrusions and misuse. The protection is based on models capable of providing a framework for the description and correlation of attacks. Research has focused on the development of techniques, mechanisms, approaches, and WIDS architectures.

It is evident that networks of all types and sizes will benefit greatly from the use of WIDSs which have the capability to be used as an effective policy monitoring tool. Open source wireless IDS programs allow a viable alternative to some of the extremely expensive commercial tools available, and thus wireless its adoption is not going to be cost prohibitive.

## REFERENCES

AirMagnet. The Top Seven Security Problems of 802.11 Wireless. AirMagnet Technical White Paper. http://www.airmagnet.com/products/ wp-index.htm

Albers, P., O. Camp, O.J. Percher, B. Jouga and M. Puttini. 2002. Security in *Ad Hoc* Networks: A General Intrusion Detection Architecture Enhancing Trust Based Approaches. Retrieved October 5, 2007, from http://www.rennes.supelec.fr/ren/rd/ssir/publis/wis02_albers_camp_percher_als.pdf

Borisov, N., I. Goldberg and D. Wagner. 2001. Intercepting Mobile Communications: The Insecurity of 802.11. Proceedings of the 7th Annual International Conference on Mobile Computing and Networking, pp. 180–189.

Brutch, P. and C. Ko. 2003. Challenges in Intrusion Detection for Wireless *Ad hoc* Networks. Retrieved October 7, 2007, from http://csdl2.computer.org/persagen/DLAbsToc.jsp?resourcePath=/dl/proceedings/&toc=comp/

proceeding s/saint-w/2003/1873/00/1873toc.xml&DOI=10.1109/SAINTW.2003.1210188

Deng, J., R. Han and S. Mishra. 2004. June 28–July 1. Intrusion tolerance and anti-traffic analysis strategies for wireless sensor networks. In Proceedings of the 2004 International Conference on Dependable Systems and Networks (DSN'04) (pp. 637–646). Italy.

Deng, J., R. Han and S. Mishra. 2003. May. INSENS: Intrusion-tolerant routing in wireless sensor networks. In The 23rd IEEE International Conference on Distributed Computing Systems (ICDCS). Providence.

Deng, J., R. Han and S. Mishra. 2005. INSENS: Intrusion-tolerant routing for wireless sensor networks. [Special issue]. *Computer Communications Journal.* **29(2):**216–230.

Edward, Amoroso. 1999. Intrusion Detection: An Introduction to Internet Surveillance, Correlation, Trace Back, Traps, and Response, First Edition, Intrusion.Net Books, New Jersey.

Gorodetsky, V., O. Karsaev, V. Samoilov. 2004. On-Line Update of Situation Assessment Based on Asynchronous Data Streams. Proceedings of the 8th International Conference on Knowledge-based Intelligent Information and Engineering Systems (KES 2004), LNAI vol. 3213, Springer Verlag. pp. 1136–1142.

Hutchison, K. 2004. Wireless intrusion detection systems. Retrieved October 18, 2004 from http://www.sans.org/reading_room/whitepapers/ wireless/

IEEE Standards Association. IEEE 802.11i Standard. http://standards.ieee.org/getieee802/ download/802.11i-2004.pdf

Kachirski, O. and R. Guha. 2003. Effective Intrusion Detection Using Multiple Sensors in Wireless *Ad Hoc* Networks. Retrieved September 22, 2007, from https://www.cs.tcd.ie/publications/techreports/reports.05/TCD-CS-2005-49.pdf

Low, C. 2005. Understanding wireless attacks & detection. Retrieved April 2005, from http://www. hackerscenter.com/public/Library/782_wireattacks. Pdf

Mateli, P. 2006. Hacking techniques in wireless networks. In H. Bidgoli [ed.], Handbook of information security (pp. 83–93). John Wiley & Sons.

NetStumbler. http://www.netstumbler.com

Nichols, R.K. and P.C. Lekkas. 2002. Telephone system vulnerabilities. McGraw Hill.

Pleskonjić, D. 2003. Wireless Intrusion Detection Systems. Proceedings of the 19th Annual Computer Security Applications Conference, Las Vegas, USA, December 8–12.

Slobodan Petrović. 2005. "Vulnerabilities in wireless networks and intrusion detection", Telektronikk 1.

Sun, B., K. Wu and U.W. Pooch. 2003. Alert Aggregation in Mobile *Ad Hoc* Networks. Retrieved September 14, 2007, from http://portal.acm.org/ft_gateway.cfm?id=941323&type=pdf&coll=GUIDE&dl=GUIDE&CFID=2659735&CFTOKEN=45841533

Sterne, D., P. Balasubramanyam, D. Carman, B. Wilson, R. Talpade, C. Ko, R. Balupari, C.Y. Tseng, T. Bowen, K. Levitt and J. Rowe. 2005. A General

Cooperative Intrusion Detection Architecture for MANETs. Retrieved October 1, 2007 from http://seclab.cs.ucdavis.edu/papers/manet_ids.pdf

The Advanced Encryption Standard (AES). http://csrc.nist.gov/publications/fips/fips197/fips-197.pdf

Vladimirov, A.A., K.V. Gavrilenko and A.A. Mikhailovsky. 2004. Counterintelligence: Wireless IDS systems. In WI-Foo: The secrets of wireless hacking (pp. 435–456). Pearson/Addison-Wesley.

Zhang, L. and W. Lee. 2003. Intrusion Detection Techniques for Mobile Wireless Networks. Retrieved September 15, 2007, from http://www.cc.gatech.edu/~yian/Zhang_03.pdf

# 5

# Using Bio-inspired Algorithms to Provide Privacy on Wireless Sensor Networks

*María de los Ángeles Cosio León, Juan Iván Nieto Hipólito, Jesús Luna García and Jetzabel M. Serna-Olvera*

**ABSTRACT**

Wireless Sensor Networks (WSNs), often referred to as ubiquitous computing, are the basis for "Internet of Things and Services" (IoTaS) applications. WSNs are usually employed in a non-intrusive way and have the potential of providing real-time information in key areas of concern e.g., e-Health. Unfortunately, due to their distributed and ubiquitous nature, WSNs are vulnerable to several security and privacy attacks. Privacy issues represent an interesting challenge, in particular with potentially unauthorized access to personal information e.g., user's geo-location. This chapter explores the use of Computational Intelligence (CI) paradigms in WSNs and proposes a bio-inspired routing algorithm that provides location privacy, while optimizing overall network performance. Empirical results and a set of recommendations for using CI algorithms in WSNs are also given. These recommendations emphasize the trade-offs between privacy and performance, when bio-inspired approaches are adopted.

## Introduction

Nowadays people share their physical spaces with a great diversity of heterogeneous devices, so-called "things", most of which have processing, sensing, and communications capabilities; therefore, they are capable of providing Internet connectivity to couple together the real world with cyberspace (e.g., by linking their personal devices to similar objects and services around them). Their communication features allow devices with small pieces of code to react as their environment changes. Consequently, "things" become smarter by providing learning and self-adapting capabilities, which are the basis of the Internet of Things and Services (IoTaS). The wide variety of IoTaS-enabled devices support not only simple services for people (via distributed tasks), but also make it possible to use "things' services" to develop more complex environments (e.g., digital hospitals), which resembles the way Wireless Sensor Networks (WSNs) are being used in e-Health applications(Alemdar and Ersoy 2010).

The deployment of IoTaS has given rise to a wide number of ubiquitous services that not only surround us, but also cohabit our personal spaces (e.g., home and personal area networks). These ubiquitous services include, for example, patient tracking, service discovery, and many similar services that can be accessed through a large number of "things" via different channel-based communications, employing technologies such as 4G, WiMax, WiFi, Bluetooth and IEEE802.15.4 (WSNs).

Despite its benefits, the IoTaS has also introduced new privacy and security concerns, in particular those related with the unauthorized access of user personal data (e.g., her geo-location). Additional considerations should be taken into account when things are used in critical environments. For example, in e-Health, when patients must be monitored, each and every data sampled must be protected and personal information and physiological measures should be transmitted in compliance with the applicable data protection legislation such as The *Health Insurance Portability and Accountability Act* (HIPAA) (Senate & Congress assembled 1996) for the USA. In México, the legal framework is provided by the "*Ley Federal de Protección de Datos Personales en Posesión de los Particulares*" (LFPDPPP) (*The Federal Law for the Protection of Personal Data in Possession of Private Interests*, in English) (United Mexican States 2010) and in the report *Legally eHealth. Putting eHealth in its European Legal Context* (Doosselare et al., 2008), which presents an overview of related legislation in Europe. Furthermore, for organizations managing data via IoTaS-enabled devices, the aforementioned legal restrictions must be met to protect their clients' personal space and thus avoid legal penalties. As a result, users of these technologies feel more comfortable using this technology knowing that their personal data are being protected, which ultimately increases trust and confidence that might help avoid life-endangering situations.

Currently, a broad range of solutions has been proposed to achieve *security and privacy in traditional distributed systems*, (Freier et al., 1996), (Kelly 2009); (Dierks and Allen 1999). However, for WSNs, this is a particularly challenging task because of the use of constrained devices and the performance overhead inherent to these privacy solutions. State of the art proposals are primarily based on privacy mechanisms that have not been fully optimized for these constrained devices (e.g., using flooding-based routing algorithms which consume a lot of energy in sensing nodes). Therefore, there is a need for research to improve user privacy while being resource efficient. With this need in mind, this chapter focuses on biological systems, specifically ants, which are able to learn, evolve, and self-organize in a fully distributed manner. Despite harsh environmental conditions, they have been able to survive for millions of years by means of evolution and adaptation, giving rise to the "Computational Intelligence" (CI) paradigm, where "intelligence" is defined as the capacity to adapt and self-organize to reach a steady state on an assigned task. Therefore, the research presented in this chapter aims to demonstrate that CI algorithms represent a suitable option for providing privacy-location in WSNs.

Summarizing, the main objectives of this chapter are:

- To advance the state-of-the-art on the use of Computational Intelligence solutions; and the problem of geo-location privacy on WSNs.
- To propose a bio-inspired WSN routing algorithm (based on ant colony behavior), that provides location privacy while optimizing overall network resources, and
- To empirically prove our hypotheses through a set of quantitative results related to the evaluation of the proposed routing protocol.

Ubiquitous devices created a technological infrastructure that can support digital services employed, for example, in e-Health (Alemdar and Ersoy 2010). One part of the IoTaS technologies is comprised of WSNs. By definition, WSNs possess a tiered structure (Man et al., 2006) that includes three different types of devices (see Figure 1): Sink device, Coordinator device (Full Function Device, FFD) and End-device (Reduced Function Device, RFD). A sink device must be robust and have continuous energy support so that they can constantly forward traffic from the WSN to the Internet. A coordinator device has routing capabilities that depend on the network. Finally, a subset of so-called End devices exists, where their physical characteristics are similar to other devices comprising WSNs. However, they also possess additional software restrictions related to the specific network topology.

WSN nodes (also known as "motes") have many advantages when used for long-term monitoring, including low maintenance, low cost and low energy consumption. However, security and privacy issues become more critical when e-Health services are supported (Cho et al., 2004; Steele

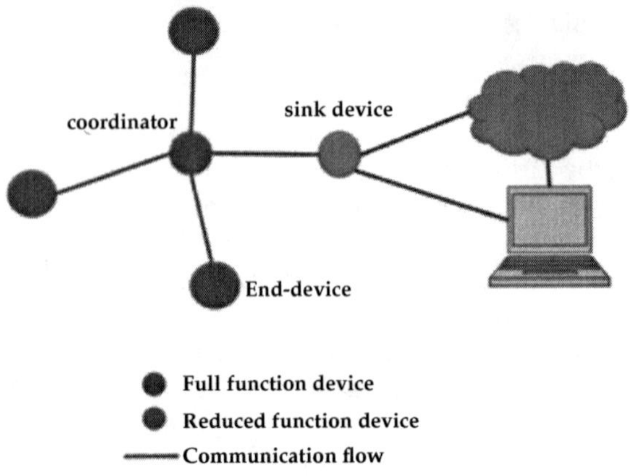


**Figure 1.** The WSN represented as a tiered structure.

*(Color image of this figure appears in the color plate section at the end of the book.)*

et al., 2009), especially where networks of distributed and autonomous devices have to provide sufficient privacy guarantees when cooperatively sensing physical or physiological parameters. In these scenarios, each sensor node must transport data originating at a source node (e.g., the patient) all the way to a sink node.

Nowadays, tracking a source of information allows potential attackers to infer habits, geo-location preference, and real-time location information of WSN users. Although it is clear that unauthorized users with that knowledge can cause damage in the real world, location tracking has no legal framework in developed countries such as the USA or in the rest of the world. Many people using mobile devices refer to privacy and the lack of legal mechanisms to protect the privacy of people using them as two of the major challenges mobile computing faces. As a result, a lot of personal information is being mishandled and misused, potentially endangering the most vulnerable population—children, the elderly and the handicapped.

State-of-the-art proposals that provide location-privacy in WSNs mostly focus on routing algorithms with the goal of keeping potential attackers away from source nodes. These proposals use different approaches, ranging from some type of flooding (Kamat et al., 2005), to "opportunistic routing" (Spachos et al., 2010), Despite their particular features, most of these proposals pursue a fair trade-off between privacy and other operational parameters of the WSNs (e.g., energy consumption and packet loss rate, among others).

Take, for example, GROW, a routing proposal that uses a two-way greedy random walk as described in (Schwiebert and Shi 2006), where the first stage of a route is generated at the sink node and transmitted to another node *n* hops away from it via a "greedy" random walk. Once this path exists, each node in the WSN will try to send packets through it. Even if this procedure is safe for privacy, the arrival of a packet is not guaranteed and it will depend of routes it needs to traverse. In addition, authors propose using fake energy-aware packets to improve privacy. In this practice, nodes will send packets only if the energy threshold has not been reached yet.

In 2010, Li and Ren (Li and Ren 2010) proposed a set of different techniques to protect privacy through intermediate node selection. In their first technique, the intermediate node is selected using a threshold distance; in the second technique, the selection of the intermediate node is totally random. The first technique insures privacy by selecting an intermediate node before it sends a packet. The latter, in turn, guarantees privacy by producing long routes, while keeping a low relation between the delivered packets and loss packets. As the main conclusion authors highlight is that routing through a single-intermediate node is more suitable for small-scale sensor networks. For large-scale networks, the authors propose a technique based on angles: Prior to data transmission, the source node defines an angle $\beta$ between the sink and the last intermediate node; this angle helps them define a new intermediate node location. Larger angles provide a higher level of location privacy.

Finally, it is worth mentioning that different approaches have adopted bio-inspired algorithms to overcome common security threats, including Denial of Service (DoS) attacks.

Ghanekar et al., 2011, presented an overview of different bio-inspired approaches aimed at preventing DoS security attacks and proposed an Intruder Detection System (IDS) approach (bio-inspired), based on a multi-cell adaptive immune system. They presented experimental results obtained via simulations to prove empirically the feasibility of the adoption of biological inspired systems.

Saleem et al. (Saleem et al. 2011) proposed an autonomously secure routing mechanism to overcome common network layer attacks. This is a hybrid approach, which adopts and combines two bio-inspired mechanisms: The ant-based ACO algorithm and an Artificial Immune System (AIS). The former algorithm provides route discovery and the latter works as a self-security mechanism.

Their results demonstrated not only the feasibility in terms of security, but it also provided an insight onto the performance, which is a key challenge in WSNs. In summary, bio-inspired algorithms are considered a promising field of research to be adopted by many areas of WSNs, especially in security-related solutions. However, until now, they have not been applied in any privacy location solution.

## Computational Intelligence Paradigms

Bio-inspired algorithms have been proven useful for solving problems related to networking and envisioned information systems (e.g., smart cities and e-health services based on distributed processes). These types of algorithms mimic the dynamics of many biological systems and laws governing them (a small number of simple and generic rules) without the need for any centralized entity. Based on the bio-inspired paradigm, computational intelligence (CI) algorithms have been successfully used in recent years to address various challenges such as: (a) data aggregation and fusion, (b) energy aware routing, (c) task scheduling, (d) security, (e) optimal deployment and (f) localization. These features make them perfect for creating novel routing algorithms that are able to preserve location privacy in WSNs.

In the following paragraphs, we will define two key issues about CI:

1. Adaptability: a process executed by an object using its environmental interactions by means of a set of basic rules.
2. Self-organization: a process without a central entity in charge of managing an entire system, but only parts of it to make restrictions and relations between them to fulfill a task.

### Adaptability

The article "Design for a brain: the origin of adaptive behavior" (Ashby 1960), mentions that adaptation leads to a survival rule, thus, an adaptive system follows two feedback loops. The first feedback loop helps the system to learn a pattern, which is called the "soft-change". The second feedback loop operates when a new pattern of behavior needs to be learned and the entire system has to change its behavior ("hard-change"); therefore, three types of "adaptation algorithms" (Cass and DePietro 1998), have been suggested:

- *Supervised Adaptation*—This kind of algorithm is supervised by an entity called "teacher" that provides a set of input data and expected results. The actual system is fed with previously generated data and the system's expected results are used to compute the error rate by comparing expected results proposed by "the teacher" versus the real result set offered by the algorithm to be analyzed.
- *Unsupervised Adaptation*—Its input is a dataset where there is no fitness information (an optimal solution), either qualitative or quantitative, within it. Results of these algorithms are clusters of data that must be analyzed by humans to understand the reason behind such groups.

- *Reinforcement Adaptation*—The system achieves adaptation through its interaction with a "critic" that provides input variables and fitness and heuristic reinforcement information, but there is insufficient information to assure that an optimal solution exists. There are different stages, each providing information to the next, as a way to improve previous solutions. This adaptation system resembles the behavior of biological entities. For example, the task carried out by the ants to find food: routes with heavy ant traffic will attract even more ants, whereas routes with low ant traffic will become increasingly less attractive for them (Kulkarni et al., 2011).

## Self-organization

The meaning of self-organization (Ashby 2004) is simple, it refers to systems that start with its constituting parts as separate entities (the behavior of each one is independent of the other parts' states), and then the parts begin to reach a steady state by enforcing restrictions and connections among them.

In summary, beyond the WSNs' design challenges related with energy consumption, and physical restrictions, it has become *critical to harmonize the entire network in order to work for a common objective*, just as what happens in self-organizing and adaptive systems. This is precisely the rationale behind our research, taking into account that CI paradigms mimic groups (e.g. mammals, insects, cells and particles), where an entire group works to reach a common objective by "adapting" its behavior.

The proposal presented in the rest of this chapter explores the use of "ant colony optimization" (ACO) algorithms (Dorigo et al., 1996) and the behavior of ant colonies to develop collective intelligence with the common goal of providing location-privacy to the WSN source node.

## Ant Colony Optimization

ACO algorithms have been inspired by the behavior of ants searching for food, leaving a trial of a chemical substance called a pheromone; the ants create several routes, selecting an optimum route between the ant hill and the food source, according to an optimization criterion. Usually, ants build colonies ranging from small groups to large and highly organized territories. Ant organization is based on specialized groups doing single tasks, e.g., searching for food. Ant-based algorithms mimic this behavior to build routes from a source node to a destination node located on "a common food preferences area" (the sink node itself).

Ant-based algorithms were modeled taking into consideration the behavior of ant colonies in the real world. Figures 2, 3, and 4 show

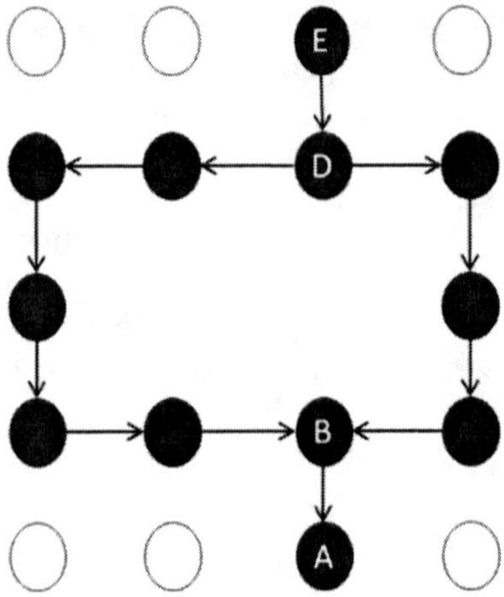


**Figure 2.** The scenario at time $t_0$; an ant in node$_i$ tries to reach the node A. At this time the ant might choose both routes.

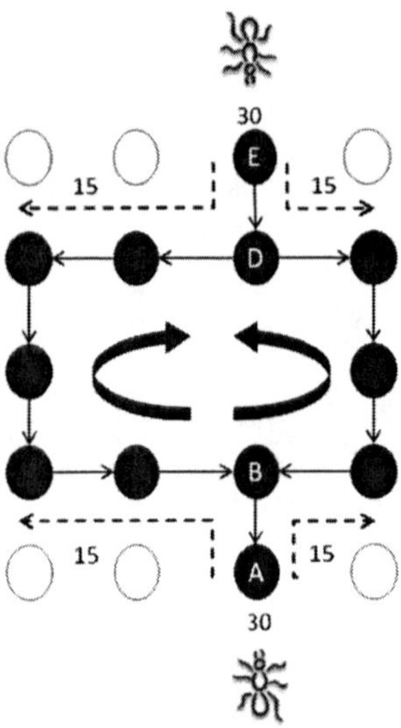


**Figure 3.** On $t_1$, there is not yet still a pheromone trail.

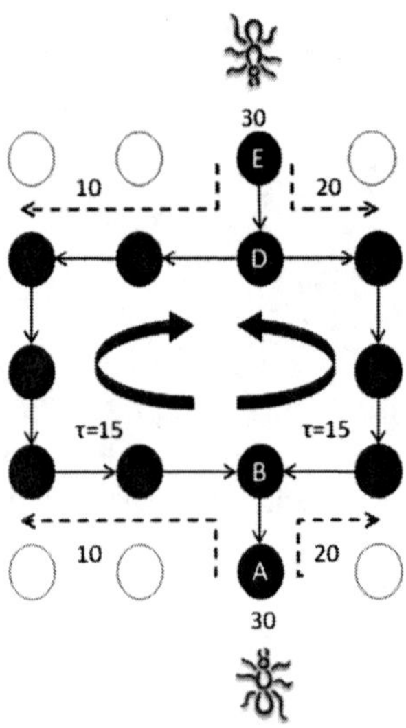


**Figure 4.** On $t_i$ there is a route with different pheromone levels.

experiments that describe the "positive feedback mechanism". This mechanism is basic for the ant decision function when its aim is to improve the routes: When a group of ants starts to search for food, it will walk randomly along different routes and not create an established pheromone trail and will transverse the area after a time $t_0$. The decision process for taking a route is as shown in Fig. 2.

On the other hand, the same group of ants at time $t_1$ will follow a pheromone trail from the source node to a destination node, just as shown in Fig. 3. The likelihood of taking a route depends on the pheromone level along the route. Finally, at time $t_i$ (Fig. 4), the pheromone level will begin to increase, becoming more attractive for other ants. Pheromone trails will guide ants to the food source using the shortest available path when following the routes with the highest pheromone levels (until reaching its destination at a time *t*).

In the real world, each ant might be able to find a route to the food zone. Part of the selected path, or even the entire path, can be shared among the ants due, in part, to the "positive feedback mechanism" and its food preferences (Deneubourg and Goss 1989). For example, a delta pattern will be produced on destination routes if the ant's food preferences are broad. On the other hand, narrow food preferences will produce single routes near the "destination node neighborhood." Ants are social insects: they exchange information about food location or contextual information such as if there are adversaries or dangers in the route. In our proposal, delta-form patterns are important because they will become key in our privacy structure proposal.

### Transition and Evaporation Functions on Bio-inspired Algorithms

Dorigo et al. (1996) proposed a formal model of the ACO algorithm. The authors used the classical optimization problem, "Travel Salesman Problem" (TSP) to explain their model. Equation 1 shows the proposed model's transition function to select next node.

$$P_{ij}^{k}(t) = \begin{Bmatrix} \dfrac{\left[\tau_{ij}(t)\right]^{\alpha} * \left[\eta_{ij}\right]^{\beta}}{\sum_k \varepsilon allowed_k \left[\tau_{ik}(t)\right]^{\alpha} * \left[\eta_{ik}\right]^{\beta}} \\ 0 \end{Bmatrix} \; if\, j\, \varepsilon\, allowed_k \; otherwise$$

**Equation 1:** The transition probability function from $location_i$ to $location_j$ for the $ant_k$.

Where:

- $P_{ij}^{k}$ is the likelihood that $node_j$ will be selected by $ant_K$ on $node_i$.
- $Allowed_k$ is the list of nodes that have not been visited yet. Therefore, $ant_k$ can avoid visiting nodes by checking this list.
- $d_{ij}$ is the path length between nodes i and j.
- $\eta_{ij}$ is the visibility that $node_i$ has of $node_j$
- $\tau_{ij}$ is the pheromone level (also called "trail intensity") on $edge_{ij}$
- $\alpha$ and $\beta$ are the tune up parameters that control the relative weight of the trail intensity $\tau_{ij}$ and the visibility $\eta_{ij}$ for ant decisions.

An *iteration* of the algorithm is refered to the travel of every ants in the interval (t,t+1) from $node_{jk}$ to $node_{(j+1,k)}$.

n iterations of the algorithm means that, an ant had finished a tour using a time equal to t+n

The trail intensity $\tau_{ij}(t+n)$ on the $edge_{ij}$ at time t+n is represented by:

$$\tau_{ij}(t+n) = \rho * \tau_{ij}(t) + \Delta\tau_{ij}$$

**Equation 2:** Intensity of trail on $edge_{ij}$.

Where:

- ρ is a coefficient such that (1 - ρ) represents the evaporation of the pheromone trail between time t and t+n (a *tour)*, with $0 \le \rho < 1$.

Increased pheromone levels between times t and $t+1$ is represented by:

$$\Delta\tau_{ij} = \sum_{k=1}^{m} \Delta\tau_{ij}^{k}$$

**Equation 3:** Pheromone level increased by iteration, $time_t$ to $time_{(t+1)}$.

Where:

- m is ant population.
- $\Delta\tau_{ij}^{k}$ is the quantity of pheromone trial laid on $edge_{ij}$ by the $ant_k$

The increase of pheromone levels for a particular $ant_k$ is represented by:

$$\Delta\tau_{ij}^{k} = \begin{cases} \frac{Q}{L_k} & \textit{if ant k use the link ij in its tour (betwen time t and t +n)} \\ 0 & \textit{otherwise} \end{cases}$$

**Equation 4:** Quantity of pheromone levels increased by $ant_k$ at $edge_{i,j}$.

Where:

- Q is a constant value that stores the maximum pheromone level, which a problem approach will define.
- $L_k$ is the *tour* length of $ant_k$.

**Ant-based Routing**

In WSNs, the transmission of packets from source nodes to the sink node is performed following a common pattern usually referred to as "hop-by-hop communication". Nevertheless, if nodes are deployed in large areas, routing tasks should be implemented with specialized algorithms mainly based on two well-known routing techniques: proactive and reactive. In proactive routing protocols, due to the use of constantly updated routing tables, the incurred cost is usually unsuitable in WSNs. These costs are related to (i) the number of messages to be transmitted (ii) the cost of maintaining these routing tables and, (iii) the space to store the whole routing information. On the other hand, reactive routing protocols only calculate a new route when it is necessary (e.g., if previously selected paths become unfeasible). Consequently, they start a new search procedure is and the cost of updating routing tables is eliminated. However, the need to provide a balance between functionality and performance has given birth to a hybrid approach, combining proactive and reactive techniques, where the decision of using any of them is based on performance measurements.

Usually, it is also true that ants find routes between their nest and the food zone after a certain amount of time, by using simple rules to take local decisions. However, they are also able to make a global effort, by building a short path from the source to the destination node. Using a reactive routing process, ants will search for food, but unlike other reactive processes, the route offered by the ant it is not the shortest path. Therefore, an optimization process that considers all of the non-optimized solutions offered by ants is needed. In order to search for the optimal path, a discrimination process applied to factors that affect the decisions becomes necessary; leaving only critical factors with the aim to reduce processing cost. Hence, this process must be carefully designed beforehand. Otherwise the processes involved will become too expensive for embedded devices. Once the discrimination process has finished, two mechanisms will be triggered: first a "building" process to create routes to food zones and second, an "improving" process that improves the initial set of solutions until (i) a shortest path has been selected or (ii) any other solution as near as possible to the ideal solution is obtained (global effort). In addition, a set of paths leading to the same destination will be offered, enabling "the ants" to send information through any of them if the currently selected *shortest path* fails. In the end, we are not interested in that shortest path but in the provisional set of routes to the sink node.

**Developing ACO Algorithms for Embedded Devices**

In order to deploy an ACO mechanism in embedded devices, it is important to take into consideration the device's physical capabilities and system requirements because of the close interaction between WSN software and hardware (Alkazemi and Felemban 2010); (Blumenthal et al., 2003); (Picco 2010). To develop ACO algorithms for WSNs, we defined four main subsystems: *task, skills, local processing and communication*. Fig. 5 shows the software components considered in our proposal.

The *task* subsystem is in charge of the data acquisition procedures (analogue or digital) and sensing control procedures (e.g., timers, interrupts, signal's port selection). The *skill subsystem manages* additional properties of the device (e.g., routing capabilities, security, and privacy mechanisms). Both *tasks* and *skills* subsystems refer to services supported by the mote (node); there are no additional gadgets considered to increase functionalities. Besides, there is a key structure for environment interactions, the so-called *communication subsystem,* which allows the devices to gather information from the environment as well as send commands and data to the WSNs.

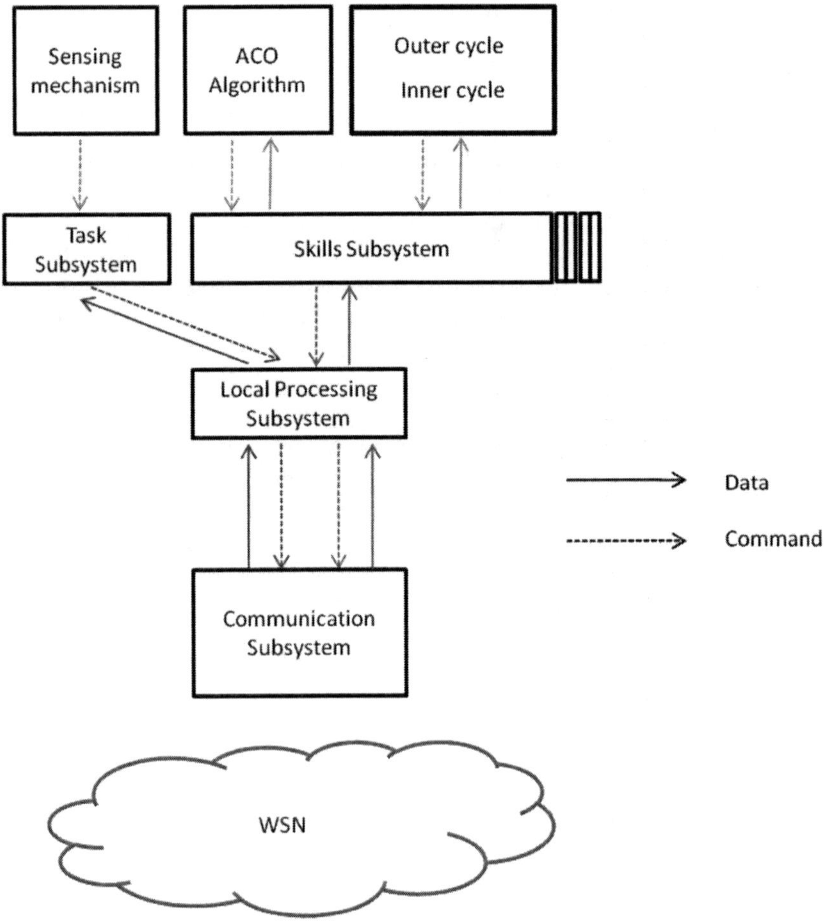


**Figure 5.** ACO software components on a WSN node.

Finally, self-control functionalities are implemented by the *local processing subsystem* (adding adaptive capabilities), which makes decisions based on two different blocks: the "outer cycle" that controls hard-changes and operates infrequently searching for routes mechanisms and the "inner cycle", controlling soft-changes (i.e., the decision function along the routes) where changes are frequent.

For a device with a reduced set of functions (RFD), there are no specialized processes; instead, a basic architecture is considered allowing interactions between RFD and FFD. The software architecture should be built taking into account the reduced capabilities of these kinds of devices.

## Basic Considerations About Memory and Energy for ACO Algorithm Implementing

Bio-inspired algorithms to optimize processes require several iterations to find the best possible element $x$ for a set **X** according to a set criteria $F=\{f_1,f_2,\ldots,f_n\}$. When the relation between a solution candidate and its "fitness" are not obvious, too complicated, or the search space is dimensionally very high, a distributed process could be a good option. Thus, work is divided among the devices forming the network with the devices acting as a team. Consequently, when one device has insufficient resources, it uses the neighborhood device resource to fulfill its task. For embedded devices, centralized solutions are not suitable because of the scarcity of resources.

Random number generation on embedded devices is an expensive process, because of their constrained resources. Therefore, the use of sensing samples could be a suitable solution to this matter. In our scenario, ECG provides an excellent source of randomness that does not add any operations and, as a result, is an excellent source for resources.

However, the sensing process and the function related to ACO are not expensive regarding memory resources because many of their variables can easily be held in 1 byte. Figure 6 shows structures used to implement the Nuku algorithm.

The ACO transition function is the most consuming resource task of the mote, due to the algorithm's kernel uses float variables, The kernel process follows a cycling behavior (inner cycle) under the control of an objective function. Inside this cycle, measures are continuously evaluated and, as a result, positive or negative feedback is applied to the context (nodes along the routes), thereby decreasing or increasing the pheromone level of the routes.

ACO algorithms take information from their surroundings and share it. (e.g., the pheromone level of the neighborhood and the nodes

```
struct node {
    unint8_t circuit[8];
    unint8_t pheromone;
    unint8_t eta;
    unint8_t links;
    unint8_t routes[9];
    unint8_t ttl[9];
    unint8_t conocov[3];
    char service;
    unint8_t evaporation;
};
```

```
struct ant {
    unint8_t recluting;
    unint8_t source[2];
    unint8_t TTL;
    char      destination[8];
    uint8_t  evaporation;
    unint8_t roundTrip;
};
```

**Figure 6.** Two data structures used to develop Nuku.

transmitting data, other nodes waiting for messages from the Internet, and link quality). However, the procedures for listening and sending data are the highest energy consumers and are consequently the most expensive. Therefore, to conserve energy resources, it is important to know the host software (e.g., Operating system software and hardware drivers) that is to be used to employ all of the data provided by the above mentioned systems, instead of computing it again or adding redundant functions.

## Implementing Location Privacy Routing with ACO

Despite no questions arising regarding the usefulness of ubiquitous services and devices, it is true that an equally large set of security and privacy concerns have arisen (e.g., Alemdar and Ersoy 2010; Li et al., 2009). Let us take, for example, WSNs, where common attack vectors include impersonating a sensor node making it possible not only to eavesdrop personal data, but to perform Man in the Middle attacks (MITM,), and even to accurately compute a user's geo-location.

Geo-location based services, as used in e-Health applications, allow patients to live by themselves in either their homes or "e-near" their families, friends, and medical services (staying in touch by means of electronic devices or Internet Services). These services are based on geo-location technologies like e.g., Global Positioning System (GPS) with outdoor coverage and Radio Frequency Identification (RFID) for short indoor or outdoor coverage. However, wireless technologies—just like WSNs—can avoid coverage restrictions by following the hop by hop transmission pattern to reach a device. Even more, with a device like a gateway WSN can joing two different network technologies.

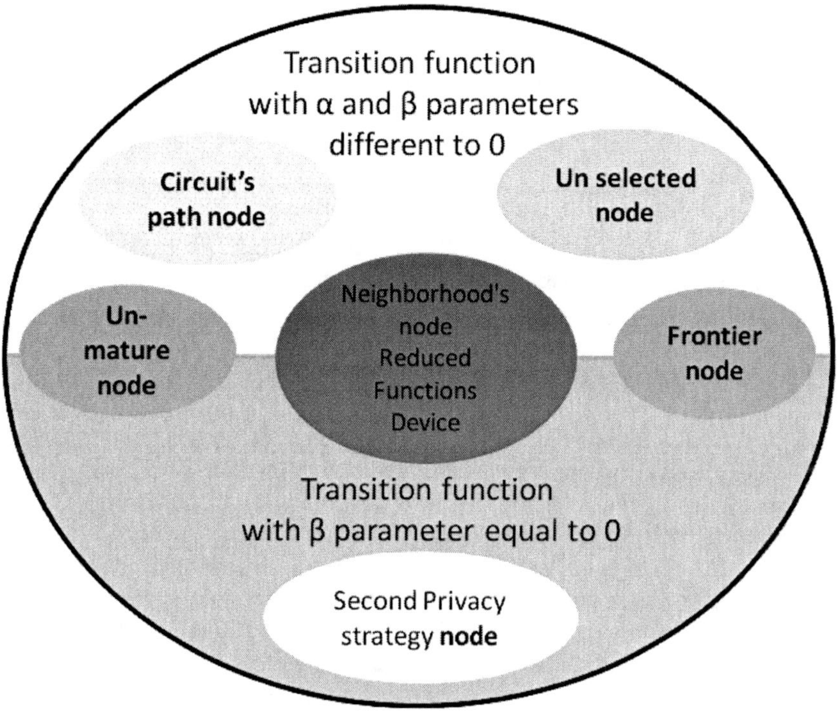


**Figure 7.** Nuku's transition function and sort of nodes.

The rest of this section presents a way to develop ACO-based routing in order to provide location privacy to WSNs used in e-Health applications.

## ACO-based Privacy for WSNs

Information Privacy is defined as an individual right to control the acquisition, use, and disclosure of his or her identifiable data as well as the right to claim these data as property. A person's geolocation is considered Personal Identifiable Information (PII). We propose an approach to protect it through a bio-inspired algorithm called Nuku (*Nu*cu *K*eep *U*n-tracking).

## Nuku Privacy Routing Algorithm

The ACO algorithm drives Nuku's kernel behavior. When a RFD must send a packet, ants start to search for routes to sink devices, traversing the FFD. Once they have searched successfully, the ants head back to the source node executing an "activation" procedure. Ants walking back through the route, or Backward-ants (Di Caro et al., 2004), will lay a maximum trail of pheromones related to the route threshold (Fig. 10). In addition, the evaporation value is computed. This value is proportional to the maximal value of pheromone along the route and the distance between the source and destination nodes affected by a privacy function called time to live private (TTLp).

Nuku employs two strategies to protect privacy (Fig. 8).

The first one computes the routes or "circuits-path". The circuits-path has a unique ID, and the packets traversing along them inherit this ID label. The route can transport more than one data flow belonged to different sources nodes; this is because of, circuits frequently share nodes. At end of the circuit, there is a node called the "middle-node," which defines the frontier between the first and second privacy strategy. Middle-nodes are defined as "immature nodes," because they do not have sufficient pheromone levels to become part of a second strategy. The "immature node" behavior is evaluated for a period "*j*" by a trust mechanism. Once the period is completed, if the node shows and demonstrates a "good behavior," it becomes a "mature node," which increases its pheromone level and takes the privacy primitives to a second phase. The expression "good behavior" means that, the immature node helped circuits-path nodes to send packet inside of "food zones", and as a result, circuit-path nodes pay for it with pheromones. The trust mechanism's aim is to enforce

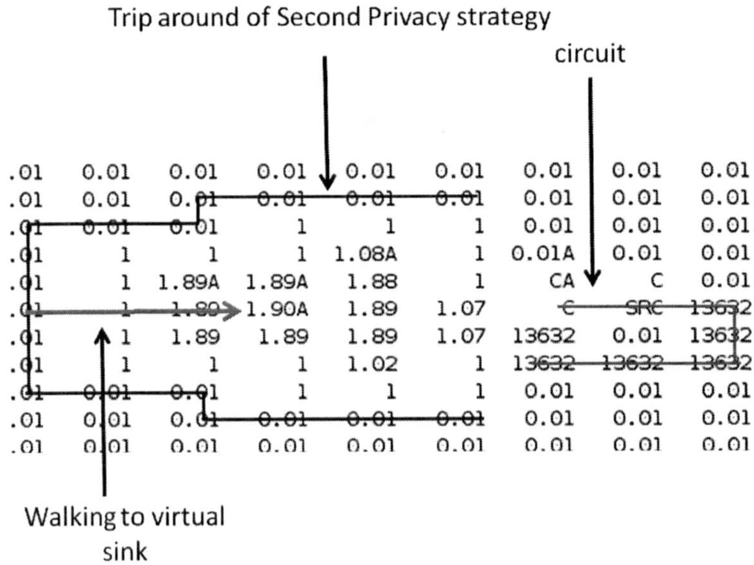


**Figure 8.** Privacy strategies by phase on transmissions.

*(Color image of this figure appears in the color plate section at the end of the book.)*

privacy, reducing possible misbehavior on the frontier of the "destination node neighborhood." If an "immature" node on this zone did not do its work, its pheromone level will be less attractive for ants and nodes with higher pheromone levels will be *"more qualified,"* and consequently more attractive.

The second strategy starts when packets enter the "destination node neighborhood." This type of structures follow a uniform distribution, along the scenario. Inside of them, a key device called a virtual sink is found; this node will have the highest pheromone level in the scenario. Virtual sinks support requests for services and data flow, they forward traffic to sink nodes. Because they are geographically distributed, they are a gate for traffic to the sink. The virtual sinks are also important because they add capabilities to the scenario to mitigate potential attacks.

**Considerations Using the Transition Probability Function by Nuku**

To perform the route selection and data transmission processes, Nuku considers six types of nodes (showed in Fig. 7). Four nodes have pheromones because they belong to structures (e.g., circuits or virtual sink, destination node neighborhood), whereas the other two do not. The decision is driven by two parameters: visibility as $\beta$ and the pheromone level as $\alpha$. Therefore, node selection depends on the physical radio characteristics (e.g., noise, reception sensibility, etc.), node current state (e.g., transmitting, waiting for routes, waiting for transmissions, etc.), as visibility parameters and the pheromone level on the circuit-path.

Once the $ant_k$ selects a neighbor node as its next hop, two variables must be updated according to the node type and phase: Nuku's "tabu buffer" and the pheromone level. In the original ACO algorithm for the former variable, each ant carried a list of unvisited nodes (tabu list) and in our proposal, visited nodes are tagged. The second variable concerns the pheromone levels on circuit-path's nodes, this value decreases while a transmission is in process; but in an immature node, the value of pheromone is increased, if its behavior is trustworthy.

**Backward-ant Mechanism**

In the process of circuit activation, ants walk back and lay pheromones until they have reached the source node, getting them ready to transport packets. The pheromone levels in the scenario correspond to the proposed thresholds shown in Fig. 10.

The return process uses the model represented in Equation 5 to set the level of pheromone affecting each of the nodes found in the trail along the

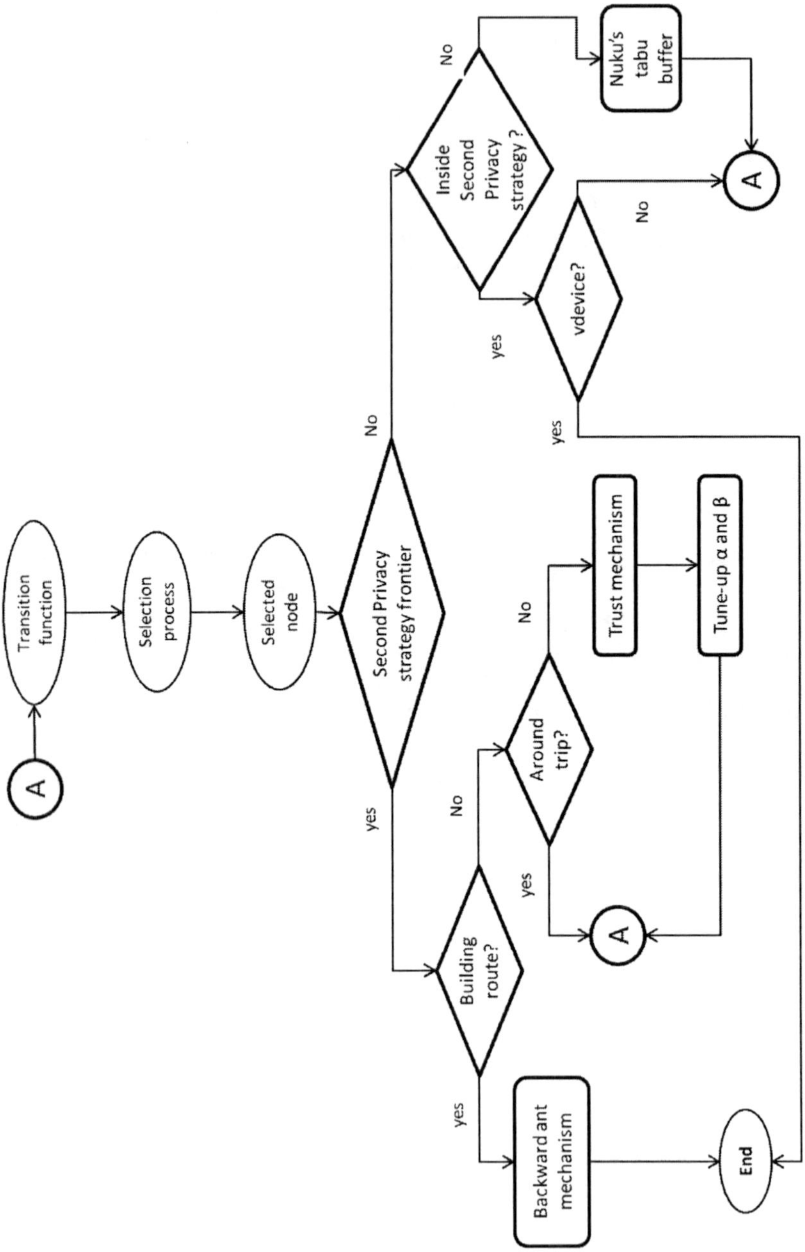


**Figure 9.** Nuku's processes.

circuit path. Each node controls its pheromone level and the evaporation quantity; they exchange information with their neighboring nodes about changes concerning the type of node (immature, mature, food zone, frontier node, etc.).

$$Ph_{it}^{r} = \left[ \left( Ph_{c} \right) / \left( TTLp \right) \right]$$

**Equation 5:** Pheromone evaporation on the route.

Where:

$Ph_{it}^{r}$ *Pheromone to be evaporate on node—i at time t and circuit—part r*

$Ph_{c}$ *Maximal pheromone level along the route*

*TTLp Length of CircuitC, hboring by privacy function*

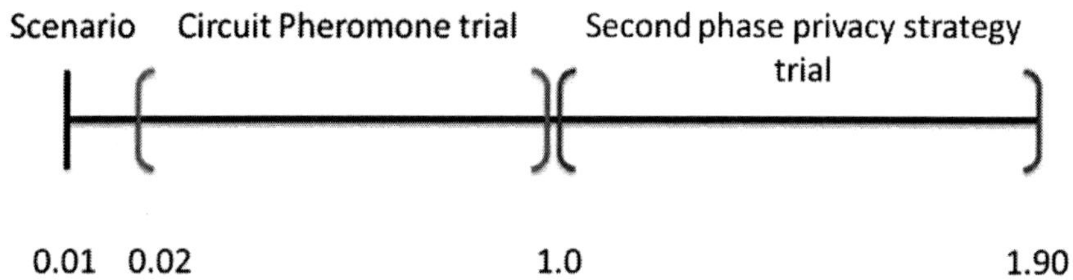


**Figure 10.** Nuku's pheromone thresholds.

## Experimental Results

As proof of concept, attack models are required to prove Nuku's performance and their adversaries' characteristics, which include the following:

### Attack Models

The *Clever adversary* attack model is considered specifically for our solution, whereas *Cautious Adversary* and *Passive Adversary* were originally proposed (Kamat et al., 2005) for the Phantom algorithm. Additionally, we also use the Phantom algorithm as a reference to compare the performance of our proposal.

The "Tracking attack" goal is to gain private information for long periods through non-intrusive mechanisms. To avoid this, mechanisms wishing to avoid this attack must be made aware of adversaries by listening to the same environmental parameters and capturing part of the network by storing and processing data that passes through them. The damage that might be inferred by static adversaries is related to the likelihood of the adversary's location and the packet's crossing route, as well as the adversary's capacity to break the encryption mechanisms. Consequently, likelihood of a successful attack is low when the network is big. This leads to analyze mobile adversaries aware of environmental perturbations because of transmissions, and propose a solution to protect the flow of source data.

## Adversaries' Description

*Passive adversary:* This type of adversary starts its attack on a sink node or two hops away of sink node ($node_i$). It waits for a packet arriving to the sink node or $node_i$; once a packet arrives, it moves to an immediate source $node_{(i+1)}$, and repeats the same procedure until it founds a source node. This adversary represented the greatest danger for the Phantom algorithm because of the probabilistic flooding being used.

*Cautious Adversary:* This kind of adversary is much like a passive adversary; the major difference, however, is that it will wait for a specified time at sink node or $node_i$. Once this time has passed, if the adversary has not sensed any transmission, it will return one hop to one visited node ($node_{(i-1)}$) and it will wait for a new transmission.

Described adversaries were initially proposed to test the performance of the Phantom algorithm (Kamat et al., 2005), mainly because of Phantom uses flooding at its second stage to transport packets, and Nuku does not. Therefore, we propose an adversary with more successful probabilities that knows when a transmission is being sent through a route near it.

*Clever adversary*: The attack can start on the *virtual sink* or two hops away of it. Once the adversary senses a packet, it starts to follow the transmitting source although the packet destination is not the adversary's node. This adversary has memory so it can remember which nodes it had visited, thus it never returns to a previously visited node, unless a set time for waiting has expired and no more packets are being transmitted by nodes in the vicinity. This attitude becomes more aggressive because the attacker will not lose its time waiting for a packet to come.

## Adversary's Knowledge of Network Protocols and Privacy Mechanisms

Based on the Kerckhoff Principle, we assumed that the adversary has ample knowledge on network protocols and privacy mechanisms. Besides, the adversary knows the sub-graph nodes distribution as well as nodes' relations but it does not know their exact location.

### Adversary's Capabilities

The adversary has software to perform traffic analysis, and to find out the type of each packet (e.g., data or control packets). However, packet payload is cipher on source node, therefore the adversary cannot access to data contained. To perform tracking, the adversary's speed is as WSNs' transmission rate.

### Performance Metrics

The considered scenario is an e-Health monitoring service in which a WSN must send physiological data over a specified area (e.g., a house or a hospital). This scenario considers a continuous power source and that transmitted data are in the form of packet flows (e.g., voice, video, or vector data—EGG-).

Considering the aforementioned conditions, the metrics selected for the performance analysis are:

- *The safety period*: For a given adversary like in Phantom routing (Kamat et al., 2005), this parameter refers to the number of safe packets being sent by the source node to the destination node where there is no adversary that knows the source.
- *Capture likelihood*: The probability of an adversary finding the transmitting source.
- *Latency*: This parameter is defined as the time elapsed since the first bit is sent to the network until a destination node receives it.
- *Loss packets*: This parameter measures packets that do not arrive on time to the destination node. Because the transmitted information is physiological data, this parameter is a critical measure used to evaluate the "quality of monitoring".

### Network Model and Performance Parameters

The adversaries and the Nuku algorithm were implemented in C code, using a computer PII with 2 Gigabytes RAM and the Ubuntu 10.10 operating system. Besides the experiment configuration follows the Phantom structure (Kamat et al., 2005) to better compare results.

To demonstrate our proposal, we configured our scenario as a 100 * 100 grid of equidistant nodes, all within a previously defined radio to insure good coverage; this parameter defines the restrictions for the number of nodes in the vicinity of each node. In addition, the scenario follows IEEE 802.15.4 (2006) standard restrictions regarding packet payload, including the header packet. The scenario had one-sink and one-source node as

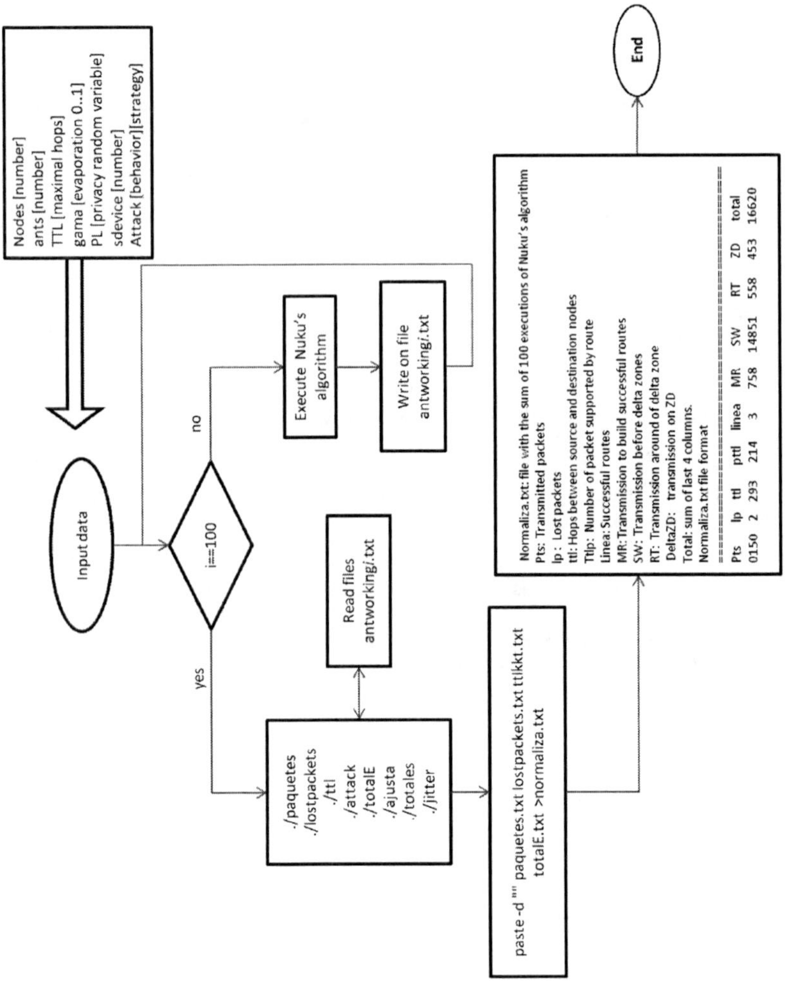


**Figure 11.** Nuku's results processing.

its Phantom structure (Kamat et al., 2005). We implemented two attack strategies and one additional adversary so that we could compare results.

Nuku's results were obtained by running the algorithm 100 times for each scenario, as shown in Fig. 9. The results were stored on a "normalize. txt" file and then used to analyze Nuku's performance.

In the first stage, the algorithm was configured considering parameters such as the number of nodes, ant population size, and Time To Live (TTL) as maximal search size in hops. We do not consider the evaporation variable ρ in our approach because Nuku's negative feedback behavior removes pheromones on the routes instead of adding them as the ACO algorithm. There is a parameter to allot privacy level, it takes values following a uniform distribution between 0 and 1, and is used to define privacy level on transmissions; where values near to one means low privacy, and values near to zero are used for transmissions with high privacy requirements. The geo-location of sink and virtual devices should be defined in the scenario. Finally, Nuku's behavior can be evaluated under 3 types of passive attacks; it is not possible to use more than one adversary at the same time; hence, the scenario must be re-initialized to use another adversary.

**Results**

Table 1 summarizes the proposal reviewed; some of them developed their own experiments based on Phantom algorithm (Kamat et al., 2005), and reproduced it; as a common characteristic, all proposals build their routes in a dynamic manner, thus allowing us to compare results under similar conditions. We selected loss packet and latency parameters because these are key issues in delay sensible communication systems; as our eHealth scenario.

Table 1 shows that latency value in proposal A is similar to result obtained on C. B has the worst perform, considering this parameter; however, it avoids flooding techniques for routing, and thus has advantage of using significantly less energy. In proposal E, the authors show that their protocol does not add latency and it performs better than A because of the way it makes the decision to advance to the next node, which is based on link quality. Authors of E did not considered latency because they select the best local node (link quality) in a specific time, however results let us to see that latency exist when we compare versus short-path solution. Also Table 1 shows that all proposals increase the loss packet parameter when the privacy mechanism is turned on, except for proposal E.

Next, we show the results obtained for Nuku. Table 2 shows *Loss packet for* Nuku. These values are at different distances in hops between the source and destination node. These results were obtained using one virtual device and one source; on the other hand, the worst case is when the destination

**Table 1. A summary of reviewed proposals.**

| | *Approach* | *Loss Packet* | *Latency* |
|---|---|---|---|
| A | (Kamat et al., 2005), Probabilistic Flooding (70% of retransmission probability) | This value is related to probabilistic flooding value, thus 30% of packets are lost. | Average message latency (as a hops number, there are not information in this article about time): 70 hops is Average Message Latency (AML) for directed walk hop-count is 0; 91 AML for 20, and 110 of AML for 40 directed walk hop-count. |
| B | (Xi et al., 2006), Greedy random walk | It is associated with $t^{-5/8}$ value that determines the likelihood of two Random Walk do not crossing. (The length of the Brownian motion path is the value of t.) Almost a 5% of packets are lost. | The sum of *t* and intersection node's distance to sink defines latency value (measures in seconds, the information about Phantom was provided by author on this paper). Their results show that 80% of packets received by sink node after 12 seconds, where Phantom spend 1 second. |
| C | (Li and Ren 2010), Single intermediate node selection | Packet length and packet rate generation and route length affect the delivery radio, thus in the worst case it is 0.35% and in the best case 0.05%. | The performance is as phantom, when the authors constrain the distance between intermediate node and the source. The rest of techniques have a worse performance than Phantom. |
| D | (Li and Ren 2010), Angle base, multi-intermediate nodes | As the angle increases, packet lost increases its value from 0.1%, to 0.4% for angles ranging from 0 to 200. | This algorithm can transmit 128 bytes through 480 meters with a delay of 0.039 seconds as the best case, while the worst case is 0.109 seconds. |
| E | (Spachos et al., 2010) Opportunistic routing | The authors define this parameter as 0, due to they consider ACK mechanism between hop by hop transmissions as a solution to this problem. | Their results shown that, a 50 hops route length can have 28 hops for latency, and a route of 35 hops, the latency is 20 hops |

node is 60 hops away from the source. Nuku's best performance is when the distance between source and destination is 4 hops, following by the 34-hop distance, which had a 1.12% packet loss ratio.

**Table 2. Loss Packet ratio for Nuku (Cosio-Leon et al., 2011).**

| *Distance on hops between S and D nodes* | 2 | 4 | 8 | 16 | 34 | 60 |
|---|---|---|---|---|---|---|
| Packets | 100 | 73806 | 10770 | 5872 | 21145 | 5017 |
| Packet loss | 0 | 606 | 278 | 66 | 214 | 996 |
| Packet loss ratio | 0 | 0.82% | 2.58% | 1.12% | 1.01% | 19.85% |

*Capture likelihood parameter* is showed in Table 3. This parameter was obtained considering each one of the three adversaries previously described. Each one of them starts at the virtual sink node. Capture likelihood for proposal A and E follow a similar condition, but E is the best when the distance between pair source-destination is of 50 hops. Nuku's likelihood capture parameter has a constant behavior and it is not related to route length.

Nuku's Safety Period (Table 3) is better than any other proposals shown in Table 1. The last two rows in Table 3 are about the Clever Adversary, but we do not have information about proposal A, because a clever adversary is a future work for their proposal.

**Table 3. Capture likelihood and safety period.**

| *Attack type* | *Nuku* | | *Phantom* | | *Source—Destination Hops distance* |
|---|---|---|---|---|---|
| | *Safety period* | *L Capture* | *Safety period* | *L Capture* | |
| 1 | 396 | 0.15 | 90 | 1 | 34 |
| 1 | 154 | 0.38 | 32 | 1 | 8 |
| 2 | 249 | 0.24 | 301 | 0.6 | 34 |
| 2 | 67 | 0.88 | 54 | 0.9 | 8 |
| 3 | 211 | 0.28 | ND | ND | 34 |
| 3 | 59 | 1 | ND | ND | 8 |

L Capture: Likelihood of capture All attacks start inside virtual sink. 1 referees to passive adversary, 2 to cautious one and 3 to the clever adversary. S-D: Source-virtual sink.

It is convenient to mention that all solutions described in Table 1 are aware of the tradeoff between privacy and latency, and with this knowledge they justify latency overhead.

However, the tradeoff between privacy and latency is not necessarily true, because of authors of proposal E proved that high latency is not a characteristic for privacy algorithms as Nuku does.

## Future Research Directions

Nuku incorporates innovations into the location-privacy issue using bio-inspired procedures. The next immediate steps include take privacy solutions from Internet to autonomous embedded devices; we are now able to move the conscience of location-privacy in WSN to gateway devices that have Internet connections. On the other hand, it is clear that Nuku needs to be changed from its "personal" space to simulators that include the IEEE802.15.4 (2006) standard to measure node performance and, therefore, WSN performance. For us, scalability has an important meaning so we need to populate the IEEE 802.15.4-described scenario with more than two destinations and source devices. Finally, circuits (path, routes) are structures that offer security and privacy conditions that need to be improved and evaluated.

### Conclusions

CI paradigms are suitable solutions for WSNs; therefore their use might be a good option when overall system performance is paramount. However, it is necessary to take into account considerations about data management, data types, and routines to develop on embedded devices.

With the purpose of knowing Nuku's privacy costs and evaluating its performance, we developed a simulator, which covers Nuku's privacy structures and processes. As a result, we have data that prove the ACO algorithm can be implemented on motes, making only minimal changes to conserve scarce resources.

## REFERENCES

Alemdar, H. and C. Ersoy. 2010. Wireless sensor networks for healthcare: A survey. Computer. Network, 54 , pp. 2688–2710. Available from http://dx.doi.org/10.1016/j.comnet.2010.05.003

Alkazemi, B.Y. and E.A. Felemban. 2010. Towards a framework for engineering software development of sensor nodes in wireless sensor networks. In Proceedings of the 2010 icse workshop on software engineering for sensor network applications (pp. 72–75). New York, NY, USA: ACM. Available from http://doi.acm.org/10.1145/1809111.1809130

Ashby, W.R. 1960. Design for a brain; the origin of adaptive behavior. New York,Wiley. Available from http://www.biodiversitylibrary.org/item/31036 (http://www.biodiversitylibrary.org/bibliography/7309)

Ashby, W.R. 2004. Principles of the self-organizing system. Principles of Selforganization, (2000), pp. 102–126. Available from http://csis.pace.edu/~marchese/CS396x/Computing/Ashby.pdf

Blumenthal, J., M. Handy, F. Golatowski, M. Haase and D. Timmermann. 2003. Wireless sensor networks—new challenges in software engineering. In Conference on emerging technologies and factory automation (Vol. 1, pp. 551–556). IEEE. Available from http://ieeexplore.ieee.org/lpdocs/epic03/wrapper.htm?arnumber=1247755

Cass, R. and J. DePietro. 1998. Computational intelligence methods for process discovery. Engineering Applications of Artificial Intelligence. **11(5)**:675–681. Available from http://linkinghub.elsevier.com/retrieve/pii/S0952197698000335

Chaum,D. 1985.Security without identification: transaction systems to make big brother obsolete. Common. ACM , 28 , 1030–1044. Available from http://doi.acm.org/10.1145/4372.4373

Cho, Y., S. Cho, D. Choi, S. Jin, K. Chung and C. Park. 2004. A location privacy protection mechanism for smart space. In K. Chae and M. Yung [eds.], Information security applications (Vol. 2908, pp. 47–61). Springer Berlin / Heidelberg. Available from http://www.springerlink.com/content/d383xrw7wty593cq/(10.1007/978-3-540-24591-9 13)

Cosio-Leon, M., J.-I. Nieto-Hipolito and J. Luna-Garcia. 2011. Nuku privacy algorithm for WSNs (Tech. Rep.). Universidad Autónoma de Baja California.

Deneubourg, J. and S. Goss. 1989. Collective patterns and decision-making. Ethnology Ecology and Evolution. **1(4)**:295–311. Available from: www.ulb.ac.be/sciences/use/publications/JLD/53.pdf

Di Caro, G., F. Ducatelle and L. Gambardella. 2004. Anthocnet: An ant based hybrid routing algorithm for mobile *ad hoc* networks. In Parallel problem solving from nature-ppsn viii (pp. 461–470). Springer. Available from: www.springerlink.com/index/GQVYDM540KF0VDL9.pdf

Dierks, T. and C. Allen. 1999. The tls protocol version 1.0 [standards process]. http://www.ietf.org/rfc/rfc2246.txt. Available from: www.ietf.org/rfc/rfc2246.txt

Doosselare, C. van, J. Herveg, D. Silber and P. Wilson. 2008. Legally ehealth; putting ehealth in its european legal context (Tech. Rep.). EHMA, CISCO and CRID. Available from http://ec.europa.eu/information society/activities/health/docs/studies/legally ehealth/legally-ehealth-report.pdf

Dorigo, M., V. Maniezzo and A. Colorni. 1996. Ant system: Optimization by a colony of cooperating agents. IEEE transactions on systems, man, and cybernetics. Part B, Cybernetics: A publication of the IEEE Systems, Man, and Cybernetics Society. **26(1)**:29–41. Available from http://www.ncbi.nlm.nih.gov/pubmed/18263004

Estados Unidos Mexicanos, C. de Los. 2010. Ley federal de protección de datos personales en posesión de los particulares. Electronic and paper document. Available from http://www.diputados.gob.mx/LeyesBiblio/pdf/LFPDPPP.pdf

Freier, A.O., P. Karlton and P.C. Kocher. 1996. Draft-ietf-tls-ssl-version3-00-the ssl protocol version 3.0. Available from: http://tools.ietf.org/html/draft-ietf-tls-ssl-version3-00

Ghanekar, S., N. Alrajei and F. Mili. 2011. Bio-inspired intrusion detection for wireless sensor networks. In A.-S.K. Pathan [ed.], (Vol. unique, pp. 347–373). Auerbach Publications. Available from http://dx.doi.org/10.1201/EBK1439819197-19

Kamat, P., Y. Zhang, W. Trappe and C. Ozturk. 2005. Enhancing source-location privacy in sensor network routing. In Distributed computing systems, 2005. icdcs 2005. Proceedings. 25th ieee international conference on (pp. 599–608). IEEE. Available from http://ieeexplore.ieee.org/xpls/abs all.jsp?arnumber=1437121

Kelly, D. 2009. Taxonomy for and analysis of anonymous communications networks. Electronic. Available from: www.freehaven.net/anonbib/cache/Douglas-thesis.pdf

Kulkarni, R.V., A. Forster and G.K. Venayagamoorthy. 2011. Computational intelligence in wireless sensor networks: A survey. IEEE Communications Surveys and Tutorials. **13(1)**:68–96. Available from http://ieeexplore.ieee.org/lpdocs/epic03/wrapper.htm?arnumber=5473889

Li, N., N. Zhang, S.K. Das and B. Thuraisingham. 2009. Privacy preservation in wireless sensor networks: A state-of-the-art survey. Ad Hoc Networks. **7 (8)**:1501–1514. Available from http://linkinghub.elsevier.com/retrieve/pii/S1570870509000407

Li, Y. and J. Ren. 2010. Source-location privacy through dynamic routing in wireless sensor networks. 2010 Proceedings IEEE INFOCOM, 1–9. Available from http://ieeexplore.ieee.org/lpdocs/epic03/wrapper.htm?arnumber=5462096

Man, L.A.N., S. Committee and I. Computer. 2006. IEEE standard for information technology telecommunications and information exchange between systems- local and metropolitan area networks—specific requirements–part 15.4: Wireless mac and phy specifications for low-rate wpans (Vol. 2006) (No. September).

Picco, G.P. 2010. Software engineering and wireless sensor networks: Happy marriage or consensual divorce? In Proceedings of the fse/sdp workshop on future of software engineering research (pp. 283–286). New York, NY, USA: ACM. Available from http://doi.acm.org/10.1145/1882362.1882421

Saleem, K., N. Fisal, S.H.S. Arinffin, S.K.S. Yusof and R.A. Rashid. 2011. Biological inspired autonomously secure mechanism for wireless sensor networks. In A.-S.K. Pathan [ed.], (Vol. unique, pp. 375–408). Auerbach Publications. Available from http://trg.fke.utm.my/members/kashif/bookcha1.pdf

Senate and Congress assembled, H. of Representatives of the United States of America in. 1996. HIPAA statute. Available from: www.hhs.gov/ocr/privacy/hipaa/administrative/statute/index.html

Spachos, P., L. Song and D. Hatzinakos. 2010. Opportunistic routing for enhanced source-location privacy in wireless sensor networks. Network, from 315 to 318. Available from http://ieeexplore.ieee.org/lpdocs/epic03/wrapper.htm?arnumber=5472946

Steele, R., A. Lo, C. Secombe and Y.K. Wong. 2009. Elderly persons' perception and acceptance of using wireless sensor networks to assist healthcare. I. J. Medical Informatics. **78(12)**:788–801. Available from http://dx.doi.org/10.1016/j.ijmedinf.2009.08.001

Xi, Y., L. Schwiebert and W. Shi. 2006. Preserving source location privacy in monitoring-based wireless sensor networks. In Ipdps. IEEE. Available from http://dblp.uni-trier.de/db/conf/ipps/ipdps2006.html#XiSS06

## KEY TERMS AND DEFINITIONS

E-Health: Term used to characterize everything related to information technologies and medicine, as well as the wide variety of services that can be developed and offered to people when they are put into Internet.

Nucu: Word from ethnical culture of Mexico, refers to *Atta fervens* female ant, this sort of insect can measure until 2.5 cm of length. Some people like to eat them.

Fitness function: This function helps to quality measure on solutions as a problem solution.

Heuristic: It is part of an optimization algorithm that uses the information gathered by the algorithm to help to decide which solution candidate should be tested, heuristics are problem class dependent.

Metaheuristic: It is a method for solving very general classes of problem. It combines objective functions or heuristics.

Positive feedback: The behavior of a system in response to a perturbation, it acts to increase the magnitude of the perturbation, unless the feedback loop is controlled by being clamped, dampened, gated, channel-limited, or otherwise physically limited.

Negative feedback: Almost all systems search for the equilibrium, this sort of behavior helps them get stabilization, hence when a perturbation increasing the system's parameters, it react decreasing the perturbation.

# 6

# Indoors and Outdoors Event Detection for Embedded Wireless Sensors

*Marco Antonio López Trinidad, Maurizio Valle and Cora Beatriz Excelente Toledo*

**ABSTRACT**

We show by experimentation that the effectiveness of acoustic event detectors is strongly influenced by the work environment, sensor signal filtering and event detection algorithm implementations. In particular, our work considers networks of wireless embedded hardware and two application scenarios. The first application is aimed to detect human activities that are developed in *indoor* environments, home or office buildings, whereas the second one detects moving motor vehicles within urban environments (*outdoors*) *sensing* their motor sounds. In each case, the event detector implementations are made up of two components, one filtering and one detection algorithm, designed considering the signals features observed in the environments of interest. In our experimentation we analyze the performances of event detectors integrated by two different implementations of the event detection modules. Particularly, in the indoors event detection scenario the filtering module issues an *adaptive low-pass filter*, whereas the event detector module implementation includes experimental results running a *signal energy change rate estimator* and a *signal energy threshold based detector*. On the other hand, in the outdoors case the detector filter is a *46 taps band-pass filter* and like in the indoors case, the event detector module considers the use of the signal energy change rate estimator and the signal energy threshold based detector. The realization of the event detectors is guided by the results of a calibration procedure, for this purpose we gather actual acoustic signals of the environment under consideration. In the case of the indoors scenario three basic human activities have been collected: *closing a door*, *dropping a plastic bottle*, and *clapping*. In the outdoors case, we register two types of car sounds: *a single car passing* and *a car stopped*. To quantify the event detection effectiveness we employ more general environmental acoustic signals such as: *people talking* for the indoors case and *a flow of cars moving down street* for the outdoors case. The indoors event detection results showed that the best events rate detection was achieved by the detector made up of the low-pass filter and the energy change rate estimator arriving up to 90% of the people talking activity detection. In the outdoors case, we find that unlike the indoors case, the event detector featuring the band-pass filter and the energy threshold detector has the best performance, accomplishing up to 85% of the vehicle signals detection.

## Introduction

To the date the detection of disperse events occurring simultaneously in large portions of land can be realized by the use of a *Wireless Sensor Network* (WSN) which features a large collection of tiny *wireless sensor nodes* or *motes*. These small motes, which are deployed in the environment, include facilities to *acquire* environmental physical data, *process* sensors data and *communicate* with other sensors in the network, to receive and transmit sensor data.

Particularly, an events detector system analyzes sensor data series searching for those subsets that contain substantial variations. These variations or *events* belong to unusual behaviours that may indicate *system bad-functions, intrusions* of individuals, *leakages* or *plume spreads* of chemical substances in the environment, *unpredicted changes in the climate conditions*, etc. (Anastasi et al., 2009; Gao et al., 2005; Johnsonet al., 2006; Ackdere et al., 2008; Gupchup et al., 2007 and Kerman et al., 2009; Kapitanova et al., 2010).

The detector design becomes more complex when the targeted hardware platform is an embedded device like the wireless motes that are extremely resource constrained such as *memory size available, processor power and speed, null floating point support, communication low rate* and very limited *electrical energy battery* (Hill et al., 2000; Polastre et al., 2005). In this manner a mote application design must consider resource efficiency instead of high performances. In fact, searching for mechanisms that enable the mote to save electrical power energy is still an open research question. The most widely used electric power saving strategy is switching the motes between the *processing* and the *low power energy states*, waking them up frequently enough to gather readings of their sensors, realize networking operations and then go back to sleep (Hill et al., 2000; Polastre et al., 2005; Wang and Yang 2007). (Polastre et al., 2005; Wang and Yang 2007) have reported that the radio communication operations, transmitting, and receiving data packets are important electric energy consumption sources that can reduce dramatically the sensor network life-time. Therefore, WSN applications must be designed to use the radio communications wisely.

Unlike centralized WSNs applications where all the motes in the network send raw sensor readings to *data collectors, sinks* or *coordinators* nodes, to be processed or forwarded. In-networking signal processing is a very efficient method to preserve the electrical energy of a wireless mote (Akdere et al., 2008; Bahrepour et al., 2010a; Bahrepour et al., 2010b; Gao et al., 2005; Gu, et al., 2005; Gupchup et al., 2007; Kapitanova et al., 2010; Kerman et al., 2009; Johnson et al., 2006; Liang and Wang 2005; Zhang et al., 2010). Under this approach, motes individually gather the surrounding environmental signals, and each mote is programmed to filter and analyse the sensor samples searching for significant events. Once the motes determine that an event is in course, a data packet is issued and sent to the data collector. In this manner, the number of packets a mote should transmit is reduced considerably.

Ideally, in an event-based application scenario, the motes are most of the time silent and only transmit packets when they observe irregularities. With the events information, WSN applications can make decisions such as switching off actuators or triggering alarms. Unfortunately, real world environments are extremely noisy, thus motes can incorrectly compute and report events that may not actually happen (*false positives*). As we have previously seen, when this problem happens frequently, the motes battery energy can be completely depleted in a very short time, thus significantly reducing the network lifetime. On the other hand, applications cannot rely on unreliable event estimations because they can lead to inaccurate inferences or erroneous predictions about a physical phenomena, machinery state or environmental situations. Therefore, the design of reliable event detectors is currently a critical WSN need (Ackdere et al., 2008; Gupchup et al., 2007 and Kerman et al., 2009; Kapitanova et al., 2010).

Generally speaking, an event detector algorithm runs a standard set of signal processing procedures such as *filtering, energy computing, and event estimating* over the sensor data in order to estimate event occurrences (Kerman et al., 2009). In our experience, the detection scenario (indoors or outdoors) and filtering stage strongly lead to the detection criteria being implemented. This article shows a design procedure and describes implementation issues regarding the realization of event detectors, based on the energy change rate and threshold approaches to detect acoustic events of indoor activities and the detection of cars presence, based on the engine sounds.

The organization of this chapter is as follows: in the methods section, we explain the filtering principles of the detector front-end, signal energy computation background, the energy change rate and energy threshold-based event detector algorithms. In the experimentation section, the detector tuning process and the detector performance tests are shown. Finally, in the conclusions and future work section, we provide a brief discussion about our achievements and provide a vision of the work that can be addressed in the area of event detection.

## Methods

### Systems and Experimentation Descriptions

Our research work is oriented to the realization of applications for embedded wireless hardware such as the Berkeley Micaz sensor motes (Hill et al., 2000). Since the Micaz has been the de facto research, development and deployment embedded platform reference. The Micaz hardware features an 8-bits microcontroller running at 8 MHz, has 128 KB memory for code, 4 KB of EEPROM, 512 KB of external memory, and a low cost, low power, short range Chipcon C2420 IEEE 802.14.15 compliant radio transceiver.

#### *Sensor Signals Description*

Our experimentation procedures consider the gathering of actual sensor signals from the environment of interest. In the indoors scenario, we employ a Micaz mote to record signal samples of human activities such as: *clapping, closing a door, dropping a plastic bottle* and *people talking*. The signal samples are 10-bits precision, sampled at a 2 KHz rate and recorded by a period of 20 seconds. The activity signals are grouped within two sets being *BHA* the set of basic signals such as closing a door *CLD*, dropping a plastic bottle *DPB*, and clapping *CLP*, see Fig. 1a), 1b), and 1c) respectively. On the other hand, *THA* is the test signal set and only considers the people talking *PT* signal, see Fig. 2.

In the cars detection scenario, we employ a desktop microphone attached to a portable computer to record motor sound samples such as: *one car moving down the street*, *one stopped car*, and *cars flow moving down the street*. The signal samples are 16-bits precision, sampled at a 44 KHz rate, and the record durations are variable because to observe a car sound state change, in some cases, there is the need to wait for longer time periods. Like in the indoors case, we group the motor sound signals within two groups, one basic signals set *BCS* and one tests signals set *CFM*. *BCS* groups the signals: *one stopped car CST*, and *car moving down the street SCM*, see Fig. 3a) and 3b) respectively. *TCS* Groups the signal *cars flow moving down the street CFM* and its duration is 120-seconds, see Fig. 4.

#### *Data analysis procedure*

Because the mote lacks of user friendly interfaces that allow tracing the algorithms execution, we process *off-line* the sensor signals in a Personal Computer (PC) system. In particular, the analysis procedure considers the design and tests of filter algorithms, and the event detector algorithm's design and tuning. This procedure is supported by the use of mathematical

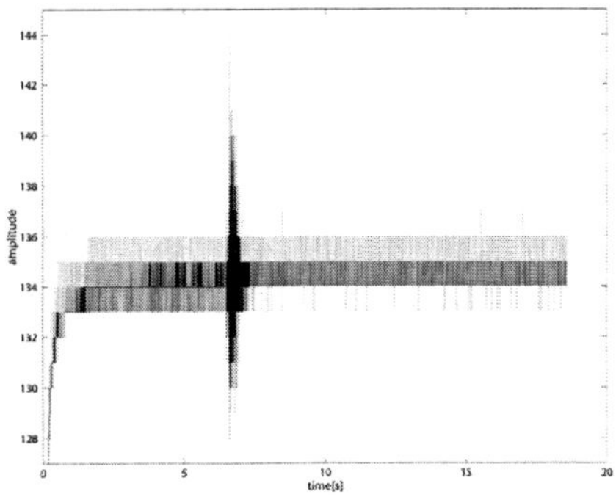

*a)*

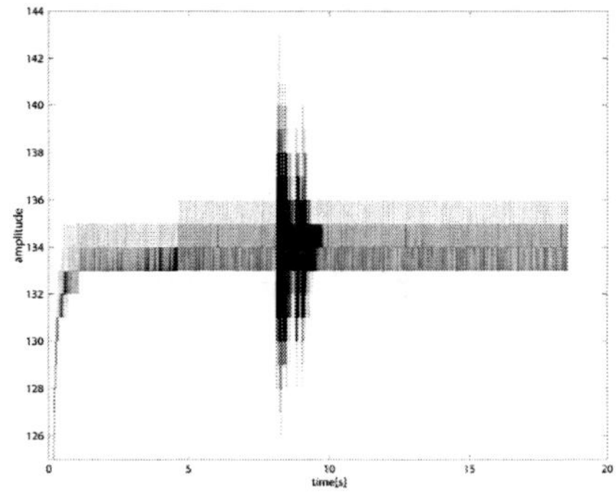

*b)*

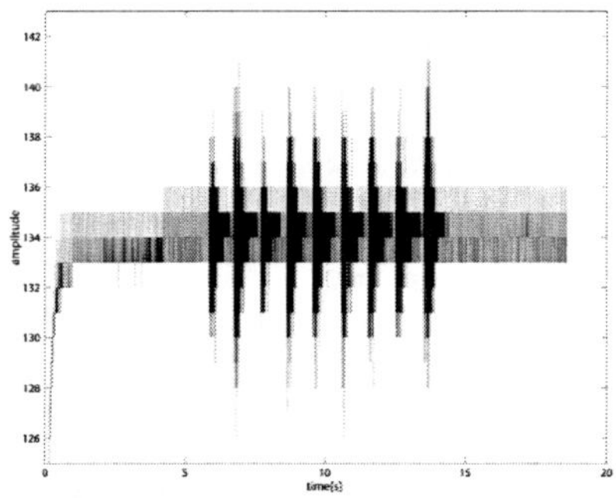

*c)*

**Figure 1.** Indoor people basic activity signals set *BHA*. (a) closing a door *CLD* (b) dropping a plastic bottle *DPB* (c) clapping *CLP*.

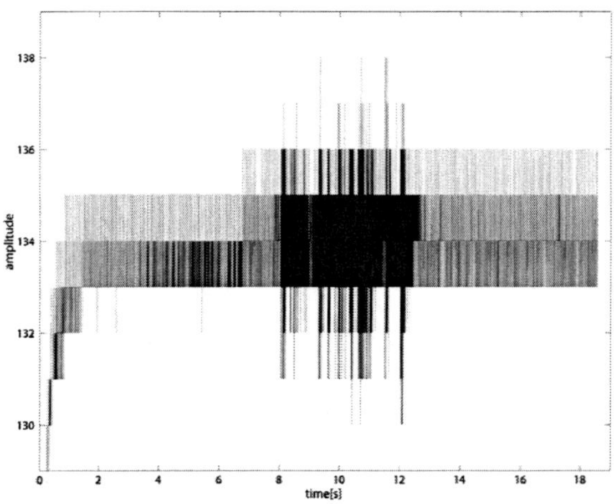


**Figure 2.** The indoor test *THA* data set, people talking *PT*.

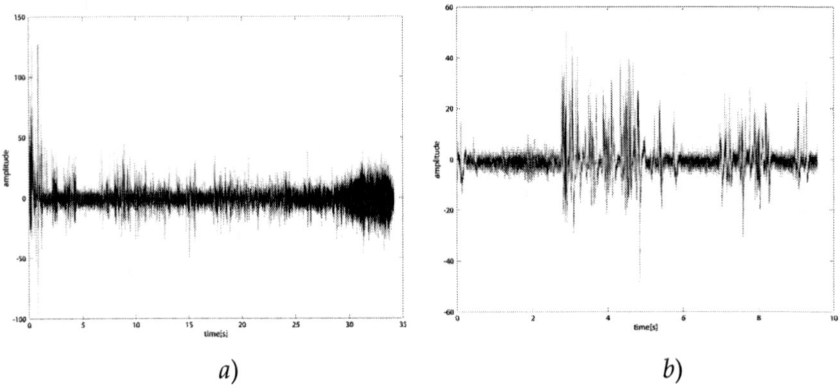


**Figure 3**. The car sounds set *BCS*. (a) signal of one stopped car *CST* (b) signal of one single car moving down *SCM*.

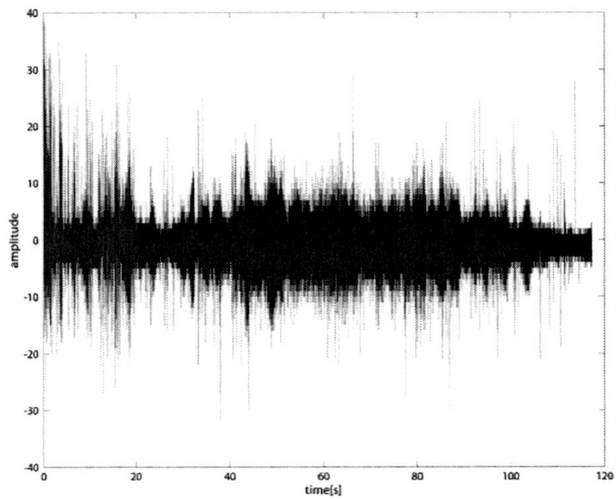


**Figure 4.** The test cars data *TCS* set, cars flow moving down the street *CFM*.

tools like Octave (Octave 1988), a Matlab type of program and AVRora (Titzer et al., 2005) a Mica and Micaz hardware emulator and sensor network simulator.

In order to observe the effects of the embedded hardware limitations, the indoor event detection performance tests are run in a PC, in a hardware emulated Micaz mote and in an actual Micaz mote. On the other hand, in the car detection scenario, the detector performance test results include only algorithm executions on a P.C.

## Signal Processing

In our approach, we are interested in adaptive in-networking signal processing algorithms that outline and accurately extract the relevant sensor signal features and do this economically in terms of memory space allocation, processing speed and electrical energy consumption, among others. This section provides details about the algorithm implementations for the filtering, computing of signal energy and detecting energy changes.

### *Indoor human activity signals filtering procedure*

In an indoor environment, we expect low noise background signal energy levels therefore we select a low pass filter as the detector filtering module. This is particularly the case of the *Infinite Impulse Response* (*IIR*) *Exponentially Weighted Moving Average* (EWMA) low pass filter (Steiner 1999, Woo and Culler, 2003), which is an economic resources algorithm. The acoustic signals frequencies we are interested in are located within the 50Hz to 2KHz range. In this manner, the filter output is computed by the relation

$$\langle s_k \rangle = \alpha \cdot s_k + (1-\alpha) \cdot \langle s_{k-1} \rangle \tag{1}$$

where $\langle s_k \rangle$ is the estimated average value of the input signal $s_k$ acquired at the time instant and $\alpha = 0.2$ (Steiner 1999) which gives us a 2KHz cut-off frequency. Fig. 5 shows the EWMA filter output $\langle s_k \rangle$ when the filter is fed with the *THA* signal.

It must be noted from Equation 1 that the filter output $\langle s_k \rangle$ is computed on the present signal sample and the last previous value of the signal average $\langle s_{k-1} \rangle$. Therefore, an implementation of the EWMA only requires about three 16 bit memory storage locations; instead of a large block of memory allocated for data and computations such as the convolution-based algorithms require, for instance the *Fast Fourier Transform* algorithm.

### *Car signals filtering procedure*

In the outdoors case, it is expected that the noise background presents higher signal energy levels. We therefore opt for a more selective filter whose implementation is still memory, computing and electrically spending economic. In this case our interest is the engine sounds that are

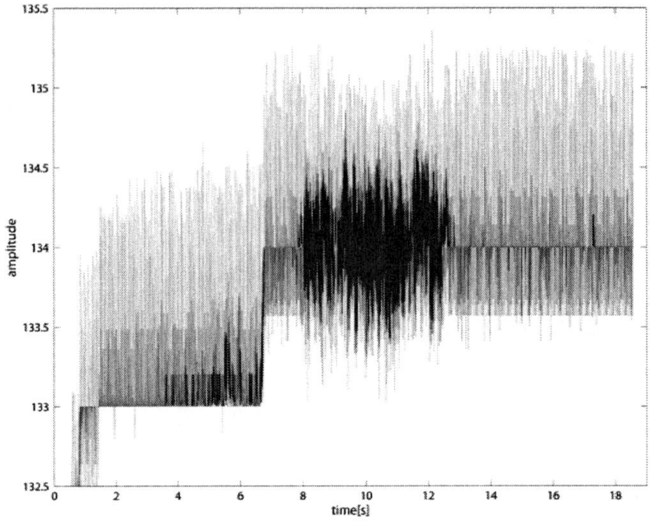


**Figure 5.** The indoor *THA* time series $s_k$ filtered by the EWMA low pass filter.

within the 500Hz-5KHz band (Varaiya 2004), therefore a 40 taps *Hamming FIR band-pass filter* is implemented. The filter taps are computed by the use of the filter design wizard tool FIWIZ (FIWIZ 2008) which employs the linear expression:

$$y_k = b_0 s_k + b_1 s_{k-1} + \ldots + b_N s_{n-N} \tag{2}$$

where $y_k$ is the filter output, $\{b_0, b_1 \ldots b_N\}$ are the filter taps and $\{s_k, s_{k-1}, \ldots, s_{k-N}\}$ is a N + 1 length vector of acoustic input sample $s_k$. In Figure 6, a plot of the Hamming low pass filter output is shown when the car *TCS* signal is fed to the filter.

Differently to EWMA implementations, we can observe that the computation of the filter output is obtained after about 40 sensor samples have been collected and the implementation requires a total of forty-one 32-bit memory locations, forty 16-bit memory locations, one 32-bit register for the filter output $y_k$, forty 32-bit registers for the filter taps and forty 16-bit registers for the acoustic input signal $s_k$.

As expected, comparing the input signal, see Fig. 4, and the filter output, see Fig. 6, we can observe that the noise levels are significantly reduced enhancing the car sound signal.

### *Indoors signal energy computation*

We compute events on the base of the changes observed in the signal energy to that end we search for economic signal energy computation algorithms. In this case we employ the energy estimator reported in (Gu et al., 2005).

$$e_k = |s_k - \langle s_k \rangle| \tag{3}$$

where $\langle s_k \rangle$ is the average value of the input signal $s_k$ computed at the $k$ intant.

We can observe that an implementation of Equation 3 requires only two 32-bits and one 16-bits memory registers: one 32-bits for the computation of $e_k$, one 32-bits for the computation of $s_k$, and one 16-bits for input signal $s_k$. Although other estimation methods are reported in the literature (Lian and Wang 2005), we experimentally have observed that Equation 3 provides better estimations energy consumption, even when sensor signal is weak. An estimation of the indoor *THA* energy $e_k$ is presented in Fig. 7.

### *Car signal energy computation*

The signal energy estimation is realized in a similar fashion as described in the previous section by the use of Equation 3. The *TCS* energy computation is plotted in Figure 8.

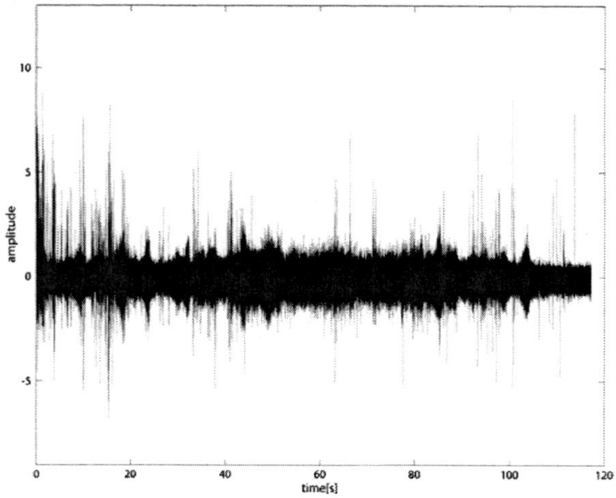


**Figure 6.** The car *TCS* time series filtered by the Hamming band-pass filter.

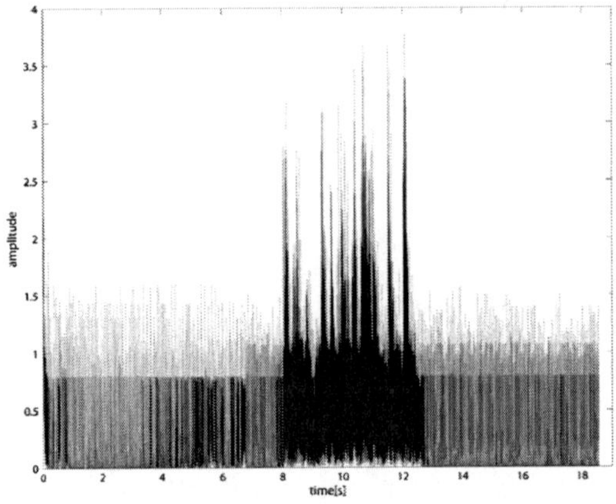


**Figure 7.** Estimation of the indoor *THA* energy $e_k$.

### *Indoor event estimation*

The threshold procedure, although common in detection literature, whose theoretical background we explain in detail in the section *Car moving estimation based on engine sounds*, is extensively cited (Gu et al., 2005; Lorincz et al. 2008). (Mhatre and Papagiannaki 2006; Kapitanova et al., 2010) and we by our own experimentation experience (Lopez and Valle 2009), have found that tuning a threshold, based on an event detector, is an extremely difficult task. After an intensive search, we noted that the energy reference values and energy change indicators values are infinite in number, and they would not be used to detect possible events because none of all the computed changes indicators are able to cross the reference value as required by the threshold criteria. For instance, Fig. 9 shows a group of energy references and energy change indicators of the indoor *THA* data set computed for different threshold parameter values.

Instead of a threshold implementation, a computationally and electrically efficient change rate-based detector, which has been successfully tested in wireless *Quality of Service* (*QoS*) systems (Mhatre and Papagiannaki 2006), is shown below:

$$ce_k = \frac{\left[\langle e_k \rangle - \langle e_{k-TC} \rangle\right]}{TC} \quad (4)$$

where $\langle e_k \rangle$ is the signal energy average computed at the $k$ time instant, and $\langle e_{k-TC} \rangle$ is the average signal energy computed at the $k$–$TC$ time instant. It must be observed that an implementation of Equation 4 requires three

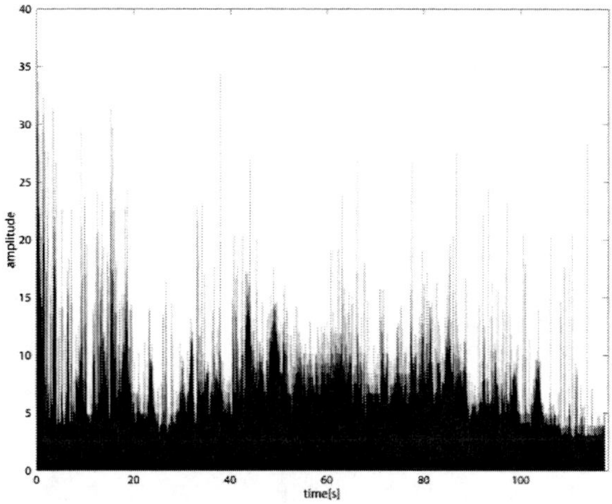


**Figure 8.** The cars *TCS* energy $e_k$ computation.

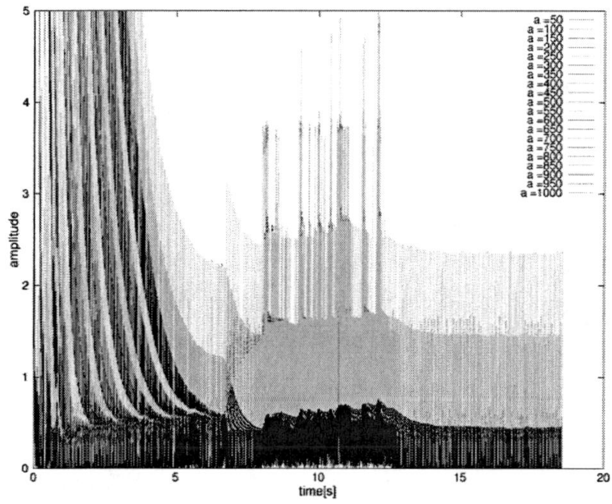


**Figure 9.** The *THA* set energy reference and energy change indicator computations.

*(Color image of this figure appears in the color plate section at the end of the book.)*

32-bit memory allocations and one 16-bit integer constant register, one 32-bit register for the $ce_k$ variable, one 32-bit register for the $\langle e_k \rangle$ variable, one 32-bit register for the $\langle e_{k-TC} \rangle$ and one 16-bit register for the *TC* constant.

An ideal event detector works on the assumption of two conditions: (a) when events are absent $\langle e_k \rangle$ settles to a semi-constant level. In this manner, it can be expected that two computations of $\langle e_k \rangle$ at different $k$ instants will be identical. This is the case when the signal component is only made of the environmental noise background component, and thus $ce_k = 0$. (b). On the contrary, when events occur $e_k$ largely bursts having at different $k$ time instants an ample range of $\langle e_k \rangle$ values, and thus getting $ce_k \neq 0$. In this manner an implementation of the events detector considers a fixed time interval that we call *detection time interval TD*. Within *TD* $ce_k$ is computed in an integer number of times *TC* with *TC* chosen as ($TC \leq TD$), counting the number of times $ce_k \neq 0$ and storing the counting number in the device memory. Then, at the *TD* end, event packets are signalled if it is found that the counting number is greater than a pre-programmed value. The pre-programmed value is adjusted on the base of the requirements of the application.

Instead, real-world indoor scenarios $ce_k$ has slight and continuous swings around the cero value, even though the sensor signal is only comprised of background noise. Therefore, a limit is set on the event counting process and the detector counts events when the following condition is achieved:

$$l < |ce_k| \tag{5}$$

where $l$ is a constant number that is experimentally set, depending the conditions of the indoor scenario of interest.

***Cars moving estimation based on the engine sounds***

As stated previously, the threshold-based event detector is an extensively used criterion where two quantities are sought: (a) a *reference or threshold* value that steadily tracks signal energy tendencies, and (b) a *change indicator* that is a reactive quantity that provides estimations of instantaneous energy changes. In order to get the threshold and change indicator, we compute signal energy statistics, *average* $\langle e_k \rangle$, *variance* $e\_var_k$ and *standard deviation* $e\_std_k$. As in the previous algorithm implementations, we seek efficient and accurate estimations. More specifically, the searched threshold should provide *steady* estimations about the instant signal energy behaviour trends. On the other hand, a changes indicator that outlines sharply changes observed in the signal energy is also required. Specifically, we find that the EWMA $\langle e_k \rangle$ provides stable estimations of the signal energy trends moving smoothly and achieving the signal energy average. On the other hand, we have observed that both $e\_var_k$ and $e\_std_k$ have shown values below $\langle e_k \rangle$ values when the signal energy component is only background noise and reacted strongly on energy changes.

To illustrate how we can establish the threshold criterion in the detection of moving cars, we run a comparison between $\langle e_k \rangle$, $e\_std_k$ and $e\_var_k$. In particular, in Fig. 10) upper and lower parts the computations of $\langle e_k \rangle$ against $e\_std_k$ and $\langle e_k \rangle$ against $e\_var_k$ are respectively shown for the *SCM* signal, see Fig. 3b). In this case we can observe two particular situations: (a) when there are not car sounds the signal energy is only due to noise background, in this case we observe that $e\_std_k$ and $e\_var_k$ stand all the time below $\langle e_k \rangle$. Furthermore, it must be noticed that the distance between $e\_var_k$ and $\langle e_k \rangle$ is larger than the distance between $e\_std_k$ and $\langle e_k \rangle$. (b) when the car sounds are present the signal energy shows large changes being the $\langle e_k \rangle$ crossed several times by $e\_std_k$ and $e\_var_k$. In particular, it is observed that $e\_var_k$ shows the largest changes around $\langle e_k \rangle$. In this manner, we choose the pair $\langle e_k \rangle$ and $e\_var_k$ as the event detector threshold and changes indicator respectively.

Finally we explain the functioning principle of the threshold-based event detector algorithm. Within a fixed time period that we call *event detection time window EDTW*, the algorithm continuously computes and compares $\langle e_k \rangle$ against $e\_var_k$ counting the occurrences when $e\_var_k$ crosses $\langle e_k \rangle$. Once *EDTW* has elapsed the detector checks the $e\_var_k$ crosses counter and signals the positive detection of an event if the crossing number is bigger than certain pre-programmed number. In particular, the pre-programmed number is searched by trial and error choosing the one that provides better detection rates and accomplishes the application requirements.

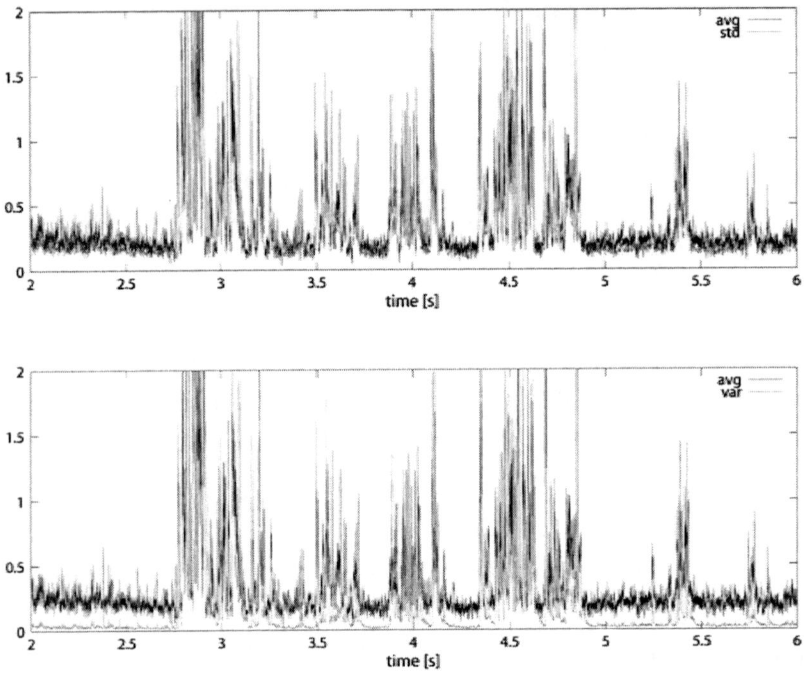


**Figure 10.** Signal energy statistics computations of the *SCM* data series. Upper part, in black color the computations of $\langle e_k \rangle$, and in red color the computations of $e\_std_k$. Lower part, in black color the computations of $\langle e_k \rangle$, and in red color the computations of $e\_var_k$. *(Color image of this figure appears in the color plate section at the end of the book.)*

## Experimental Procedure

In order to get a complete description about the limitations of the event detector we run our experimental procedure in an incremental complexity fashion, where we realize the following tasks: (i) the detector is programmed with a fixed set of detector parameters. (ii) the event detector algorithm is executed feeding it with basic sensor signals, *BHA* and *BCS* data series. (iii) the detected events by the algorithm are recorded, counted and grouped in *True Positives TP* and *False Positives FP* sets. (iv) the detector parameters are changed and (ii) is repeated. Particularly, this procedure is realized for a large range of detector parameter values. (v) it is searched the parameter intervals where the detector can sense most events. (vi) finally, we program the detector with values of the intervals found in (v) and record the detector performance execution when the algorithm is fed with the signals of the test sets, *THA* and *TCS*.

### *Tuning outcomes of the indoors human activities detection*

In Fig. 11 is shown the event detector work space, it is all the events the indoors event detector detected when the *BHA* signal set, *CLD, DPB, CLP,* is input to it. More specifically, we show *FP*, and *TP* events sensed by the detector when $0<l<0.14$.

Fig. 11 shows that events can be roughly grouped in three *l* regions that feature the following properties: *Region I*) The detector senses *FP* and *TP* events. *Region II*) the detector senses only *TP* events. *Region III*) The detector cannot signal *FP* or *TP* events, see Table 1. For instance, we can see from the top of Figure 11 that when the detector parameter is programmed with *l* values within the interval of (0,0.0005), the detector sensed *FP* and *TP* events, with *FP* events more numerous than *TP* events. When we program the detector parameter *l* with values within the (0.0005,0.00735) interval, the detector senses only *TP* events. Finally, we observe that with the *l* parameter values within the (0.00735, ∞) interval, there are no *FP* or *TP* events seen at all. In short, the detector did not observe any events for those *l* values.

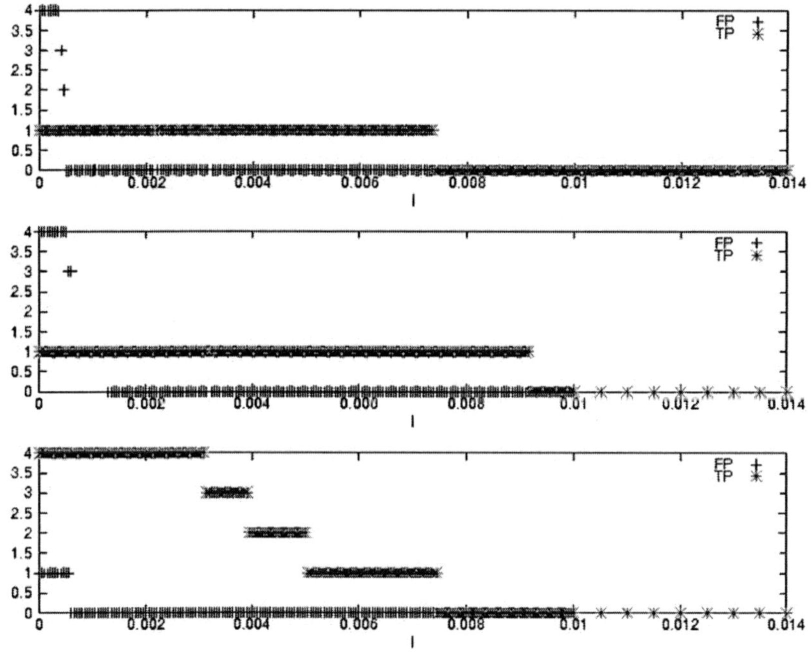


**Figure 11**. Events detected from the *BHA* signal are set for 0<l<0.014 values. *Top part, FP,* and *TP* events are sensed when the detector is fed with the *CLD* signal. *Middle part, FP,* and *TP* events are sensed when the detector is fed with the *DPB* signal. *Bottom part, FP,* and *TP* events are sensed when the detector is fed with the *CLP* signal.

*(Color image of this figure appears in the color plate section at the end of the book.)*

**Table 1. The detector work space of the *BHA* signal set.**

| *Signal* | *l* | | |
|---|---|---|---|
| | *I* | *II* | *III* |
| $BS_1$ | [0,0.0005) | [0,0.0005,0.00735] | (0.00735, ∞) |
| $BS_2$ | [0,0.0013) | [0,0.0013,0.00915] | (0.00915, ∞) |
| $BS_3$ | [0,0.0006) | [0,0.0006,0.00745] | (0.00745, ∞) |

As can be inferred from the previous discussion, we are particularly interested in *l* intervals where the detector observes most *TP* events. In this way, we find that a more general event detector algorithm can be obtained merging *l* intervals of Regions I, II, and III. In short, we apply *XOR* operations over the event detection interval intersections in the following manner:

$$l_I = \cap\{[0,0.0005),[0,0.0013),[0,0.0006)\} = \varnothing$$

$$l_{II} = \cap\{[0.0.0005,0.00735],[0.0.0013,0.00915],[0.0.0006,0.00745]\}$$
$$= [0.0013,0.00735]$$

$$l_{III} = \cap\{(0.00735,\infty),(0.00915,\infty),(0.00745,\infty)\} = \varnothing$$

where $l_I$ is the result of the disjunction operation applied on the events in region I. $l_{II}$ is the result of the disjunction operation applied on the events in region II. $l_{III}$ is the result of the disjunction operation applied on region III. In this manner, a unified region $l_F$ can be computed as

$$l_F = \cap[l_I, l_{II} l_{III}] = \varnothing \cap [0.0013,0.0075] \cap \varnothing = [0.0013,0.0075] \quad (6)$$

Equation 6 means that programming the event detector with $l_F$ values, any events from the *BHA* set can be detected by the algorithm. Furthermore, we believe that any sort of indoor human activity events can be detected whether the energy change rate-based event detector is programmed with *l* values within [0.0013,0.00735].

### *Tuning outcomes of the cars moving detection*

Table 2 shows that neither *FP* nor *TP* events were detected when the algorithm is fed with signals of the $CST \in BHA$ data set. Trials have been run with *EDTW* lengths of one, two, three, four, and five seconds, and the energy changes indicator crosses number is greater than ten. We observe that for all the *EDTW* lengths, there are no *FP* events detected at all. Also, we note that the *TP* events number is inversely proportional to the *EDTW* length. Particularly, the larger number of detections is obtained when *EDTW* is set to one second, and the smaller amount of detections is obtained when *EDTW* is set to five seconds. This is explained analyzing two cases: (a) when *EDTW* =1 second, since *CST* is a 35 second length record, the detector computes events 35 times finding that there were energy changes within 34 periods. (b) when *EDTW* = 5 seconds, and since *CST* is 35 seconds length, the detector computes events 7 times, also finding that there were energy changes within 6 periods.

**Table 2. The cars detection of the base signal set *CST*.**

| *EDTW* (seconds) | *FP* | *TP* |
|---|---|---|
| 1 | 0 | 34 |
| 2 | 0 | 17 |
| 3 | 0 | 11 |
| 4 | 0 | 8 |
| 5 | 0 | 6 |

In Table 3 we observe the detection of events belonging to the $SCM \in BHA$ that the detector sensed when *EDTW* takes values within the range of one to five seconds. In this case, *SCM* is ten seconds length. Roughly, we would see from Table 3 that events are allocated within the 2.8 to 4.8 seconds period. Unlike the case shown in Table 2, here we observe events of the two kinds, *FP* and *TP*, signalled for *EDTW* values programmed within the interval of one to three seconds, being nine the largest number of events that happened when *EDTW* = 1 second. Like in the previous case, we observe that the number of events signalled by the the detector decreases as *EDTW* gets widened. Particularly, we can observe three cases: (a) when *EDTW* = 1 the detector estimates events in ten time periods and finds nine events in total. From the nine events, it is found that the majority are miss-estimations and only two events are correct detections. We explain this result in the following manner: because *EDTW* is short the detector is highly sensitive to the motor sounds and tire noises even for cars that are far away from the sensing point. (b) when *EDTW* = 4 seconds the detector computes two times events, this *EDTW* gives the detector more robustness to the background noise and therefore two events are effectively detected. Finally, (c) when *EDTW* = 5 the algorithm computes one single event that in this case is a correct estimation.

**Table 3. The car detection of the base signal set *SCM*.**

| *Time Window* (seconds) | *FP* | *TP* |
|---|---|---|
| 1 | 7 | 2 |
| 2 | 2 | 2 |
| 3 | 1 | 2 |
| 4 | 0 | 2 |
| 5 | 0 | 1 |

In this manner, we conclude that a reliable car presence detector must be programmed with *EDTW* = 5 seconds to provide correct estimations of event occurrences from the *CST* and *SCM* sets.

Finally, it must be noted that although in both cases the algorithm correctly detected events from both data sets, *CST* and *SCM*, there is a trade-off between the detection granularities against the number of packets transmitted by the mote so that the sensor network life-time is not compromised.

**Detection Test Results**

In this section we probe our assumptions. As previously explained we obtained detector parameters and we believe that these parameters enable the detector senses more general acoustic signals related to the application work environment. Here we program the detector with test *THA* and *TCS* signal sets. Finally, the event detector outcomes are recorded then on the base of the obtained results we compare and discuss the performances observed.

***Test outcomes of the indoor human activities detection***

We program a Micaz sensor with $l = 0.0017$ that was arbitrarily chosen from the values within the interval [0.0013,0.00735], then the detector is feed with the *THA* set. It is observed that there are two events detected by the algorithm, see Fig. 12 bottom part. The first one *FP* event occurs within the first two seconds. This estimation is due to the circuit sensor starting up process and because the algorithm had not yet arrived to a settled estimation of $\langle e_k \rangle$ and $e\_var_k$, see top part of Fig. 12, therefore it must be discarded. Within the eight to ten seconds period, the detector observes a second one *TP* event which effectively corresponds to a person talking.

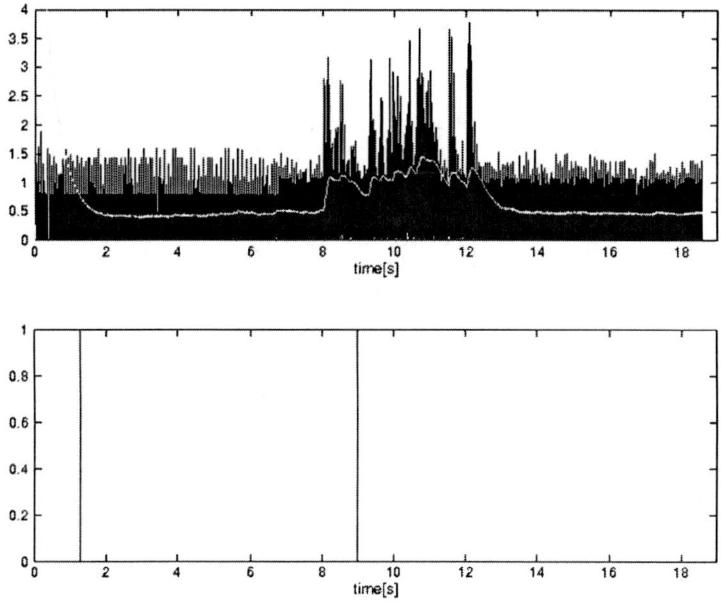


**Figure 12.** Event detection of the test *THA* set. (a) Top part, $\langle e_k \rangle$ and $e\_var_k$ estimations. (b) Lower part, event detection estimations.

*(Color image of this figure appears in the color plate section at the end of the book.)*

### *Test outcomes of the cars moving detection*

As we have noted in the tuning section, $EDTW = 5$ is the best compromise to detect both signals of the *BCS* set. Therefore, we program *EDTW* with this value and feed the detector with the *TCS* set.

In the upper part of Fig. 13 it is showed the computations of $e_k$ and $\langle e_k \rangle$ belonging to the *TCS* signal set and in the bottom part of Fig. 13 it is plotted the events effectively detected by the detector algorithm. The observed detections are explained as follows: as *TCS* is a 120 second length record, there are 24 time intervals of five seconds. Within those intervals, the detector computes events finding that within all the 120 seconds interval events are present.

Because it is not clear whether the results are correctly computed, we separately analyse *TCS* segments of ten seconds and see whether there are energy changes. For instance, the first ten second period is presented in Fig. 14. At the top $e_k$ and $\langle e_k \rangle$ computations are shown and the lower part of

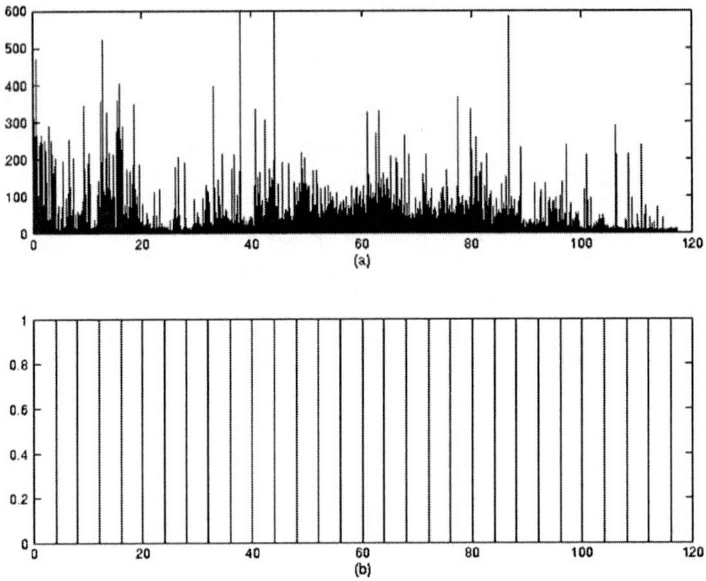


**Figure 13.** Event detections within the120 sec. of the *TCS* signal set. (a) $\langle e_k \rangle$ and $var_k$ of the signal set *TCS*. (b) cars detected by the algorithm.

*(Color image of this figure appears in the color plate section at the end of the book.)*

Fig. 14 plots the events observed within this time interval. From Fig. 14 we can see that the detector, in fact, observes continuous energy changes and therefore events are sensed. We interpret these results as follows, although the running cars are separated from each other by certain distance $e\_var_k$ is all the time over $\langle e_k \rangle$ and therefore events are always signalled. This grossly means that even if more than one car can pass with an *EDTW* period, the detector will only detect one vehicle.

***Event detection effectiveness comparison***

As a manner of comparison we ran event detectors with crossed data and (i) we observe the events sensed when the energy change rate detector is fed with the *TCS* data set, and (ii) recorded events when the energy threshold detector is fed with the *THA* data set. In both cases the detector parameters, *l* and *EDTW*, were set to a range of values within intervals that provided high *TP* detection rates and we compute an effectiveness factor $\frac{TP}{TP+FP}$.

In Table 4, the first column shows the test signal set that is fed to the detector algorithm. In the second column we show the effectiveness

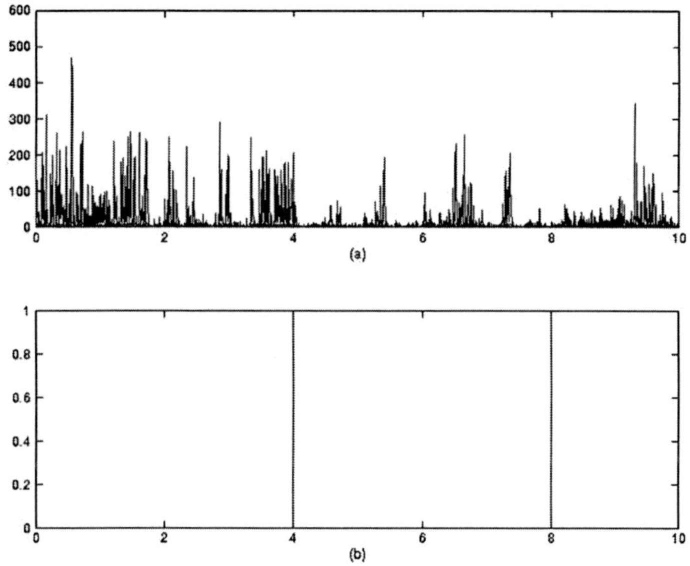

**Figure 14.** Event detections within the first 10 seconds of the *TCS* signal set. (a) $\langle e_k \rangle$ and $var_k$ estimations of the signal set *TCS*. (b) events signalled.

*(Color image of this figure appears in the color plate section at the end of the book.)*

computations of the energy change rate detector when it is fed with *THA* and *TCS* data sets. Finally, in the third column is showed the effectiveness of the energy threshold detector when it is fed with *THA* and *TCS* data sets.

**Table 4. Event detection effectiveness of the rate-based energy change against the energy threshold algorithm.**

| *Test signal set* | *Energy change rate-based Effectiveness* | *Threshold-based Effectiveness* |
|---|---|---|
| *THA* | 20 to 90% | N/A |
| *TCS* | 25 to 50% | 10 to 85% |

*Energy change rate detector effectiveness.* It is observed that with both data sets, *THA* and *TCS*, the algorithm can be adjusted to obtain a large range of effective detections, seeing that the largest effective detection range is for the *THA* data set. On the other hand, for the *TCS* data set, only half of the events can be sensed.

*Energy threshold detector effectiveness.* As previously discussed, we were not able to set an energy threshold to detect events from the *THA* set; therefore, we cannot show a significant effectiveness of this algorithm. On the other hand, we were able to find an ample range of energy threshold values in the *TCS* case that allowed us to detect a large range of events, *FP* and *TP*, finding that the maximum effectiveness rate is approximately in a proportion of eight to ten events correctly sensed.

### Conclusions

We have presented event detectors of acoustic indoor human activities and cars in motion. The detector designs are aware about the featured strong constraints of wireless sensors hardware. In principle, although the detection of events considers standard procedures, we show that the detector performance can be strongly determined by the signals environment, filtering process and the detection algorithm criteria. In particular, we observed that the energy change rate based algorithm performed satisfactorily in the detection of indoor acoustic events with a 90% success rate. Meanwhile, the detector performed only moderately well in detecting the presence of cars. On the other hand, in the threshold-based approach, we find that threshold tuning is a very difficult task when indoor human activity events are fed to the detector. On the contrary, in the case of detecting cars moving, a threshold-based detector can successfully detect up to 85% of the cases studied.

### Future Investigations

Our future investigations will consider the deployment of networks of acoustic detectors in both indoor and outdoor scenarios. Our principal interest is to search for correlations in activities developed in different spaces and specific hours and days. On the other hand, in the case of car moving detection, we want to realize event detections based on richer sensor platforms, like accelerometers or magnetometers, as well as the use of data fusion algorithms.

## REFERENCES

Akdere, M., U. Çetintemel and N. Tatbul. 2008. Plan-based complex event detection across distributed sources. In the Proceedings of the VLDB Endowment 1. **1**:66–77.

Anastasi, G., O. Farruggia, G. Lo Re and M. Ortolani. 2009. Monitoring high-quality wine production using wireless sensor networks, 1–7. In the Proceedings of the 42nd Hawaii International Conference on System Sciences HICSS '09, Big Island, Hawaii, IEEE Computer Society Press.

Bahrepour, M., N. Meratnia, M. Poel, Z. Taghikhaki and P.J.M. Havinga, 2010a. Distributed Event Detection in Wireless Sensor Networks for Disaster Management, 507–512. In the Proceedings of the 1st International Workshop on Computational Intelligence for Disaster Management (CIDM-2010) in conjunction with the 2nd International Conference on Intelligent Networking and Collaborative Systems (INCoS 2010), Thessaloniki, Greece. IEEE Computer Society.

Bahrepour, M., N. Meratnia and P.J.M. Havinga. 2010b. Fast and accurate residential fire detection using wireless sensor networks. In the *Journal of Environmental Engineering and Management*. **9(2):**215–221.

FIWIZ 2008.User Manual Version 2.7, Rainer Storn, International Computer Science Institute, Berkeley University of California, California, USA.

Gao T., D. Greenspan, M. Welsh, R. Juang and A. Alm. 2005. Vital signs monitoring and patient tracking over a wireless network, 102–105. In the Proceedings of the 27th Annual International Conference IEEE Engineering in Medicine and Biology Society EMBS, Shanghai.

GNU Octave. 1988. A high-level interpreted language. In http://www.gnu.org/software/octave/about.html.

Gu, L., J. Dong, P. Vicaire, T. Yan, L. Luo, A. Tirumala, Q. Cao, T. He, J.A. Stankovic, T. Abdelzaher and B.H. Krogh. 2005. Lightweight detection and classification for wireless sensor networks in realistic environments, 205–217. In the Proceedings of the 3rd international conference on Embedded networked sensor systems SenSys '05 , San Diego, California, USA, ACM.

Gupchup, J., A. Terzis, R. Burns and A. Szalay. 2007. Model-Based Event Detection in Wireless Sensor Networks. In the Proceedings of the Workshop for Data Sharing and Interoperability on the World Wide Web (DSI 2007), Cambridge, Massachusetts, USA. April 2007.

Hill, J., R. Szewczyk, A. Woo, S. Hollar, D. Culler and K. Pister. 2000. System architecture directions for networked sensors, 93–104. In the Proceedings of the 9th International Conference on Architectural Support for Programming Languages and Operating Systems ASPLOS '00, Cambridge, Mass, USA, ACM SIGPLAN Notices.

Kapitanova, K., S.H. Son and K.-D.Kang. 2010. Event Detection in Wireless Sensor Networks—Can Fuzzy Values Be Accurate? 168–184. In Jun Zheng, David Simplot-Ryl and Victor C.M. Leung [eds.] *Ad Hoc* Networks: Second International Conference, ADHOCNETS 2010. Victoria BC Canada. Springer.

Kerman M.C., W. Jiang, A.F. Blumberg and S.E. Buttrey. 2009. Event Detection Challenges, Methods, and Applications in Natural and Artificial Systems, 1–19. In the Proceedings of the 14th International Command and Control Research and Technology Symposium (ICCRTS), Washington, DC. USA.

Johnson, J., J. Lees, G. Werner-Allen, K. Lorincz and M. Welsh. 2006. Fidelity and yield in a volcano monitoring sensor network, 381–396. In the Proceedings of the 7th USENIX Symposium on Operating Systems Design and Implementation (OSDI), Seattle.

Liang, O. and L. Wang. 2005. Event detection in wireless sensor networks using fuzzy logic system, 52–55. In the Proceedings of the IEEE International Conference on Computational Intelligence for Homeland Security and Personal Safety CIHSPS'05, Orlando, Fla, USA, IEEE.

Lopez T.M.A. and M. Valle. 2009. Reliable event detectors for constrained resources wireless sensor node hardware. In the *EURASIP Journal on Embedded Systems*, 2009, doi:10.1155/2009/474903.

Lorincz, K., B. Chen, J. Waterman, G. Werner-Allen and M. Welsh. 2008. Resource aware programming in the pixie OS, 211–224. In the Proceedings of the 6th ACM Conference on Embedded Networked Sensor Systems SenSys'08, Raleigh, NC, USA, ACM.

Mhatre, V. and K. Papagiannaki. 2006. Using smart triggers for improved user performance in 802.11 wireless networks, 246–259. In the Proceedings of the 4th International Conference on Mobile Systems MobiSys '06, Applications and Services, Uppsala, Sweden, ACM.

Polastre, J., R. Szewczyk and D. Culler. 2005. Telos: Enabling ultra-low power wireless research, 364–369. In the Proceedings of the 4th International Symposium on Information Processing in Sensor Networks IPSN '05, Los Angeles, Calif, USA, ACM.

Steiner, S.H. 1999, Exponentially weighted moving average control charts with time varying control limits and fast initial response in *Journal of Quality Technology*. **31**:75–86.

Titzer, B.L., D.K. Lee and J. Palsberg. 2005. Avrora: Scalable sensor network simulation with precise timing, 477–482. In the Proceedings of the Fourth International Symposium on Information Processing in Sensor Networks, IPSN 2005. Los Angeles, Calif, USA, ACM.

Varaiya, P. 2004. Assessment of MeMS Sensors in an Urban Traffic Environment. Research Report, California Partners for Advanced Transit and Highways (PATH) UC Berkeley, California, USA.

Wang, Q. and W. Yang. 2007. Energy Consumption Model for Power Management in Wireless Sensor Networks, 142–151. In the Proceedings of the 4th Annual IEEE Communications Society Conference on Sensor, Mesh and *Ad Hoc* Communications and Networks, SECON '07. San Diego, CA, USA. IEEE.

Woo, A. and D. Culler. 2003. Evaluation of efficient link reliability estimators for low-power wireless networks. In Technical Report UCB//CSD-03-1270, U.C. Berkeley Computer Science Division, September 2003.

Zhang, Y., N. Meratnia, P.J.M. Havinga. 2010. Outlier Detection Techniques For Wireless Sensor Networks: A Survey. In the *Journal of IEEE Communications Survey & Tutorials*. **12(2)**:159–170.

# 7

# Armed Mobile Platform Group Swarming and Networking Simulation

*Radomir Janković*

**ABSTRACT**

Armoured units, in spite of almost 100 years since their introduction in active service in armies throughout the world, still are one of the key resources of the contemporary states ground armed forces, due to their great capabilities and versatility. The problem is that unconventional and asymmetric warfare, so prevailing nowadays, has questioned usability of traditional armoured units. The solution could be in two directions of further activities: development of new, essentially different generation of armed mobile platforms and research of new tactics of their use, among which swarming is the most promising one.

Some of the results of author's research of armoured units swarming by means of discrete events simulation have been presented in the chapter. Two simulation models have been proposed and realized: first for armed mobile platform group swarming, intended for research of its variable defended territory occupancy density impact to swarming success, and the second for swarming information system network, intended for research of various data rates and operation algorithms impact to its mobile computer network successful operation.

The motivation for such a research is the fact that adaptation of existing armoured units for swarming, by means of embedding the necessary information system equipment into every armed mobile platform, could be the best investment in small and medium sized armies' modernization.

## 1. Introduction

Originating from biology, a *swarm* represents a large group of insects or small organisms, particularly in motion. Such a group tends to organize itself in order to solve a common problem, like searching a source of food, building nests, crossing obstacles, hunting, or defending against an overwhelming threat. In spite of the fact that swarmers usually are not very powerful individuals, by using relatively simple rules of actions, communication and local interactions, they manage to build up some kind of collective decentralized intelligence (Bonabeau 2003), capable to solve complex problems. Some examples of swarming in nature can be found in (Parunak and Brueckner 2004), related to ants, bees, fishes, birds, wolves, etc. Such collective behaviour attracted attention not only of biologists, but also of scientists, engineers and experts from many other areas of contemporary human activities, who have been involved in research for the last three decades to find new methods and algorithms inspired by swarming in bio systems. A suitable definition of swarming, sufficiently general to include both biological and artificial systems, has also been given in the same text: "Swarming is useful self-organization of multiple entities through local interactions".

*Swarm* and derived concepts like *swarm intelligence* and *swarming* have quickly become words *à la mode*, due to their extensive use as an analogy or the basis of other uses of the terms, in various areas of science and technology, starting from the early works in swarm robotics (Beni 1988), through solving problems in mathematics, social and information science (Liu et al., 2011), and recently to military applications (Arquilla and Ronfeldt 1999; Edwards 2003; Sauter et al., 2005; Janković 2011).

This chapter deals with military application of swarming tactics, especially by armoured units, being still one among the most important resources of modern ground armed forces. In section 2, the military aspect of swarming has been analysed, along with some recent achievements in its application as tactical procedure. In sections 3 and 4 simulation models of two most important aspects of armoured units' swarming have been presented: armed mobile platforms group swarming and its command information system mobile communication network operation. Both sections 3 and 4 are organized as follows: statement of related problem and goal of simulation, simulated system description, simulation model, system parameters and performance measures, and experiments with analyses of results. Section 5 deals with the main directions of our future research, and in section 6 general conclusions have been given.

The main objective of this chapter is to present to readers some of the author's research results in military swarming, particularly that of the armed mobile platforms (Janković 1998, 2001; Janković and Nikolić 2009; Janković 2010b) as an expensive resource in armed forces, being one the most critical sectors of every contemporary state.

The second objective is to present some of discrete events simulation models developed during the research, relating to both important aspects of the matter: simulating of swarm participants motion, and of their wireless mobile communication during the swarming action. We believe that it could be useful for both target audience groups: academics and practitioners.

The last and not the least important objective is to inform the reading audience on our future main research directions.

## 2. Military Swarming

In military sense of the word, *swarming* is a tactics by which military forces attack an adversary from many different directions, and then regroup (Arquilla and Ronfeldt 1999). Repeated actions of many small, manoeuvrable units are going on, circling constantly through the phases of swarming:

- Dispersion of units in battlefield;
- Gathering (concentration) of many units on common target;
- Action (strike or fire) at a target from all directions;
- Dispersion of units.

The way of military swarming application is depicted in Fig. 1, and its basic characteristics have been given in Table 1.

**Table 1. Military swarming: basic characteristics.**

| *No.* | *Characteristic* |
|---|---|
| 1. | Autonomous/semiautonomous units, engaged in concentrated attack at common target. |
| 2. | Amorphous, coordinated attack from all directions by continuous fire/shock assaults . |
| 3. | Many small, space dispersed mutually networked units. |
| 4. | Integrated surveillance, sensors, and $C^4$ISR systems for upper level situation assessment. |
| 5. | Units' action capabilities, from distance as well as in direct contact. |
| 6. | Continuous attacks aiming to break adversary's cohesion. |

Although numerous examples of successful swarming application have been recorded in history (Edwards 2000 and 2003; Inbody 2003) the significance of this tactics did not reach its full measure until our days, due to brisk development of information technologies and merging of embedded computing devices and mobile telecommunications. The swarming tactics is applied by units much smaller than threat/target, but

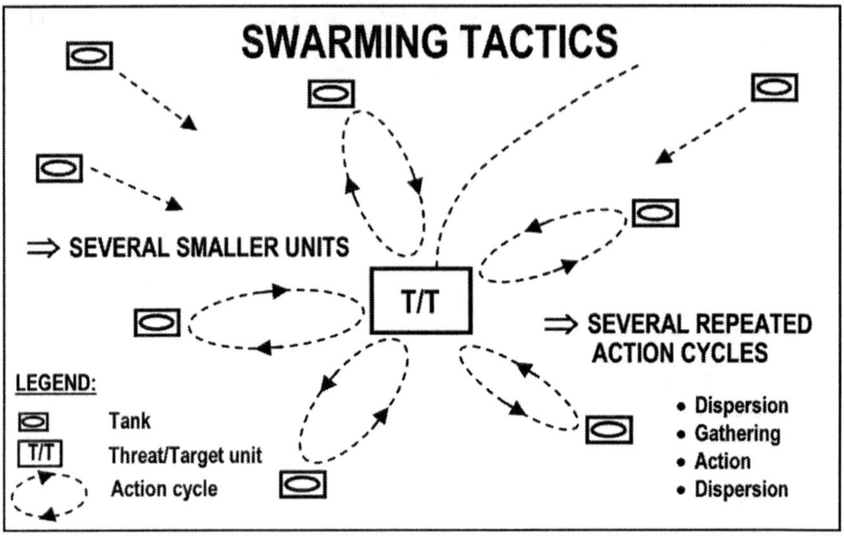


**Figure 1.** Swarming of armoured and mechanized units.

their use is far more efficient, so in their actions as a whole, they can often defeat many times superior adversary (Edwards 2005).

More intensive military swarming research began after 2000, and attained first results mainly in the areas of unmanned vehicles (Price et al., 2006; Nowak et al., 2007; Pohl et al., 2008; Singer 2009), in the air, underwater and on the ground, air force, navy and some special ground forces units. A common point to all those research efforts is that they have emphasized the need for changes of contemporary armies' organization and tactical principles.

As for military "hardware", it seems that the existing air force and navy equipment could be used for swarming on the "as is" principle, at least in the near and foreseeable future.

To our best knowledge, until now there are not many published results of the armoured and mechanized units swarming research, especially those on mainbattle tanks as their representative paradigm, placing the research results in (Janković 2010c, 2011a, b) among the first and rare exceptions. Tanks have been mentioned in other publications on military swarming systems rather as the objects of swarming attack than being the swarmers themselves.

The main reason for that lies in fact that the best known contemporary tanks in service have been conceived to be dominant over classical opponents on the ground. That means heavy, large-calibre main armament and a huge amount of extra equipment for protection, both active and passive. The consequences are huge mass, large silhouette, less capability of fast manoeuvring and extensive logistic requirements.

All these factors have impact to prevailing use of existing tanks according to former well known tactical principles, meaning that there will still be tendency to engage tanks in great joint formations, and that the main swarmers in ground forces will probably be some other kind of units, perhaps cooperating with medium and light armoured combat vehicles.

A huge mass represents particular problem for the existing mainbattle tanks attacking swarming extremely limiting their fast deployment in the combat action territory by means of air transport. That problem could be somewhat lessen in defence, because in many cases the defender's tanks are supposed to be already deployed in the defended territory.

Because of all aforementioned reasons, armies of more developed countries still rely mostly on firepower supremacy achievement and massive force concentration, along with airspace superiority, conducting the so called air-land battle.

However, attack to the units of such heavy armed and technologically extraordinary equipped army, from all directions and close distances could result in extremely good effects. Numerically strong units and their fire and strike power can be successfully opposed by psychological effect, achieved on adversary, attacked suddenly and from all directions. Besides, small units' manoeuvrability and their capability to disperse after sudden attack prevent technologically superior adversary to react on time.

Such an approach brings major changes, both in modern armies organization and equipment, and has a particular influence on the principles of use and the bare prospects for tanks survival on the 21st century battlefields. New generations of tanks will surely differ from those developed up to now, intended for totally different style of warfare. They should be smaller, faster and more agile, better armed and with incorporated elements of the command information systems.

A development new generation of tanks is going on for quite some time now. However, their introduction in service of the producer countries armed forces is expected not until the middle of next decade, and for the rest of the world even later. Besides, such procurements require enormous financial costs, not to mention actual industry problems, for there is a huge number of contemporary tanks, conceptually quite different, still in service, and nobody would like to question the existing production and logistic sectors, being the important economy factors of the most developed countries.

One of the approaches to that problem solving is to investigate adaptation possibilities of the existing tanks for swarming, so that their active life time can be prolonged until the new generation comes into service, and they could in the best way be used in the changing conditions of modern warfare (Janković 2001, 2007).

## 3. Armed Mobile Platforms Group Swarming Simulation

### 3.1 Simulation Goal

The problem with application of swarming as a new and until recently not common armoured units' tactics against superior threat is that it is generally not known how many AMPs (armed mobile platforms) in various sized defended territories should be engaged as swarmers against various types of adversary, so that his mission in the defended territory fails.

The chosen approach to the problem solution in the research has been to build a discrete events simulation model of an armed mobile platforms group swarming and to implement it in computer program-simulator form, as a tool for conducting experiments.

The goal of simulation experiments has been to explore impact of the variable defended territory occupancy density by the AMP group to success of its swarming. It has been achieved through experiments with the realised simulator, by changing the dimensions of the territory in experiments, along with retaining fixed number of AMPs in the group.

### 3.2 Simulated System

The system being simulated consists of: one group of N armed mobile platforms (AMP-*i*, where $i = 1, ..., N$), one threat/target unit (T/T), command information system and the defended territory—the theatre of combat actions (swarming).

Armed mobile platforms AMP-*i*, can move at velocities $V_i$. Armed mobile platform group (GAMP) moves at mean velocity $V_{AMP}$. Every AMP-*i* has main weapon (MW-*i*), characterized by range ($R_{MW\text{-}i}$) and compatibility with threat/target unit T/T-*j* ($K_{ij}$), which determines its total effect in swarming against every threat/target unit ($U_{ij}$).

Every AMP is active member in the command information system (CIS GAMP) and disposes appropriate equipment: GPS receiver (GPS), computer (AMPC) and radio transceiver (AMPR) for digital data transmission.

Threat/target unit can move at maximal velocity $V_{t/t}$. It is superior to every single AMP of the group, so for the successful swarming it is necessary that the critical cumulative effect of several AMPs, specific to such threat/target, be exceeded.

The primary function of the command information system CIS GAMP is regular informing of every user on its actual position and other data of interest:

- about the threat/target unit, based on the data acquired by the territory multisensor surveillance network (Lamont et al., 2011), individual AMPs of the group and other friendly forces;
- about every single AMP of the group, based on the regular reports of the AMP-*i* about changes in its own position.

CIS GAMP works successfully if it dispatches regular reports about threat/target unit and every single AMP-*i* motion before their expirations. Time of expiration (Janković 2005a, b, c) of the report about every moving entity (threat/target unit or AMP of the group) is defined by the expression:

$$t_z(i) = \frac{P_{CIS}}{V_i} \tag{1}$$

Where $P_{CIS}$ [m] is the preciseness of the CIS GAMP system (prescribed relative distance from the previous known position, due to further motion, which can be tolerated as no motion at all), and $V_i$ [m/s] is the moving entity's velocity in the period of the report.

Combat actions take place in territory which is presented in the model by means of two-dimensional rectangular coordinate system.

### 3.3 Simulation Model

The simulation model of the AMP group swarming (Janković 2008a) is discrete and dynamic, oriented to events. In the model, system activities are represented by pure time delays.

The moving entities in the model are: units that form a group (AMP-*i*, *i* = 1, 2, ..., N), threat/target unit (T/T) and messages of the CIS GAMP command information system.

The initial deployment of the group of armed mobile platforms is random, which is the worst case, because the occurrence of threat/target is unexpected, and the AMP group is prepared for it by no deployment intended for defending territory under such specific circumstances.

In the simulation, the group of an armoured battalion size has been considered, consisting of 43 armed mobile platforms (tanks, armoured personnel carriers, etc.).

The armed mobile platforms of the group (AMP-*i*) get information about the threat/target units and other AMPs of the group motion in time expiring intervals $\Delta t$, and give reports about their own current positions.

Based on that information, the armed mobile platforms, AMP-*i*, head toward threat/target unit, aiming to reach, as soon as possible, the position enabling them to perform successful swarming, for the sake of destroying, disabling, or preventing the adversary in accomplishment of its mission.

The graphical presentation of the motion of a part of the system, consisting of 1 threat/target unit (T/T) and 4 AMPs of the group (1, 2, 3 and 4), manoeuvring to accomplish the swarming is depicted in Fig. 2.

The threat/target unit is accomplishing its own mission and, contrary to AMP-*i*, has no access to information of the CIS GAMP command information system, so its primary goal is to fulfil its own task, which has been presented by motion on the given trajectory between points A and B in the model, according to functional dependencies of its coordinates on time, $x_{t/t}(t)$ and $y_{t/t}(t)$.

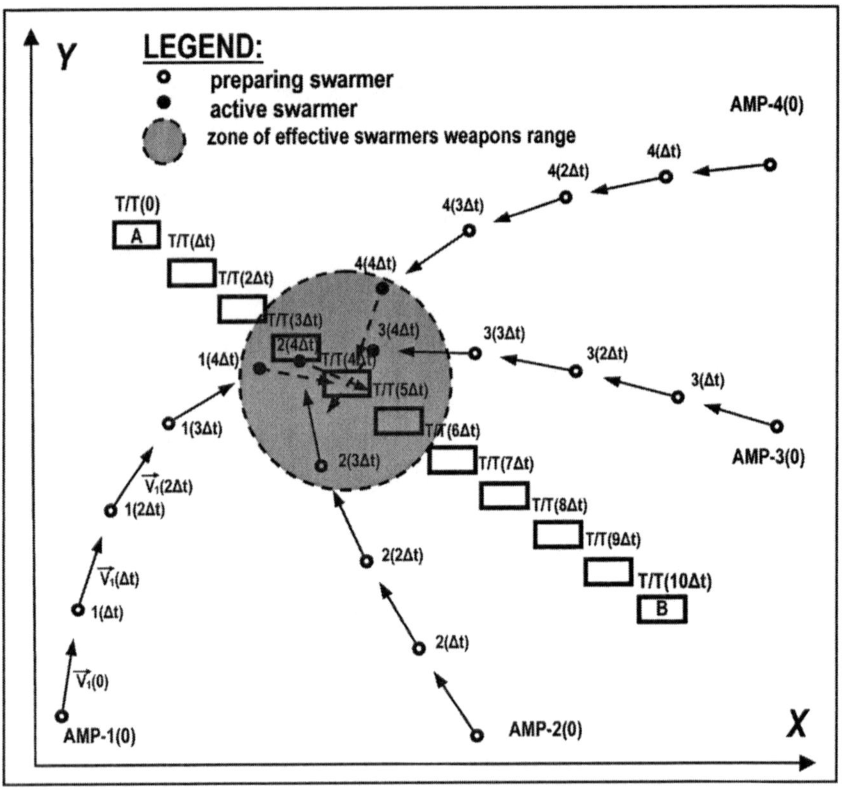


**Figure 2.** Motion of threat/target unit and AMP-1, 2, 3 and 4.

When the period of expiration elapses and CIS GAMP dispatches the report of the new threat/target position ($t = \Delta t$), threat/target has moved to its new position, T/T ($\Delta t$). Until then, platforms AMP-*i* have also moved to their new positions, 1($\Delta t$), 2($\Delta t$), 3($\Delta t$) and 4($\Delta t$), following the directions from the previous time interval, directed their velocity vectors towards new position of the threat/target, and then the process continues.

In order that individual swarming participating AMP-*i* could act upon threat/target unit, the following three conditions (Janković 2008a, c), must be fulfilled:

a. AMP-i must dispose of the main weapon MW-i, compatible with the adversary unit.
b. The distance between AMP-*i* and the adversary unit *j* must be in the main weapon MW-*i* range limit:

$$D_{ij} = \sqrt{(y_j(t) - y_i(t))^2 + (x_j(t) - x_i(t))^2} \le R_{MW-i} \tag{2}$$

c. The a. and b. conditions must be satisfied by enough AMPs from the group, so that their total cumulative effect on threat/target, $KU_j$ should be equal or greater than critical threshold of the multiple AMPs cumulative effect, $PKU_j$, specific to threat/target unit T/T-*j* :

$$KU_j = \sum_{i-1}^{N} A_{ij} \cdot K_{ij} \cdot U_{ij} \ge PKU_j \tag{3}$$

Where:

- $A_{ij}$ is the assignment coefficient (0/1), which is intended to assignment of the T/T-*j* threat/target unit to AMP-*i*, in the multitarget swarming models;
- $K_{ij}$ is the main weapon MW-*i* compatibility coefficient with the T/T-*j* threat/target unit(0/1);
- $U_{ij}$ is the possible effect of the main weapon MW-*i* on the T/T-*j* threat/target unit.

The simulation model and the algorithm for the armed mobile platform group swarming (Fig. 3), implemented by means of the discrete events system simulation language GPSSWorld (Minuteman Software) have been presented in the paper (Janković 2008b).

Two issues of swarming are possible:

a. SUCCESS: AMPs of the group have succeeded to fulfil all necessary conditions and by means of swarming have prevented the threat/

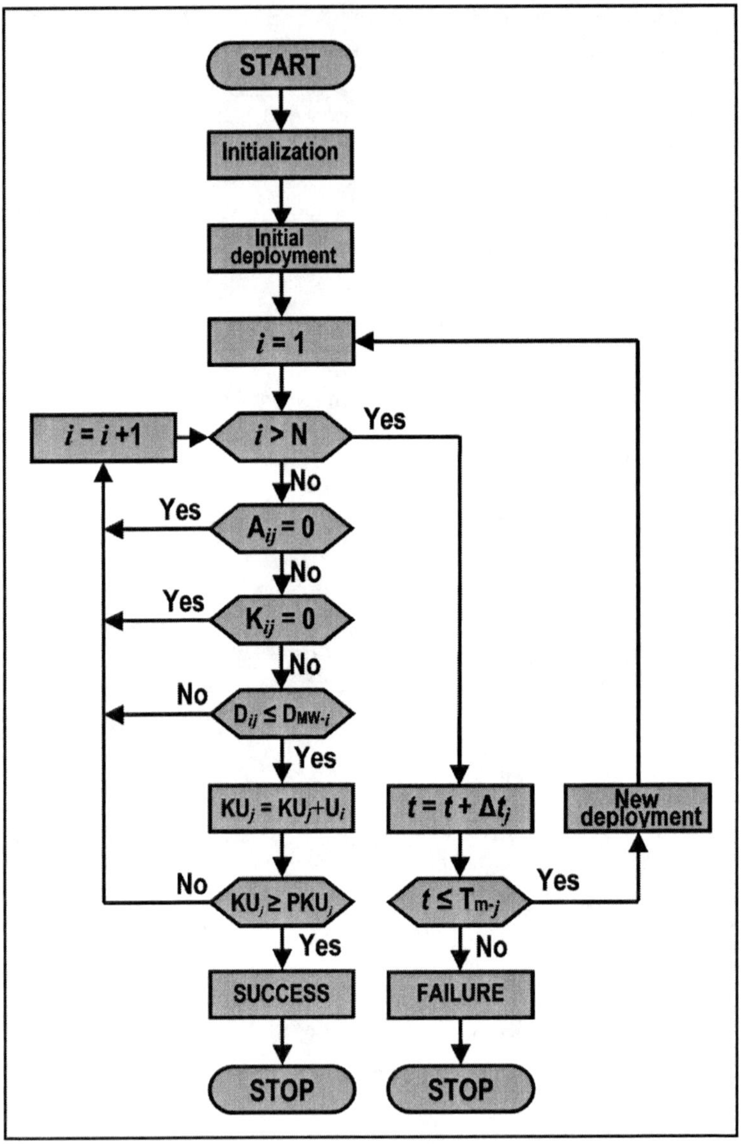


**Figure 3.** Basic swarming simulation algorithm (monotarget/threat).

target from accomplishing its mission (it has not reached the point B, in the example in Fig. 2).

b. FAILURE: AMPs of the group have not succeeded to fulfil all necessary conditions, so the threat/target has accomplished its mission (it has reached the point B).

The realized simulator of swarming, which members of an armoured battalion, as a group of armed mobile platforms, can apply in conflict with numerically and technically superior adversary units (threat/target), has been presented in (Janković 2011a, b). For testing the simulator, the experiment has been performed, with a system which characteristics are provided in Table 2.

**Table 2. Test simulated system characteristics.**

| No. | *Characteristic* | *Value* |
|---|---|---|
| **1.** | Defended territory area | 20 × 20 [km] |
| **2.** | Threat/target unit velocity | $V_t$/t = 15 m/s |
| **3.** | Threat/target unit initial position | (0, 0) [m] |
| **4.** | Threat/target unit final position | (20000, 20000) [m] |
| **5.** | Threat/target unit law of motion | Uniform, rectilinear, $V_{t/t}$ |
| **6.** | Number of AMPs | N = 43 |
| **7.** | AMPs initial deployment | Random, whole territory |
| **8.** | AMP-*i* velocity | $V_{AMP\text{-}i}$ = 15 m/s |
| **9.** | MW-*i* main weapon | $A = 1$, $R = 2500$ m, $U = 0.15$ |
| **10.** | MW-*i* cumulative effect threshold | PKU = 3 (20 AMPs needed) |
| **11.** | CIS GAMP system information period | Δt = 10 s |

The simulator output in some characteristic moments of the swarming dynamic development has been graphically presented in Fig. 4.

One can see that AMPs of armoured battalion attained successful swarming (20 among them simultaneously were inside the curve representing boundary distance to threat/target of 2500 m) in t = 480 s, i.e., 8 minutes of simulated time after first detection of the adversary in the territory—theatre of the combat actions. As the mission time of the threat/target unit—to reach the point with coordinates (20000 m, 20000 m)—has been $T_m$ = 1885.618 s in the case considered, the armoured battalion units in the presented experiment realization have successfully applied the swarming tactics in defending territory.

### 3.4 System Parameters and Performance Measures

The parameters of the simulated system (Janković 2011a), consisting of the group of $N = 43$ AMPs, which applies swarming in defending territory during the conflict with $M = 1$ threat/target unit, involve the following:

a) Threat/target unit parameters:
   - Number of threat/target units: $M = 1$

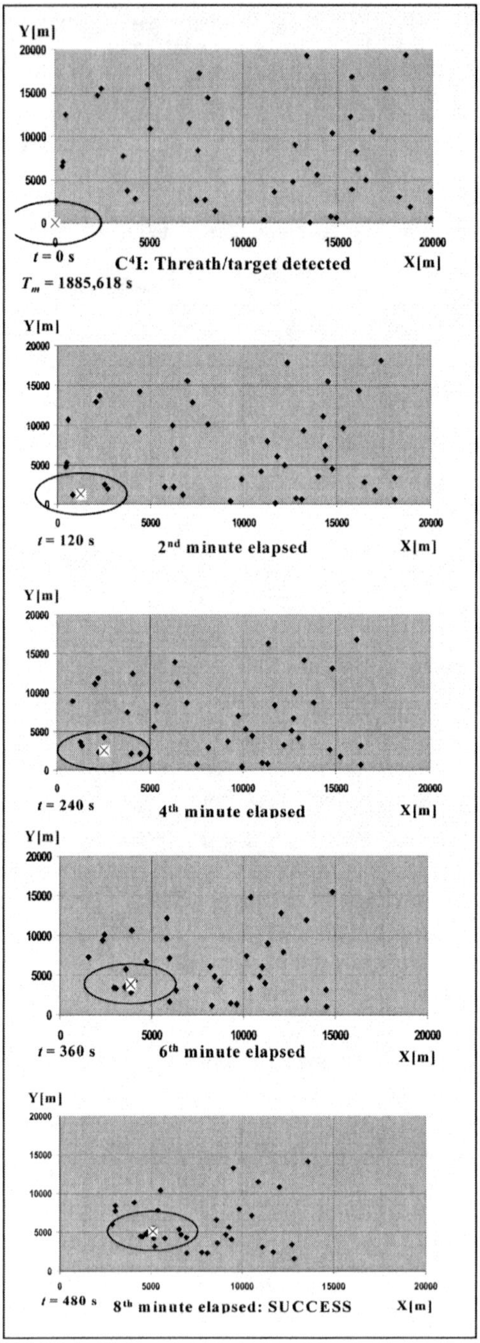


**Figure 4.** Armed mobile platforms swarming dynamic development.

- Initial position of the threat/target unit: A ($x_j(0)$, $y_j(0)$)
- Final position of the threat/target unit: B ($a$, $a$), a $\in$ {500, 1000,..., 10000 m}
- Law of motion of the threat/target unit P-j:
  $x$-axis: uniform motion, $x = X_o + V_c \cos\alpha t$
  $y$-axis: uniform motion, $y = Y_o + V_c \sin\alpha t$
- Maximal threat/target unit velocity: $V_c = 15$ m/s
- Threshold of the critical cumulative effect on the threat/target unit: $PKU_j \in \{1.5, 3.0, 4.5\}$

b) Parameters of the AMPs group:
- Number of AMPs in the group: $N = 43$
- Maximal AMP-$i$ velocity: $V_i = 15$ m/s
- Range of the main weapon MW-$i$: $R_{MV\text{-}i} = 2500$ m
- Effect of individual AMP-$i$ against the threat/target unit: $U_i$ = 0.15
- Compatibility coefficient of main weapon MW-$i$ with the threat/target unit: K-$i$ = 1

c) Parameters of the territory of swarming:
- Shape: quadratic
- Dimension: a $\in$ {500, 1000, ... , 10000 m}

d) Parameters of the CIS GAMP command information system:
- Preciseness (prescribed relative distance from the previous, known position of every threat/target unit T/T-$j$ or AMP-$j$, for which the previous position data are not obsolete): $P_{CIS} = 150$ m
- Existing communication equipment data transfer rate: $V_{RU} = 16$ kb/s.

The primary performance measure of the simulated system consisting of the armed mobile platforms group (AMP-$i$, $i$ = 1, 2, ..., $N$) which, in conflict with the threat/target unit, applies swarming in defending territory, is the probability of successful swarming, defined by the expression:

$$p_{ssw} = \frac{N_{ssw}}{N_{ssw} + N_{usw}} \qquad (4)$$

Where: $N_{ssw}$ is the number of successful, and $N_{usw}$ is the number of unsuccessful swarmings in the sample of L simulated conflicts.

### 3.5 Experiments and Analysis of the Results

Total of 60 experiments have been executed by means of the realized simulator (Janković 2011a). In each experiment $L$ = 10,000 conflicts of 1 target/threat unit and the group of $N$ = 43 AMPs, so total of 600,000 conflicts have been simulated. The CIS GAMP system preciseness has been set to $P_{CIS}$ = 150 m.

The experimental factors involved:

- Kind of threat/target unit, expressed through the critical threshold of successful swarming value, $PKU_j$ {1.5, 3.0, 4.5}
- Dimension of the quadratic defended territory, a ∈ {500, 1000, … , 10000 m}

The results of the experiments have been presented in Figs. 5, 6 and 7. For given size of the group of $N$ = 43 AMPs, successful swarming probability has been considered, depending on the defended territory area $S$ [km$^2$], the front width $a$ [m] and the defended territory occupancy density by the AMPs of the group, $G_{AMP}$ [AMP/km$^2$], defined by the expression:

$$G_{AMP} = \frac{N}{S} \tag{5}$$

Where $N$ is the number of AMPs in the group, and S is the defended territory area in km$^2$.

The successful swarming probability curves, $p_{ssw}$ [%], as function of the defended territory area, $S$ [km$^2$], have been presented in Fig. 5.

The simulation has been executed for 3 kinds of threat/targets, attacking the territory defended by the group of $N$=43 AMPs, applying the swarming tactics: T/T-1, T/T-2 and T/T-3, which need 10, 20 and 30 AMPs necessary for successful swarming, respectively.

The simulated system behaves as expected: the successful swarming probability in smaller defended territories is initially 100 percent, and later it decreases as the territory area increases, until becoming negligible. The fastest change is for the T/T-3 threat/target (which needs at least 30 AMPs for successful swarming), slower for T/T-2 (20 AMPs), and the slowest for T/T-1 (10 AMPs). If the probability $p_{ssw}$ greater than 90 percent is adopted as the criterion for the successful system, one can see that the group of 43 AMPs can be satisfactory in the territory of about 16 km$^2$ for the threat/target unit T/T-1, 9 km$^2$ for T/T-2 and 6.5 km$^2$ for T/T-3.

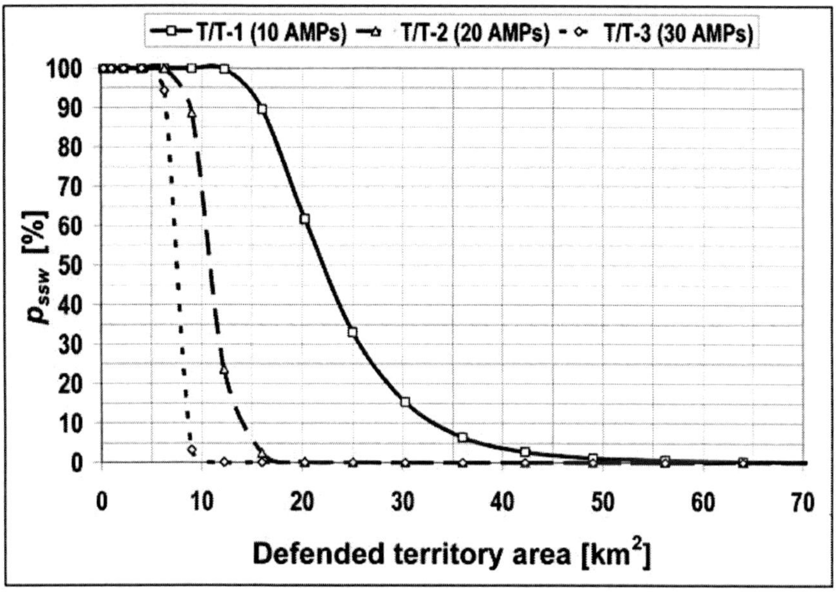


**Figure 5.** Successful swarming probability as function of the defended territory area.

One of the usual ways of dealing with the territory defending issues is by considering the defending front width, presented in the Fig. 6.

The curves are similar in shape to those in Fig. 5, with somewhat more moderate change, due to one-dimensional independent variable in the later case, expressed in linear metres. If the same criterion is considered, i.e. that $p_{ssw}$ must be greater than 90 percent, it can be seen that the group of 43 randomly deployed AMPs can satisfy it at the front width of about 4000 m for the threat/target unit T/T-1, 3000 m for T/T-2 and 2600 m for T/T-3.

Both ways of the results interpreting considered so far require knowledge of the exact number of AMPs in the group ($N = 43$).

Somewhat different way has been presented in Fig. 7, where the swarming success probability has been considered as function of the defended territory occupancy density by AMPs of the group, i.e., of the number of AMPs per km$^2$. That probability starts from 0 percent for empty territory, and increases with the occupancy density up to 100 percent. Such change is the fastest for the threat/target unit T/T-1 which requires at least 10 AMPs for successful swarming, slower for T/T-2 (20 AMPs), and the slowest for T/T-3 (30 AMPs).

It can be seen that the AMP group satisfies the criterion of the $p_{ssw}$ probability being greater than 90 percent at the defended territory occupancy density $G_{AMP} = 2.7$ AMP/km$^2$ for the threat/target unit T/T-1, $G_{AMP} = 4.8$ AMP/km$^2$ for T/T-2 and $G_{AMP} = 7$ AMP/km$^2$ for T/T-3.

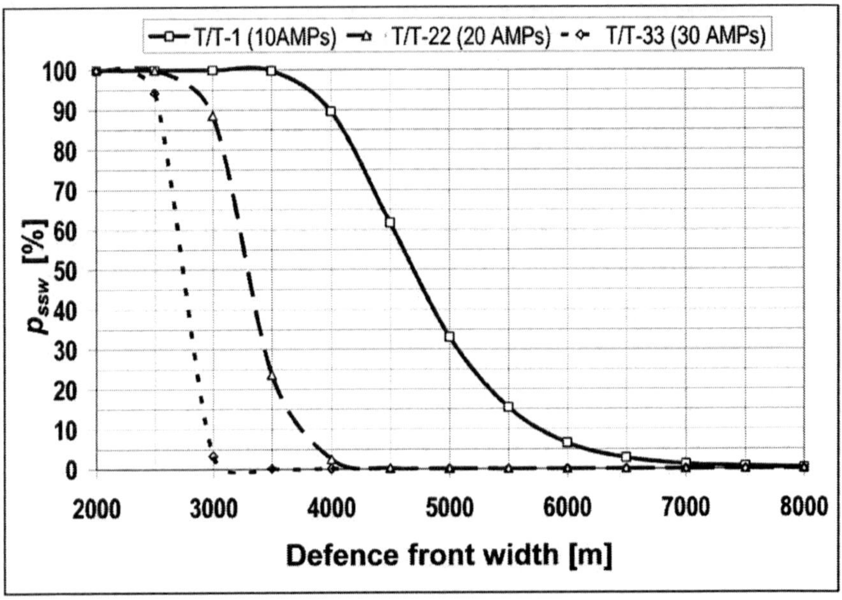


**Figure 6.** Successful swarming probability as function of defence front width.

The results produced by means of the realized armed mobile platforms group swarming simulator could be used for armoured and mechanized units' engagement in territory defence planning. If the characteristics of the threat/target unit are known, the appropriate density achievement, i.e., number of AMPs per unit of territory area, will result in the value of successful swarming combat action issue.

## 4. Swarming Command Information System Network Simulation

Two basic requirements should be met for successful implementation of swarming in combat (Janković 2005a, b, Janković and Nikolić 2009):

a. In order to attack the adversary from all directions, one must dispose with many small agile networked units, so that they can easily communicate and cooperate, which is expected from them in swarming.

b. Forces applying swarming tactics must engage themselves not only in direct combat activities (strike and fire actions) but have also to be an important part of "sensor organization", which enables observation at synoptic level, necessary for generating and maintaining the acceptable insight in general situation.

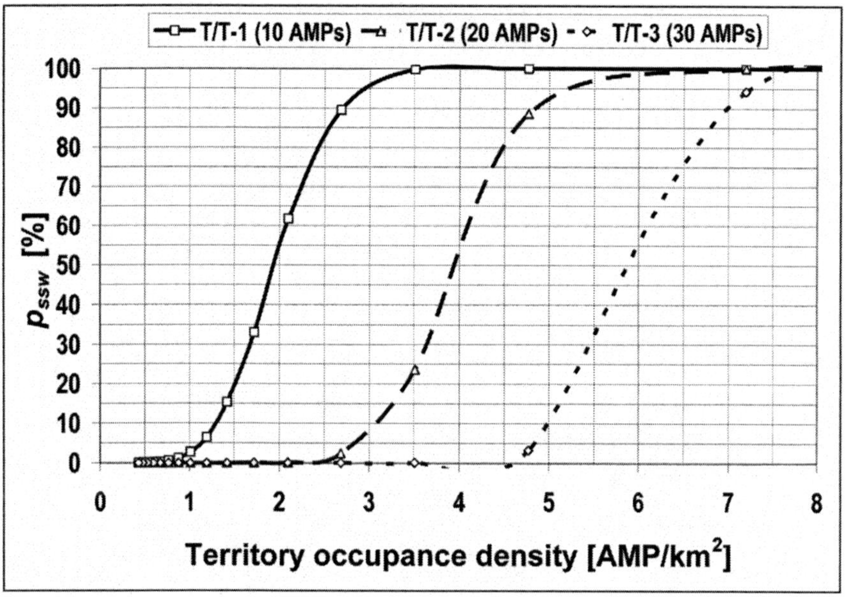


**Figure 7.** Successful swarming probability as function of territory occupancy density.

As for any other military forces, for those applying swarming tactics a higher level command element must exist, having insight in general situation, but which seldom intervene in case of swarming, issuing only general directives, in order to prevent irrational expenditures of resources for less important activities.

On the other hand, greater liberty in decision making, according to on-spot circumstances, is left to lower lever units, down to single tank crews, requiring different organization of command information system, as well as the very way of command and control combat operations, and training of all participants.

In Fig. 8 one can see that armoured and mechanized units are capable of performing both tactical and operational swarming (Janković 2007). Tactical swarming can be applied up to armoured battalion level, with the possible formation of a swarming group from all lower units, down to a single tank. On the other hand, armoured and mechanized units so organized could also perform operational swarming, when more battalion sized units take part in it.

For information support of the armoured and mechanized units' combat actions in contemporary warfare it is necessary to develop command information systems of armed mobile platforms groups (CIS GAMP). The CIS GAMP primary function is regular reporting of every AMP of the group of its current position and other information of interest,

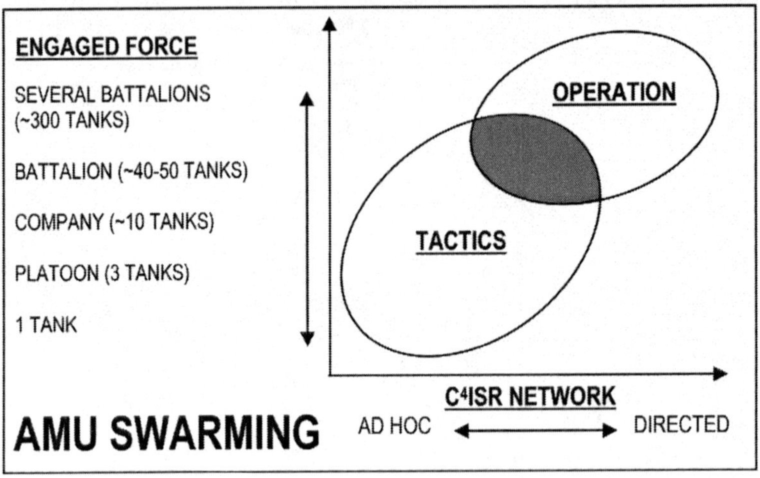


**Figure 8.** Armed and mechanized units tactical and operational swarming.

for the sake of updating the electronic maps and the situation assessment in real time. On the basis of such information, decisions on further actions should be made for the AMPs group as a whole, as well as for every single AMP.

### 4.1 Simulation Goal

The problem with swarming command information system is that participants are mobile, so its mobile local area network (RLAN) relies on relatively low data transmission rate radio communication devices. With the increase of swarmers' velocities, which is one of the principal requirements for successful fast swarming incursions, the need for more frequent reporting on their positions changes is also increasing; the same applies to the increase of number of AMPs in the swarming group. Both mean intensive traffic on RLAN radio channel, with possibility of its overloading.

The chosen approach to the problem solution in the research has been to build swarming command information system network discrete events simulation model and to implement it in computer program-simulator form, as a tool for conducting experiments.

The simulation goal has been to explore impact of three proposed algorithms to success of the swarming command information system network operation. It has been achieved through experiments with the simulators, realised for each algorithm and by changing the mean velocity of AMP group in the given range of possible values.

## 4.2 Simulated System

Every AMP has to be equipped with (Fig. 9):

- Computer (AMPC) with an electronic map on its display, for situation assessment;
- GPS receiver as a sensor for determining its own position;
- VHF radio communication device (AMPR), so that all computers of the AMPs could be connected in mobile radio local area network (RLAN).

The mobile radio local area computer network represents one of the most important CIS GAMP subsystems, because it severely influences the overall performances of CIS GAMP, AMP group and every single AMP. During the RLAN research (Janković 2006a, b, c, 2007) computer simulation has been used to explore that influence, so that the adequate decisions could be made on technical solutions of the AMP group command information system, its subsystems and components (sensors, computers, and data transmission devices).

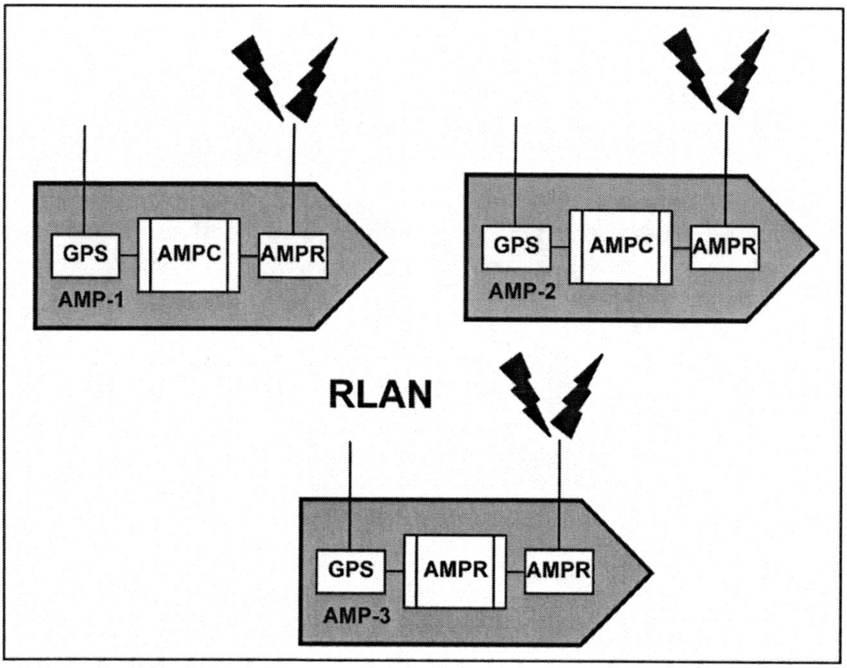


**Figure 9.** Radio LAN for armed mobile platform group command information system support.

Three different algorithms of the RLAN operation has been explored: *ad hoc* RLAN network with CSMA/CA algorithm (Khattab et al. 2002, Kurose and Ross 2003), suitable for small groups of AMPs, and two algorithms of directed RLAN, suitable for larger groups of AMPS: network with roll-call of participants and network with roll-call of participants and two types of messages—shorter ones for commands and longer ones for reports.

The radio local area computer network (RLAN) is a discrete dynamic system in which different activities are represented by pure time delays. The basic time unit in the simulation model is $T_0 = 1$ ms. It has been derived by considering characteristic times in the model, listed in Table 3.

**Table 3. Radio LAN simulation model characteristic times.***

| *Time* | *Name* | *[ms]* |
|---|---|---|
| $t_b$ | 1 bit transfer | 0.061 |
| $t_B$ | 1 byte transfer | 0.488 |
| $t_{JAM}$ | Jam signal (48 bits) | 2.929 |
| $t_{96b}$ | Radio channel in occupation signal (96 bits) | 5.859 |
| $t_{512b}$ | Back-off time (512 bits) | 31.25 |
| $t_p$ | Long message transfer (576 bits) | 35.16 |
| $t_{iapmin}$ | Minimal mean time between messages | 23.25 |
| $t_{iapmax}$ | Maximal mean time between messages | 465.1 |

*Radio communication device data transfer rate $V_{RU} = 16$ kb/s

Moving entities of the model are AMPs regular reports messages that represent requests for a shared resource—RLAN radio channel.

### 4.3 Simulation Model

Basic assumptions of all so far implemented simulation models of RLAN for CIS GAMP support were:

a) Armoured battalion is a group that consists of $N$ armed mobile platforms. It acts as a whole, and is moving at mean velocity $V_{GAMP}$ [m/s].
b) Every AMP (tank, armoured personnel carrier, self-propelled gun etc.) has GPS, receiver, computer (AMPC) and VVF radio communication device (AMPR), capable to transfer digital data at $V_{RU}$ [b/s] data transfer rate.
c) CIS GAMP basic function is real time reporting of every single AMP current position and other information of interest.

d) CIS GAMP operates efficiently if it delivers regular AMP reports before their obsolescence. The CIS GAMP time of message obsolescence $t_z$ [s] has been defined by the expression:

$$t_z = \frac{P_{CIS}}{V_{AMP}} \tag{6}$$

Where $P_{CIS}$ [m] is CIS GAMP preciseness (assigned displacement of the AMP due to further motion relative to previous position which could be tolerated as no movement at all), and $V_{AMP}$ [m/s] is the platform speed of movement within the group.

The command information system preciseness, chosen for armoured and mechanized units, has been $P_{CIS}$ = 20m, which corresponds to 3 lengths of contemporary tank.

e) Three different algorithms of the RLAN operation have been explored:
   - *Ad hoc* RLAN network with CSMA/CA protocol (Janković 2005c);
   - Directed RLAN network with roll-call of participants (Janković 2006a);
   - Directed RLAN network with roll-call of participants and two kinds of messages: command and reporting (Janković 2006c).

f) Message formats have been given in Table 4:

**Table 4. CIS GAMP message formats (in bytes).**

| *Field* | *Command* | *Report* |
|---|---|---|
| Preamble | 8 | 8 |
| Message destination address | 6 | 6 |
| Message source address | 6 | 6 |
| Type of protocol | 2 | 2 |
| Content | 1 | 46 |
| CRC | 4 | 4 |

### *4.3.1 Ad hoc Radio LAN*

The algorithm of *ad hoc* radio LAN (RLAN) for CIS GAMP support has been represented in Fig. 11.

The arrival of single AMPs regular report messages for transmission over RLAN radio channel is effected by generating pseudorandom number

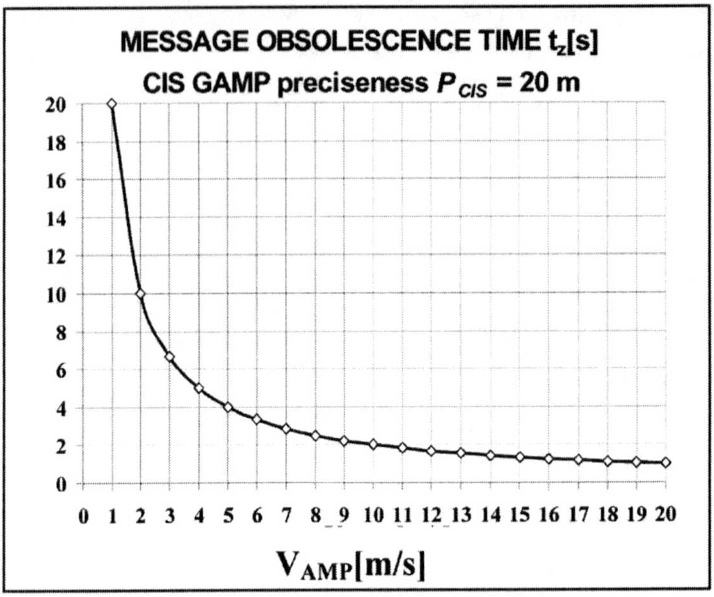


**Figure 10.** CIS GAMP message time of obsolescence as function of AMP motion velocity.

and sampling the exponential distribution. The mean time between two consecutive simulated messages, $t_{iap}$, has been defined by expression:

$$t_{iap} = \frac{P_{CIS}}{V_{GAMP} \cdot N} \tag{7}$$

Where $P_{CIS}$ is CIS GAMP preciseness, $V_{GAMP}$ is AMP group mean velocity, and $N$ is the number of AMPs in the group.

On arrival of message for transfer, the measurement of its total time delay in the system, $t_{CIS}$, begins. After the recognition of free RLAN radio channel, by finding lack of transmission of any signal for a period of 96 times of 1 bit transfer $t_{96b}$ the RLAN radio channel is seized and the message transfer begins.

If during the message transfer no collision comes up with another message that appeared in the meantime, on transfer completion the radio channel is released, the counter of successful message transmissions ($N_s$) is increased by 1 and the message is removed from the simulation.

If during the message transfer a collision with another message on the radio channel comes up, the transmission stops, the message releases radio channel and its collision counter ($N_c$) is increased by 1. If the limit

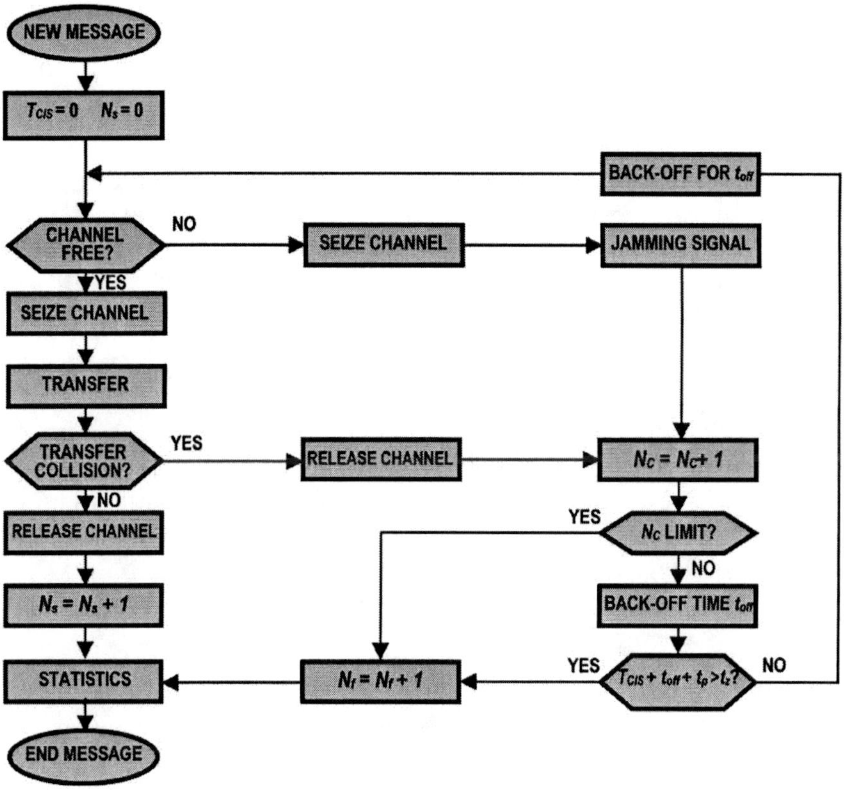


**Figure 11.** *Ad hoc* radio LAN simulator algorithm.

of the possible number of collisions for that message is reached, the failed message transfer counter ($N_f$) is increased by 1, and such message is removed from the simulation. Otherwise, that message is delayed for back-of time ($t_{off}$), which is calculated by the expression:

$$t_{off} = (1 + RN \bmod(2^{\min(n,10)} - 1)) \cdot t_{512b} \tag{8}$$

Where *RN* is generated random number (0 to 999), *n* is collision ordinal number for that message, and $t_{512b}$ is transfer time for 512 bits.

If the value of message back-of time $t_{off}$ is such that the inequality (5) is valid, the total time that the message spends in the system is so great that it will be obsolete on arrival at destination, even if there are no further delays and back-offs:

$$t_{CIS} + t_{off} + t_p > t_z \tag{9}$$

Where $t_{CIS}$ is a time that the message spent in the system until then, $t_{off}$ is back-of time, $t_p$ is the message transfer time over the RLAN radio channel and $t_z$ is the time of message obsolescence.

Otherwise, the message is delayed for the back-of time, $t_{off}$, and then a new attempt of its transfer is made.

If the RLAN radio channel is not free, the collision occurs with another message whose transmission is in course. That message transmission is then interrupted, and it is directed to simulator section for back-off time generating, while the message that caused the collision seizes the radio channel. Radio device that pre-empted the channel then transmits the jamming signal in $t_{JAM}$ = 48 times of 1 bit transfer, and after that the message is directed to afore mentioned simulator section for back-off time generating and the collision counter for that message is increased by 1.

The possible number of collisions for that message and its obsolescence are revaluated, and the message is either removed from simulation

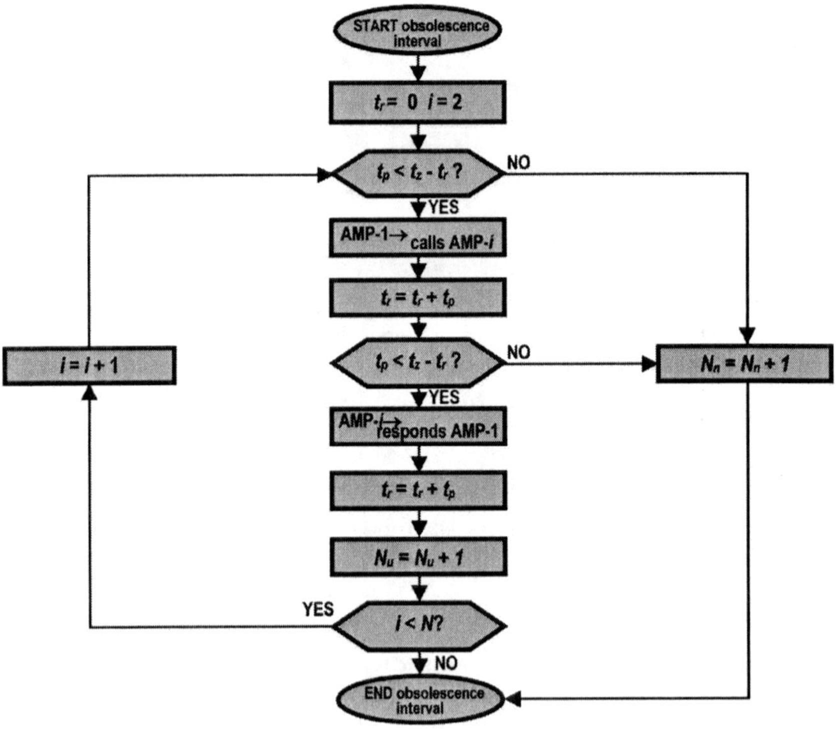


**Figure 12.** Directed Radio LAN with Roll-Call of Participants simulator algorithm.

or, after the back-off time elapses, a new attempt of its transmission is made.

### *4.3.2 Directed Radio LAN with Roll-Call of Participants*

Instead of an *ad hoc* network, in (Janković 2006b) a new RLAN operation algorithm, as directed radio LAN with roll-call of all participants has been introduced (Fig. 12).

The essence of the new algorithm can be reduced to the following:

1. The network director (battalion commander) is located in one of the armed mobile platforms (AMP-1) and he determines the time interval of message obsolescence, based on armoured battalion mean motion velocity. In those time intervals, he executes the other network participants (AMP-*i*, $i = 2, 3, \dots, N$) roll-call (according to their order, or at his discretion).
2. When (and only when) the RLAN director (AMP-1) calls some other participant (NMP-*i*), AMP-*i* can respond by its message, reporting the commander of its current position and other information of interest.
3. If until the end of the interval of obsolescence ($t_z$) AMP-1 fails to perform mutual communication with some of the remaining AMPs from the group, the director executes new roll-call, and failed communications are considered CIS GAMP losses, thus affecting the RLAN performance.

The new algorithm has been introduced with an aim to explore the possibilities of keeping the existing radio devices, with data transfer rate of $V_{RU} = 16$ kb/s.

### *4.3.3 Directed Radio LAN with Roll-call of Participants and Two Types of Messages*

The results of the directed radio LAN with improvement of the participants' roll-call algorithm, achieved by introduction of short command message as a new type of CIS GAMP message, have been presented in (Janković 2006c).

The CIS GAMP command message has data field of only 1 byte, coding a type of command. Such a message takes much less time in every communication between RLAN director and called network participant, resulting in significantly better performance when the participants—AMPs from the group—move faster, and the message obsolescence interval shortens.

### 4.4 System Parameters and Performance Measures

System parameters of simulated RLAN for CIS GAMP support are the following:

- Number of AMPs in the group: $N = 43$
- Maximal AMP-*i* velocity: $V_i = 15$ m/s
- CIS GAMP preciseness (prescribed relative distance from the previous, known position of every AMP-*j*, for which the previous position data are not obsolete): $P_{CIS} = 20$ m
- Radio device data transfer rate: $V_{RU} \in \{16, 32, 64, 115\}$ kb/s.
- RLAN operation algorithm: *ad hoc*, directed with roll-call, directed with roll-call and 2 message types.

Primary measure of performance is the RLAN operation efficiency, $U_{RLAN}$ [%], defined as percentage of CIS GAMP messages successfully transferred over RLAN during the simulated period of time.

### 4.5 Experiments and Analysis of the Results

The analysis of CSMA/CA algorithm results of 20 experiments executed by means of the realized simulator, in which 1 hour of the *ad hoc* RLAN operation has been simulated, can be found in (Janković 2005c). The *ad hoc* RLAN was intended for the CIS GAMP information support of an armoured battalion of 43 AMPs swarming operation. The CIS GAMP precision was $P_{CIS} = 20$m. The experimental factor was AMPs mean velocity, taking values $V_{AMP} = \{1, 2, \ldots, 20\}$ [m/s], and the existing radio communication devices transfer rate of $V_{RU} = 16$ kb/s.

The *ad hoc* RLAN efficiency is presented by the lowest curve in Fig. 13. Performance of such a network is not satisfactory, so in (Janković 2006a) the results of 80 experiments are published, executed with an aim to determine the impact of radio communication device choice (with data transfer rates of 16, 32, 64 and 115 kb/s), i.e., to justify the replacement of existing communication equipment.

The results of those experiments have shown that for the mean velocities of the AMPs motion up to 1 m/s, *ad hoc* radio LAN for CIS GAMP support has an efficiency of 100 percent for all 4 data transfer rates considered.

With the increase of AMPs group motion velocity, the *ad hoc* radio LAN efficiency decreases. If the minimum value of RLAN efficiency of $U_{RLAN} = 50$ percent is adopted as the acceptance criterion, then the radio communication device with data transfer rate of $V_{RU} = 16$ kb/s does not satisfy, so it should not be used in an *ad hoc* RLAN for CIS GAMP information support of armoured battalion swarming operation. The radio

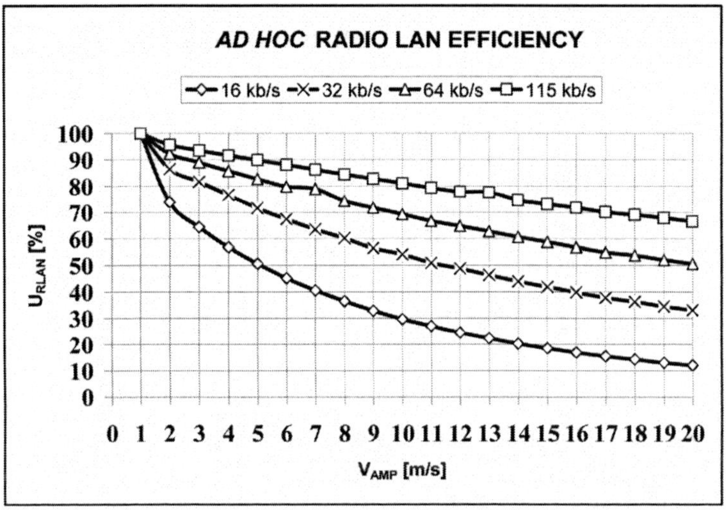


**Figure 13.** *Ad hoc* radio LAN efficiency as function of armed mobile platforms motion velocity.

communication device with $V_{RU}$ = 115 kb/s satisfies the needs of *ad hoc* radio LAN, and the remaining two considered devices have a performance between the two afore mentioned.

The new algorithm of directed radio LAN with roll-call of network participants has been introduced in (Janković 2006b), with an aim of exploring the possibilities of keeping the existing communication equipment, with data transfer rate of 16 kb/s. Simulation experiments have been executed in order to evaluate performance of that algorithm, compared to *ad hoc* radio LAN, and the results have been presented in Fig. 14.

Retaining the same, existing communication equipment, directed radio LAN with roll-call of network participants performs with 100 percent efficiency up to the armoured battalion mean velocity of motion of $V_{AMP} = 6$ m/s. Its efficiency also decreases with the increase of AMPs velocity, except that the boundary of acceptable network operation ($U_{RLAN} = 50$ percent) has moved to $V_{AMP} = 13$ m/s.

At maximum velocity of the AMPs group ($V_{AMP} = 20$ m/s), directed radio LAN with roll-call still operates with an efficiency of $U_{RLAN}$ = 33 percent, which means that under such conditions every third communication of CIS GAMP is successful. So, by changing the algorithm of radio LAN operation, its performance can be improved, proving the approach much more cost-effective compared to *ad hoc* radio LAN and procurement of new communication equipment.

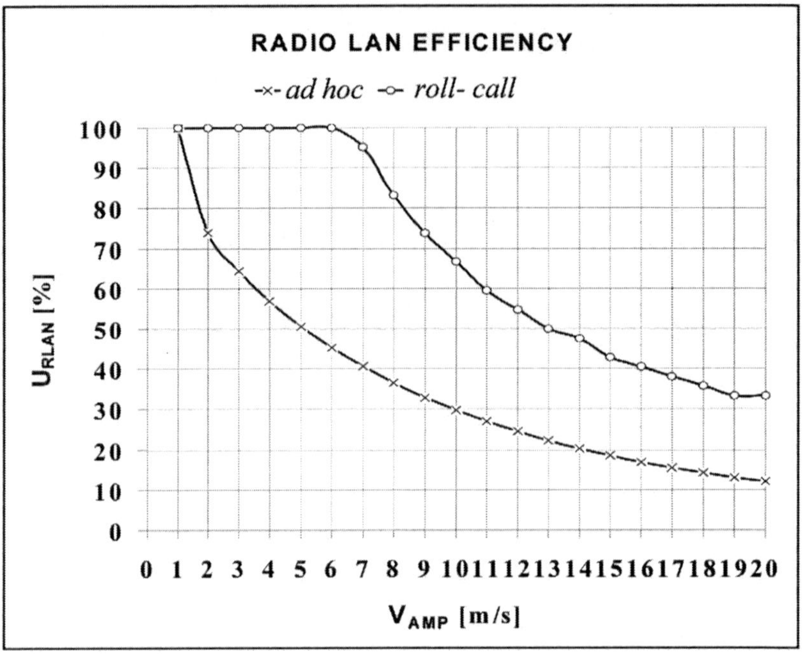


**Figure 14.** Radio LAN efficiency as function of its operation algorithm.

An improved algorithm of directed radio LAN with roll-call of participants and two types of CIS GAMP messages (shorter ones for commands and longer ones for reports) have been introduced in (Janković 2006c). The results of simulation experiments have been presented in Fig. 15, along with results for two aforementioned algorithms.

The results of simulation experiments have shown that by introducing special, short command messages to the CIS GAMP, significant network performance improvement can be achieved, compared to that one operating by algorithm of roll-call of participants and unique message format.

Directed radio LAN with roll-call and short command messages has efficiency of $U_{RLAN}$ = 100 percent up to the mean AMPs group motion velocity of $V_{AMP}$ = 9.85 m/s (35.46 km/hr). For greater velocities, the network efficiency also decreases, but slower than in the case of directed radio LAN with unique format of the message, and especially compared to *ad hoc* network. The network efficiency has been greater than 50 percent up to AMPs velocity of 19 m/s, and only for the maximum velocity of $V_{AMP}$ = 20 m/s, it has dropped to 47.62 percent.

If we analyze the relative increase in network performance as a function of the network operation algorithm for two typical AMPs motion velocities, the results of experiments have shown that:

- For $V_{AMP} = 10$ m/s, directed radio LAN with roll-call and 2 message types has 1.464 times greater efficiency than directed radio LAN with unique message format, and 3.26 times greater efficiency than *ad hoc* radio LAN with CSMA/CA protocol;
- For $V_{AMP} = 20$ m/s, directed radio LAN with roll-call and 2 message types has 1.429 times greater efficiency than directed radio LAN with unique message format, and 3.946 times greater efficiency than *ad hoc* radio LAN with CSMA/CA protocol.

The results of experiments indicate that CSMA/CA protocol is not suitable for the battalion-sized group of AMPs *ad hoc* radio LAN, especially at higher velocities of AMP movement. That is consistent with conclusions in (Khattab 2002): "CSMA/CA is a protocol suitable for low number of stations and low arrival rates (i.e., low offered traffic) which is expected for a contentions-based access scheme." and "The protocol has a large delay value that makes it unsuitable for time-bounded applications."

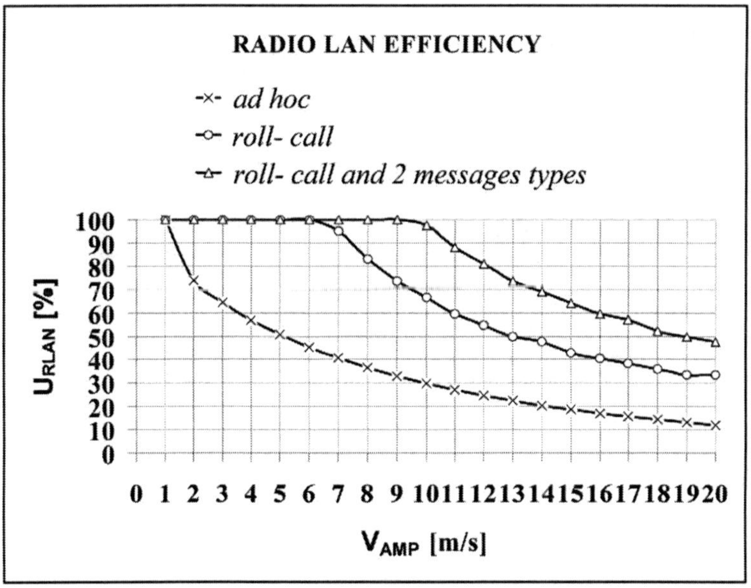


**Figure 15.** Radio LAN efficiency as function of its operation algorithm and message types.

However, in making decisions on mode of operation of radio LAN supporting command information system of armed mobile platforms, one should additionally have in mind the following:

- Radio LAN with roll-call of participants is directed network, so the problem of control take-over in case of disabling the network director has to be solved.
- Such an algorithm is not the standard one, which requires additional effort in communication equipment programming;
- Contrary to the case of *ad hoc* network with CSMA/CA protocol, where the crews of AMPs make decisions on information obsolescence, based on their own positions and motion velocities, in radio LAN with roll-call, it is the network director who calls other participants according to the information obsolescence interval, determined on the mean motion velocity of the AMPs group as a whole.

## 5. Future Research Main Directions

In course of the research of armoured and mechanized units so far, several simulators have been developed, among which the most important are:

- Simulator of an AMP group's monotarget swarming;
- Simulator of an AMP group's local area mobile computer network operation.

The developed simulators have served as tools for executing numerous experiments, of which some characteristic results have been presented in this chapter. Although practically usable even in the simulator's developing phase achieved so far, the results of those experiments raised some new issues, which marked the directions of future research work.

Presently, the most important among them are the following:

- AMP group swarming synchronization;
- Special purpose AMP group initial deployments;
- Multitarget swarming systems;
- Heterogeneous swarming systems.

### 5.1 AMP Group Swarming Synchronization

The synchronization problem arises if AMPs applying swarming against target/threat don't reach the distance from it within the range of their main weapons in approximately same time and in a number large enough to achieve critical threshold of their cumulative effect. In that case there is a probability of swarming failure, because target/threat could destroy AMPs one by one, as they appear within the range of its own weapons.

To solve this problem, we have proposed in the paper (Janković 2010a) introducing of synchronization zone (SZ), circular ring around target/threat (Fig. 16), defined by the expression:

$$S_Z = R_{WTT-j}\,(1 \pm q) \tag{10}$$

Where $R_{WT/T-j}$ is the effective range of weapon of T/T-*j*, and $0 < q < 1$.

Such a mechanism enables, while there are not enough AMPs for successful swarming, that:

- AMPs far from target/threat continue (AMP-1, 2, 3, 4 and 6 in Fig. 16) to approach it by the shortest possible way;
- AMPs in the *SZ* (AMP-5 and 7 in Fig. 15), move parallel with target/threat at relative safe distance, ready to move towards target/threat again, when enough AMPs gather for successful swarming cumulative effect;
- AMPs near target/threat (AMP-8 in Fig. 15) move away from it towards *SZ*, in order to avoid individual destruction and swarming failure.

### 5.2 Special Purpose AMP Group Initial Deployments

Certain attention is going to be paid to different initial deployments of the AMPs of the group. One of significant aspect of the research is the exploring of initial deployments intended for specific purposes of different situations, when the AMP group has to apply the swarming tactics.

The results obtained in simulation of swarming of the group of AMPs with random initial deployment will be the reference (the worst case), whereas the results obtained for the chosen specific initial deployments should be compared to, for the sake of getting knowledge of advantages that could be achieved by alternative solutions considered.

### 5.3 Multitarget Swarming Systems

The AMP group swarming simulator realized so far is limited to the case of a single target/threat and no synchronization procedure among the swarming AMPs. Those limitations are not a problem when considering relatively smaller AMP groups, which are often engaged in combat action against one target/threat, and/or when the number of

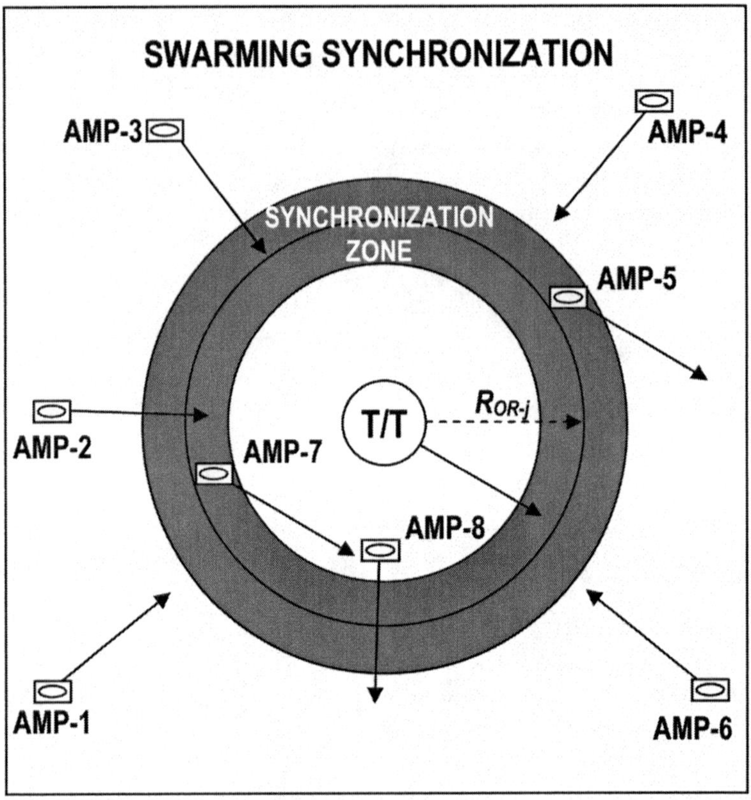


**Figure 16.** Armed mobile platforms swarming synchronization.

AMPs in the group is large enough to let the swarming develop in an unsynchronized way.

However, even an armoured battalion, as the basic tactical unit, has opportunities to apply swarming against target/threat which is near, or even beyond its capabilities, or simultaneously against 2, 3 and even more different targets/threats. This imposes some new issues related to target/threat selection, priorities, and the capabilities of the group for the self-organizing in such situations.

Possibility of multiple target/threat introductions in the swarming simulation model is one of the most important directions of the future researching work. An improved simulator algorithm and corresponding implementation mechanisms have been proposed in the paper (Janković 2010b). It has been done, having in mind the following initial assumptions:

- Threats/targets appear in the swarming territory independently, in arbitrary spatial and temporal disposition.

- Every new threat/target is on its own mission, characterized by its own law of motion and point in space which it has to reach by such motion.
- Every threat/target is defined by its critical threshold of cumulative effect of several armed mobile platforms which perform swarming against it.
- AMP group engages its armed mobile platforms assuming that all target/threats appearing in the defended territory are of equal importance, and differ only in sequence of arrival, i.e., deals with them on the FCFS (first come-first served) basis, unless some other priority policy is introduced.
- AMP group engages sufficient number of AMPs against every target/threat, providing potentially successful swarming. AMPs are assigned against some particular target/threat by least distance criterion. Armoured battalion engages the remaining AMPs against other target/threats, if there are some in the defended territory.
- After successful swarming, target/threat is removed from simulation; the AMPs engaged in swarming against that target/threat are released, in order that all those resources could be assigned to new target/threats entering the simulation.

The proposed multitarget algorithm is in fact upgraded basic monotarget algorithm, enhanced by introduction necessary new data structures and mechanisms for events management in models for simulation of swarming against multiple targets/threats simultaneously present in the battlefield.

## 5.4 Heterogeneous Swarming Systems

Swarming is a very promising cost-effective tactical concept, because its main advantage is that it gives good prospects to successful conflict issue, engaging much less military resources than it was necessary in more conventional warfare up to now.

It is, therefore, of prime interest to widen the swarming tactics research, from armoured and mechanized units, where mostly homogenous swarming systems models were developed, to other, more complex military systems, where heterogeneous models are more advantageous.

That would enable the research of complex tactical procedures of armoured and mechanized unit's cooperation with other arms of service as possible swarming participants, from modern infantry, self-propelled artillery and missile launchers, to flying AMPs (aircrafts and helicopters) and/or armed unmanned armed robotic platforms.

## 6. Conclusions

A review of several years' research in the area of armoured units swarming has been presented in this chapter. The research methodology is based on discrete events simulation.

The research basic motivation is that armoured units are still one of the most important resources of ground military forces. The problem is that contemporary and future warfare conditions impose the need for new solutions of the next generation of tanks technical features, as well as the new tactical procedures, swarming being among the most promising ones.

The armed mobile platform concept, introduced in the research, proved itself suitable for research and development of military systems characterized by autonomous propulsion, man crew, armament and significant need for logistic support. Armoured combat vehicles are the most representative examples of the kind, but the concept could also be used for researching other military and/or non-military systems.

In course of the research, several simulators have been realized, notably those of armed mobile platforms group monotarget swarming and of mobile local area computer network operation. All these simulators have been realised by means of basic mechanisms, developed as building blocs and implemented in the GPSS World simulation language.

Numerous experiments have been executed with realized simulators, in order to explore several factors impact to armed mobile platform group monotarget swarming success and to effectiveness of the armed mobile platforms group local area network operation. Among others, those factors involved: data transfer rate, algorithm of local area network operation and defended territory occupancy density.

The research results achieved so far could be used for planning of armoured and mechanized units in the territory defence, as well as useful information for future designs and new tactical procedures. To our best knowledge, the research of armoured combat vehicles swarming tactics and the results achieved so far are among the first ever published in that area.

During the research, some new issues have been raised. Presently, the research efforts are concentrated to developing the simulating mechanisms for swarming synchronization, special purpose armed mobile platforms group initial deployments, multitarget swarming and heterogeneous swarming systems.

It is expected that those mechanisms enable simulations which will yield important results to contribute the new, more effective and economical usage of armoured and mechanized units, both the upgraded existing ones and the new generations to come.

## REFERENCES


Arquilla, J. and D. Ronfeldt. 1999. Swarming and the future of conflict. Rand Corporation, Santa Monica, CA, USA.

Beni, G. 1988. The Concept of Cellular Robotic System. In: Proceedings of IEEE International Symposium on Intelligent Control, Los Alamitos, CA, USA. pp. 57–62.

Bonabeau, E. 2003. Swarm Intelligence. In: Proceedings of Swarming: Network Enabled C4ISR, Tysons Corner, ASD C3I, VA, USA, sec. C, pp. 19–25.

Edwards, S.J.A. 2000. Swarming on the Battlefield—Past, Present and Future. Rand Corporation, Santa Monica, CA, USA.

Edwards, S.J.A. 2003. Military History of Swarming. In: Proceedings of Swarming: Network Enabled C4ISR, Tysons Corner, ASD C3I, VA, USA, sec. C. pp. 2–11.

Edwards, S.J.A. 2005. Swarming and the Future of Warfare. Rand Corporation, Santa Monica, CA, USA.

Inbody, D. 2003. Swarming: Historical Observations and Conclusions. In: Proceedings of Swarming: Network Enabled C4ISR, Tysons Corner, ASD $C^3I$, VA, USA, sec. C. pp. 12–18.

Janković, R. 1998. Concept of the Armed Mobile Platform. *Vojno Delo*. **3**:81–96.

Janković, R. 2001. Methodology for Embedded $C^3I$ Systems Research and Development. *Vojno Delo*. pp. 4–5.

Janković, R. 2005a. Simulation Model of *Ad Hoc* Network for the Group $C^3I$ System Support. The Proceedings of the 11th Scientific-Professional Symposium on Information and Communication Technologies YUINFO 2005, Kopaonik, Serbia.

Janković, R. 2005b. Computer Simulation of the Group $C^3I$ System *Ad Hoc* Network Reliability. In: Proceedings of the 8th DQM International Conference DQM-2005, Belgrade, Serbia. pp. 168–176.

Janković, R. 2005c. Simulation of *Ad Hoc* Computer Network for an Armoured Battalion $C^3I$ System. The Proceedings of the 32nd Symposium on Operations Research SYM-OP-IS 2005. Vrnjačka Banja, Serbia.

Janković, R. 2006a. Approaches to the Group $C^3I$ System Mobile Network Reliability Improvement. The Proceedings of the 9th DQM International Conference DQM-2006. Belgrade, Serbia, pp. 534–541.

Janković, R. 2006b. Simulation of Two Group $C^3I$ System Mobile Computer Network Algorithms. The Proceedings of the 12th Scientific-Professional Symposium on Information and Communication Technologies YUINFO 2006. Kopaonik, Serbia.

Janković, R. 2006c. Computer Simulation of Improved Algorithm for Computer Network of the Armoured Battalion $C^3I$ System. The Proceedings of the 33rd Symposium on Operations Research SYM-OP-IS 2006. Banja Koviljača, Serbia.

Janković, R. 2007. Research of some tank modernization aspects to armoured battalion swarming effectiveness. The Proceedings of 2nd Scientific-Professional Conference on Defence Technologies—OTEH 2007. Belgrade, Serbia.

Janković, R. 2008a. An Approach to Computer Simulation of an Armoured Battalion Swarming. The Proceedings of 14th Scientific-Professional Symposium on Information and Communication Technologies YUINFO 2008. Kopaonik, Serbia.

Janković, R. 2008b. GPSS Realization of the Armed Mobile Platform Group's Swarming Algorithm. The Proceedings of 11th Annual International Conference on Dependability and Quality Management ICDQM-2008. Belgrade, Serbia. pp. 883–889.

Janković, R. 2008c. Simulator of Armoured Battalion Swarming. The Proceedings of the 35th Symposium on Operations Research SYM-OP-IS 2008. Soko Banja, Serbia. pp. 463–466.

Janković, R. and N. Nikolić. 2009. Computer Simulation Applications in Contemporary War Physiognomy Research. Institute for Strategic Research, Belgrade, Serbia.

Janković, R. 2010a. On Some Basic Control Mechanisms for Multitarget Swarming Software Simulators. The Proceedings of 37th Symposium on Operations Research SYM-OP-IS 2010. Tara, Serbia. pp. 565–568.

Janković, R. 2010b. On Armed Mobile Platform Group Multitarget/Threat Swarming Simulator Algorithm. The Proceedings of the 16th Scientific-Professional Symposium on Information and Communication Technologies YUINFO 2010. Kopaonik, Serbia.

Janković, R. 2010c. Discrete Event Simulation in Armoured and Mechanized Units research. The Proceedings of the 1st DQM International Conference ICDQM-2010. Belgrade, Serbia. pp. 97–114.

Janković, R. 2011a. Computer Simulation of an Armoured Battalion Swarming. *Defence Science Journal*, Vol. 61, No. 1. pp. 36–43.

Janković, R. 2011b. Computer Simulation of Armed Mobile Platform Group Swarming. The Proceedings of the 5th International Conference on Information Technology ICIT 2011. Amman, Jordan, paper no. 541.

Khattab, T.M.S., M.T. El-Hadidi and H-A.M. Mourad. 2002. Analysis of Wireless CSMA/CA Network Using Single Station Superposition (SSS). *International Journal of Electronics and Communication (AEÜ)*, Vol. 56, No. 2. pp. 71–81.

Kurose, J.F. and K.W. Ross. 2003. Computer Networking: A Top-Down Approach Featuring the Internet. Second Edition, Pearson Education, USA.

Lamont, L., M. Toulgoat, M. Déziel and G. Patterson. 2011. Tiered Wireless Sensor Network Architecture for Military Surveillance Applications. The Proceedings of the Fifth International Conference on Sensor Technologies and Applications SENSORCOMM 2011. August 21–27, 2011. French Riviera, Nice/Saint Laurent du Var, France. pp. 288–294.

Liu, B., P. Terlecky, A. Bar-Noy, R. Govindan and M. J. Neely. 2011. Optimizing Information Credibility in Social Swarming Applications. In: Proceedings of IEEE INFOCOM 2011 conference, Shanghai, China.

Minuteman Software: GPSSWorld Reference Manual. http://www.minutemansoftware.com/reference/reference_manual.htm

Nowak, D.J., I.C. Price and G.B. Lamont. 2007. Self Organized UAV Swarm Planning Optimization for Search and Destroy Using SWARMFARE Simulation.

The Proceedings of the 2007 Winter Simulation Conference. Washington, DC, USA.

Parunak, H.V.D. and Engineering Swarming Systems. In: F. Bergenti and M.P. Gleizes [eds.] 2004. Methodologies and Software Engineering for Agent Systems. Kluwer Academic Publishers, USA. pp. 341–375.

Price, I.C. and G.B. Lamont. 2006. GA Directed Self-Organized Search and Attack UAV Swarms. The Proceedings of the 2006 Winter Simulation Conference. Monterey, CA, USA.

Pohl, A.J. and G.B. Lamont. 2008. Multi-Objective UAV Mission Planning Using Evolutionary Computation. The Proceedings of the 2008 Winter Simulation Conference. Miami, FL, USA.

Sauter, A.J., R. Matthews, H.V.D. Parunak and S.A. Brueckner. 2005. Performance of Digital Pheromones for Swarming Vehicle Control. In: Proceedings of the Fourth International Joint Conference on Autonomous Agents and Multiagent Systems AAMAS 05. ACM, NY, USA, pp. 903–910.

Singer, P.W. 2009. Wired for war? Robots and military doctrine. *Joint Force Quarterly*, issue 52, 1st quarter 2009, pp. 105–110.

# 8

# Image Processing Applied in Agriculture

*Jesús R. Martínez-Sandoval, Eduardo A. Murillo-Bracamontes, Miguel E. Martínez-Rosas, Manuel M. Miranda Velasco and Humberto Cervantes De Ávila*

**ABSTRACT**

The purpose of this chapter is to provide an overview of recent developments in image processing in agriculture, and more specifically for aquaponic and vineyard applications. In order to satisfy the demand for cost efficiency, low-cost multipurpose image processing embedded systems with Wireless Sensor Network (WSN) capability are included. The material presented should be useful for students starting out in the field of image processing, as well as for researchers looking for state-of-the-art studies and practical applications of image processing.

## Introduction

With the advances in electronic and information technologies, various sensing systems have been developed for specialty crop production around the world. Accurate information concerning the spatial variability within fields is very important for precision farming of specialty crops.

In the evaluation of fruit and vegetable quality, producers and consumers rely on the external inspection of produce: color, shape, bruising and firmness. Therefore, in the food industry, quality inspections are usually carried out by hand. Nevertheless, judging a fruit on its external parameters is a subjective task, and human decision is inconsistent and may vary for each individual. By using a computer vision system that is able to grade automatically, and extract useful information with a degree of sensitivity closer to that of a human, the margin of error can be considerably minimized. Nowadays, with digital cameras and computers becoming ever more affordable, a computer vision system may be used for inspection tasks. A computer vision system is made of an optical setup to capture an image, and this image is then sent to a computer, first to be pre-processed and segmented, and then to be analyzed. In some cases, it is necessary to process the image in an embedded system to perform dedicated functions like fruit sorting, or to send only processed data over a very limited bandwidth channel (such as WSN) to a graphical user interface (GUI), in order to build a database concerning particular crop characteristics.

Because of its non-contact and non-destructive properties, computer vision systems are highly suitable for the fruit crop industry (Chherawala et al., 2006).

Specialty crops are living biological products (fruits and vegetables, tree nuts, dried fruits and horticulture and nursery crops) and consumers expect them to be safe and of the highest quality. Non-destructive (ND) testing for properties and characteristics of specialty crops is critical for monitoring and controlling product quality and safety because traditional manual expertise for quality inspection is no longer adequate (Lee 1995).

The quality of agricultural farming products is affected by several factors, including crop yield, soil properties and nutrients, crop nutrients, crop canopy volume and biomass, water content, and pest conditions (disease, weeds, and insects). These factors can be measured using a range of sensors and instruments such as field-based electronic sensors, spectroradiometers, machine vision, airborne multi spectral and hyper-spectral remote sensing, satellite imagery, thermal imaging, radio frequency identification (RFID), and machine olfaction system, among others (Lee et al., 2010). Still, despite this wealth of information, farmers find its on-site application complicated, as the average farmer cannot be expected to be an expert in computer science or electrical engineering just as they are not experts in agricultural pathology or nutrition (Lee 1995). Because of these issues, developments need to be as simple as possible without sacrificing operation, in order to avoid non-acceptance from farmers or perhaps abandonment of the technology, even though this is a high-performance system like wireless sensor networks (WSN).

Computer vision is a non-destructive method of inspection, a rapid, economic, consistent and objective inspection technique that has found applications in the agricultural and food industry, including the inspection and grading of fruit and vegetables (Brosnan 2002). The recent development in both hardware platforms and software frameworks has led to many significant advances in imaging and spectroscopic technology (Valous et al., 2010), so the extra help required by farmers has been made possible.

## Fundamentals of Computer Vision

An image is a 2D representation of a three dimensional scene. A digital image is basically a numerical representation of an object. Digital image processing, also referred to as computer imaging, can be defined as the acquisition and processing of visual information by computer. The importance of digital image processing is derived from the fact that our primary sense is sight. Our vision system allows us to gather information without the need for physical interaction.

Image analysis involves investigation of the image data for a specific application. Typically, it starts with a set of images and the purpose is to look beyond the raw image data, to gain insight into what is happening with the images and determine how they can be used to extract the required information (Umbaugh 2010).

## Digital Image Representation

An important aspect in digital image processing is image representation. Any monochrome image can be represented by means of a two dimensional light intensity function $f(x,y)$, where $x$ and $y$ denote spatial coordinates and $f(x,y)$ is the value of the gray level, or brightness, of the image at the point $(x,y)$. The axis convention used to represent the image is shown in Figure 1. The origin is taken at the top left corner and the horizontal line and vertical line through the origin are taken as $y$ and $x$ axes, respectively (Annadurai and Shanmugalakshmi 2007).

A digital image can be represented as a matrix whose rows and columns are used to locate a point in the image and the corresponding element values give the gray level at that point. Each element in this matrix is called a pixel. A typical digital image of size $MxN$ is represented by the function showed in the equation (Eq 1).

$$f(x,y) = \begin{bmatrix} f(0,0) & f(0,1) & \cdots & f(0,M-1) \\ f(1,0) & f(1,1) & \cdots & f(1,M-1) \\ \vdots & \vdots & & \vdots \\ f(N-1,0) & f(N-1,1) & \cdots & f(N-1,M-1) \end{bmatrix} \quad \text{(Eq 1)}$$

**Figure 1.** Pixel representation on a digital image.

An image of size *NxM* consists of *NM* pixels. The number of pixels that can be accommodated in a unit area is called the resolution of an image and it depends strongly on:

1. The number of values for *N* and *M*
2. The number of bits *m* used to represent the gray level

## Image Acquisition Systems (Components of a Machine Vision System)

The goal of a machine vision system is to create a model of the real world from images. A machine vision system recovers useful information about a scene from its two-dimensional projections (Jain et al., 1995).

### Hardware and Software Requirements

Budget constraints are the major obstacle to applying image techniques to agriculture. To satisfy the demand for cost efficiency, developing low-cost multipurpose image processing systems is especially important to develop efficient and accurate image processing algorithms as well as to integrate such image-processing algorithms onto specialized hardware (Du 2004).

The hardware requirements for vision machine systems basically include (see Figure 2) a camera, a light source (for homogeneous illumination), a background and a sample to generate an image for the

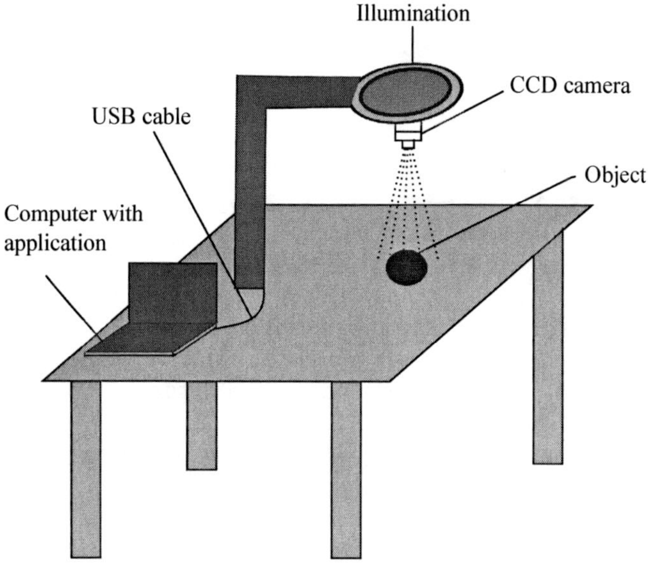


**Figure 2.** Basic vision machine (computer) system.

acquisition system and software analysis. For embedded systems a power management system must also be considered.

## Lighting (Illumination)

Illumination is a very important part of a vision system, for example, objects under an incandescent source produce an image with a much lower signal to noise ratio. This is not acceptable in some cases, especially those that are concerned with color image processing (Daley et al., 1993). In contrast, fluorescent lamps are inherently more efficient, and produce more intense illumination at specific wavelengths. Moreover, fluorescent light provides a more uniform dispersion of light from the emitting surface, and hence does not require the use of diffusing optics to disseminate the light source over the field of view, as is the case with incandescent lamps. For these reasons, a fluorescent bulb, particularly the cool white type, is a popular choice for many machine visions (Tao et al., 1995) (Abdullah et al., 2005) (Pedreschi et al., 2006).

Care must be taken when using the fluorescent light, as this source is normally AC-driven. The 50 Hz fluorescent lamp usually introduces artifacts in the image, resulting from the oversampling of the analog to digital converter. In order to reduce flickering, high frequency fluorescent bulbs, operating at a frequency in the range of a few tens of kilohertz, are preferred to low frequency ones.

### Power Management Systems

For embedded systems used in agriculture, power management systems are required to supply them, especially when they transmit continuously. Embedded systems as sensor nodes are typically powered by batteries with a limited lifetime and, even when additional energy can be harvested from the external environment (e.g., through solar cells or piezoelectric mechanisms), it remains a limited resource to be consumed judiciously. Effective energy management strategies should include policies for an efficient use of energy-hungry sensors, which become one of the main components affecting the network lifetime (Alippi et al., 2009). In order to provide a better, hassle-free node operation, it is advisable to use a power management system with dual batteries to support WSN operation 24 hours a day. This would be solar-powered via a separate battery system when charging and consuming at the same time (see Figure 3) (Miskam 2009).

### Image Processing and Analysis Software

This requirement is accomplished by a personal computer or microprocessor system, which provides disk storage of images and computational capability with vendor-supplied software and specific application programs (Jain et al., 1995).

### Frame Grabber

The features of a frame grabber required for machine vision applications include image acquisition, camera control, and image data pre-processing.

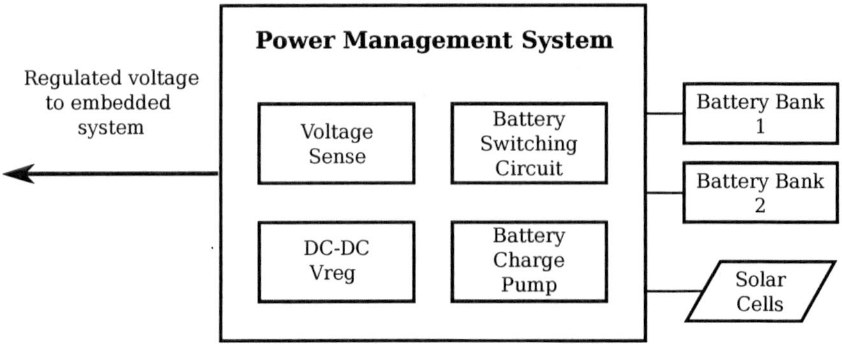


**Figure 3.** Power management system.

The frame grabber can acquire either digital or analog images depending on the camera used. For camera control, a minimum requirement is accurate analog-to-digital conversion (A/D) circuitry and precise camera timing. Input signal conditioning, such as the ability to control gain and offset, is important to minimize effects from camera variability or lighting fluctuations (Chen et al., 2002). For image processing algorithms, software implementation on a PC allows for rapid development, debugging, and testing. However, as image sizes grow larger and algorithms become more complex, the speed will be slower and will be unable to satisfy the high speed requirement in real-time systems; for example, when an application requires real-time processing like video or television signal processing, the specifications are very strict and are better met when implemented in hardware (Kalomiros and Lygouras 2008). Hardware implementation offers much greater speed than with software. There are several viable options for hardware implementation of image processing algorithms, such as application-specific integrated circuits (ASICs), digital signal processors (DSPs), and field programmable gate arrays (FPGAs) (Du and Sun 2006) (Bellis et al., 2005).

### CCD Cameras and Visualizing System

A solid state CCD array camera is required (see Figure 4) to acquire the image. Furthermore it is usually recommended to use a high resolution color monitor, which aids in visualizing images and effects of various image analysis routines (Sun 2007).

### Setup Constraints Imposed by the Type of Application

The basic assembly to set up an image acquisition system must consider the type of application, the physical place constraints, and the time required to analyze the data. The type of application refers to the significant information offered by the system in different areas such as: food quality (Chmiel et al., 2011), surveillance (Mendoza et al., 2006), fruit damage (Blasco et al., 2009), disease and infestation (Wang et al., 2011) among others (Kocer and Cevik 2011), (Clouard et al., 2011), (Lunadei et al., 2011).

Assembly concern (physical place constraints), refers to the final place where the system will be installed: production line (Figure 5) (Jarimopas and Jaisin 2008), post-harvesting montage (Figure 6) (Sanz-Uribe et al., 2008), manual operation station (Figure 7) (Rabatel and Guizard 2007) and remote monitoring (Figure 8) (Lee et al., 2010).

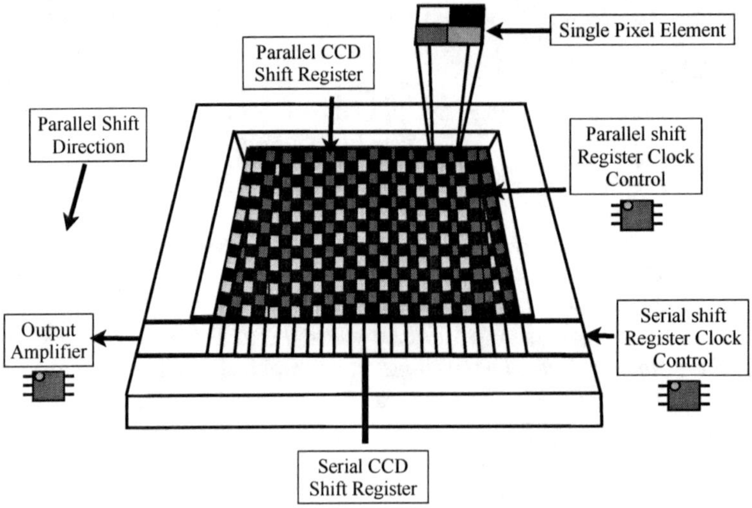


**Figure 4.** CCD array device.

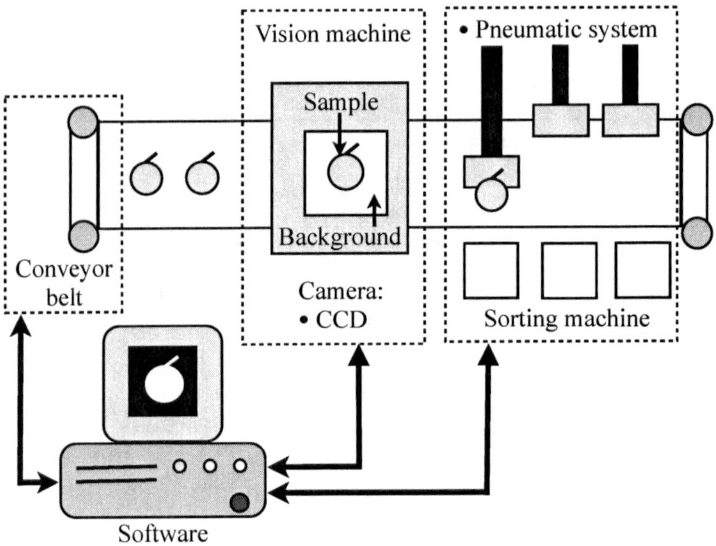


**Figure 5.** Production line setup.

## Color Image Processing in Food Science and Agriculture

The visual appearance of food, including variety, maturity, color, shape, size, and defects is a deciding factor in assessing its quality. An important part of quality control in the food industry is, therefore, based on visual

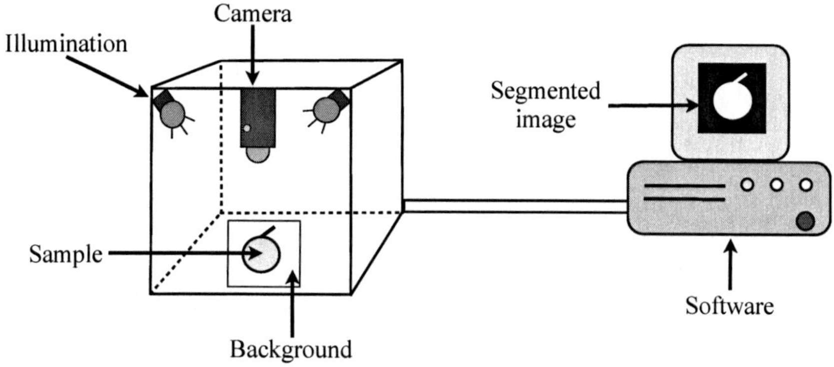


**Figure 6.** Post-harvesting setup.

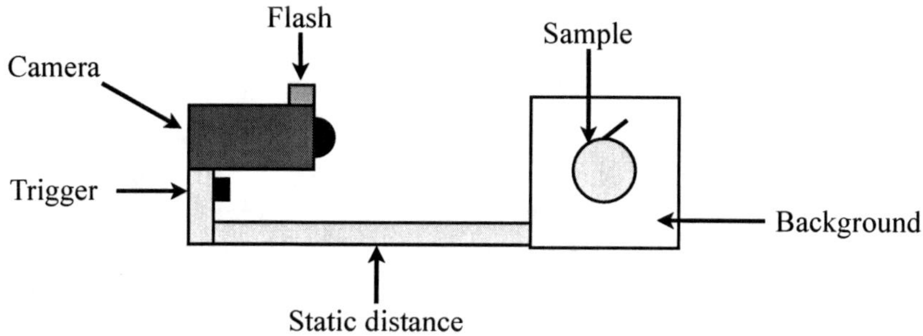


**Figure 7.** Manual station setup.

inspection. However, even if this task is traditionally carried out by the human eye, the absence of reliable quantitative assessment criteria, and the fact that visual assessment by human beings is time consuming and cost-intensive, give rise to the need for a robust, consistent, rapid, and cost-effective sorting method. For this reason, computer vision-based grading and sorting systems have been designed and implemented in order to speed up the process, as well as to maintain consistency and uniformity (Koschan and Abidi 2008) (Al Ohali 2011) (Mendoza et al., 2006). On the other hand, finding the right point of ripening in fruit from crops is not an easy task, because many parameters, like temperature, relative humidity, color and their complex behavior, are taken into account in accurately determining harvesting times. As a result, it requires particular attention from the developers of image processing systems and computer vision machines.

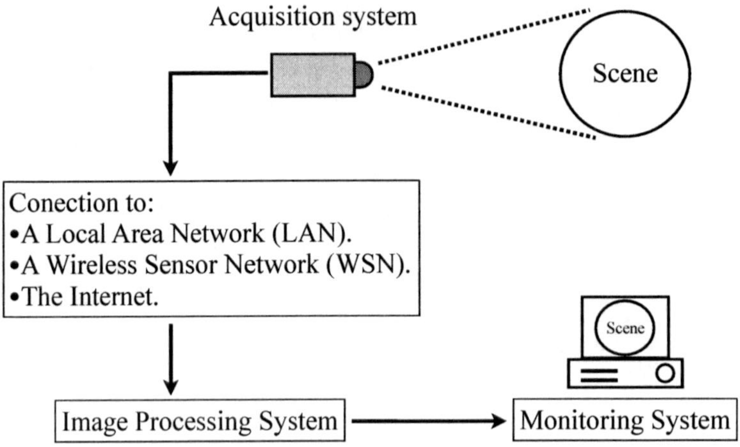


**Figure 8.** Remote monitoring setup.

**Color**

Color provides the basic information for human perception and is elementary information that is stored in pixels to constitute a digital image. Color is one of the most important object measurements for image understanding and object description (Sun 2007). Using color attributes is more discriminant than simply using grayscale since two image points with the same grayscale value can be differentiated from their color attributes (Collewet and Marchand 2009).

According to the tri-chromatic theory, color can be formed by the combination of three elementary color components, which are assigned to every pixel of a color image. Different types of values stored for the three-color components, and different color reproduction methods using these three values, lead to different color spaces. These spaces can be generally classified into three types: hardware-oriented, human-oriented, and instrumental.

Hardware-oriented spaces are developed in order to facilitate hardware processing, such as capturing, storing and displaying. The most popular hardware-oriented space is RGB (red, green blue), so called because this is the way in which cameras sense natural scenes and display them. The RGB space representation is shown in Figure 9.

A color image can be represented as the overlapping of its components (red, green and blue), as is shown in Figure 10.

The CIELAB or L* a* b* color space is a human-orientated space and this property has made it particularly attractive and useful for color analysis. The superiority of the CIELAB color space over other color spaces

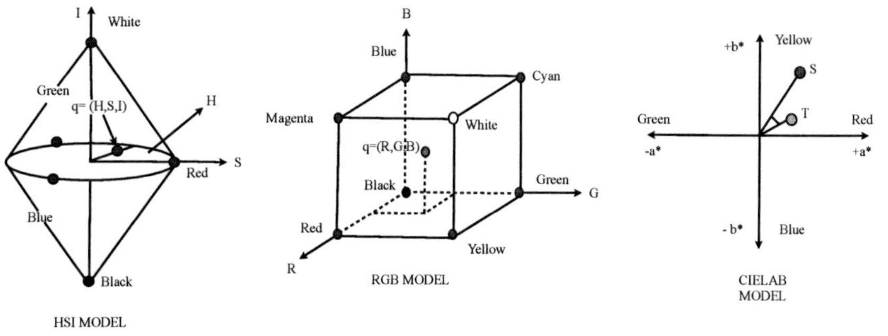


**Figure 9.** Color space representations.

has been demonstrated in many color image applications. For example, the CIELAB color space has been successfully used for color clustering. In this approach, the color difference in this space has been used in the computation of the dissimilarities between colors and this has been used for color clustering (Acharya and Ray 2005). The space representation is shown in Figure 9, where a* and b* indicate the color of the image and L* the intensity.

The HSI color system (see Figure 9) represents the colors of human perception well, because the human vision system can distinguish different hues easily, whereas the perception of a different intensity or saturation does not imply the recognition of different colors. The coordinates can be transformed from one color space to other using equations (Eq2), (Eq3) and (Eq4) (Cheng et al., 2001).

## Image Analysis

In order to perform a visual inspection of agricultural products, their color, size, shape, and texture should be characterized with a vision system.

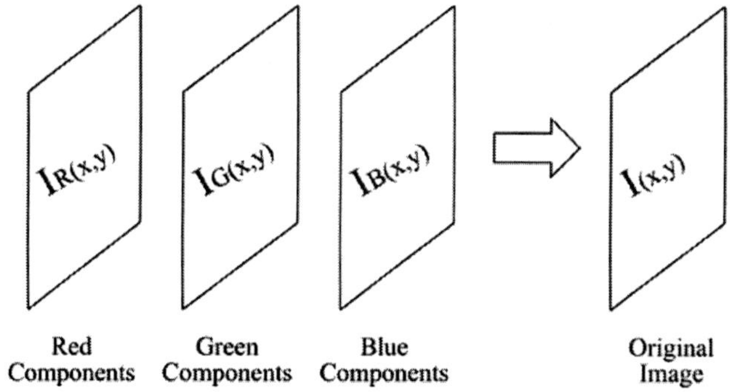


**Figure 10.** RGB components of a color image.

$$H = \arctan\left(\frac{\sqrt{3}(G-B)}{(R-G)+(R-B)}\right)$$

$$Int = \frac{(R+G+B)}{3} \qquad \text{(Eq 2)}$$

$$Sat = 1\frac{\min(R,G,B)}{I}$$

$$\begin{pmatrix} X \\ Y \\ Z \end{pmatrix} = \begin{pmatrix} 0.607 & 0.174 & 0.200 \\ 0.299 & 0.587 & 0.114 \\ 0.000 & 0.066 & 1.116 \end{pmatrix} \begin{pmatrix} R \\ G \\ B \end{pmatrix} \qquad \text{(Eq 3)}$$

$$L* = 116\left(\sqrt[3]{\frac{Y}{Y_0}}\right) - 16$$

$$a* = 500\left(\sqrt[3]{\frac{X}{X_0}} - \sqrt[3]{\frac{Y}{Y_0}}\right) \qquad \text{(Eq 4)}$$

$$b* = 200\left(\sqrt[3]{\frac{Y}{Y_0}} - \sqrt[3]{\frac{Z}{Z_0}}\right)$$

Image analysis has two phases, the training process and the application process. First, the training process refers to the training of an automatic system with the help of an expert. The expert separates the training samples into a determinate number of classes, and then the acquisition system is fed with these training samples, corresponding to each class. At this point, image analysis classifies the training samples according to the expert classification parameters.

In the image analysis process, two tasks are performed. One corresponds to the image segmentation, where the region of interest is separated from the background, over the sample image, and then is passed to the image classification process (see Figure 11).

**Image Segmentation**

Image segmentation is the process of manipulating the pixels of an entire image, analyzing them to identify and isolate a section of interest, and then creating a subimage containing the selected section (see Figure 12).

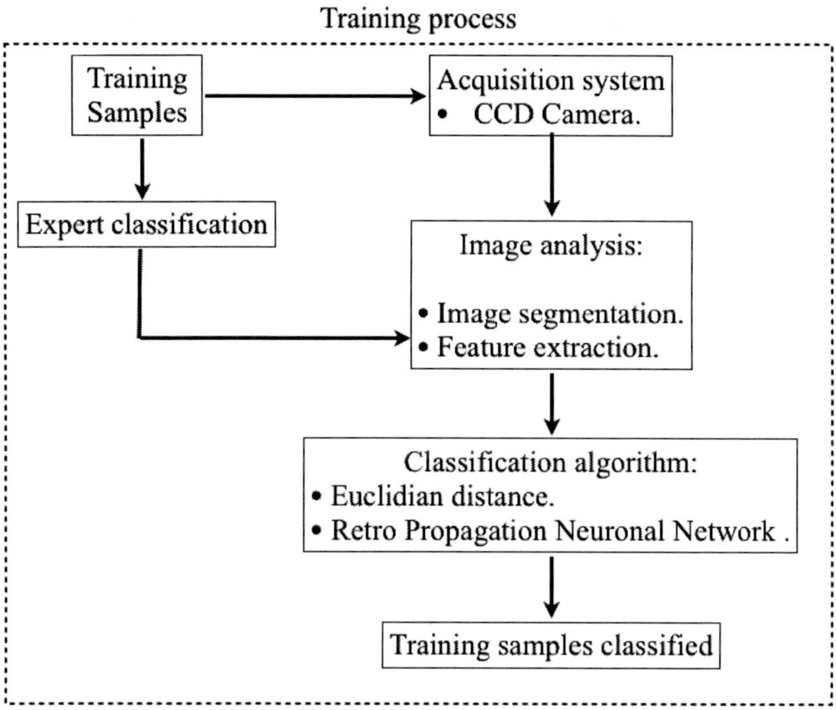


**Figure 11.** Training process for automation image analysis.

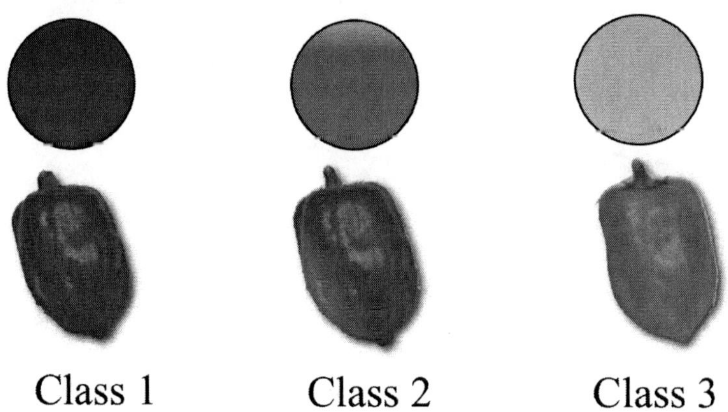


**Figure 12.** Fruit classification based on the color.

*(Color image of this figure appears in the color plate section at the end of the book.)*

The most basic attribute for segmentation is the amplitude of image luminance for a monochrome image and color components for a color image. Image edges and texture are also useful attributes for segmentation.

The time taken to segment the image is proportional to the size of the image, regardless of the segmentation method chosen.

There is no single theory of image segmentation. Therefore, there is no standard method of image segmentation. Rather, there is a collection of *ad hoc* methods that have had some degree of popularity (Pratt 2007).

To "extract" an object from an image means that pixels making up the object are identified. To express this information, an array of pixels, of the same size as the original image, is created, and a label is added to each pixel. All pixels that make up the object are given the same label and all pixels that make up the background are given a different label. The label is usually a number, but it could be anything, for example, a letter or a color. It is essentially a name and it only has a symbolic meaning. Labels, therefore, cannot be treated as numbers. Labeled images are also referred to as classified images, as they indicate the class to which each pixel belongs (Petrou and Petrou 2010).

Subsequent image interpretation tasks, such as feature extraction and object recognition, rely on the quality of the segmentation results. The difficulty arises when the segmentation performance needs to be adapted to the changes in image quality. Image quality is usually affected by variations in lighting conditions, imaging devices, density and the type of product. Despite the large number of segmentation techniques presently available, no general methods have been found to perform adequately across a diverse set of imagery under variable conditions (Abbasgholipour et al., 2011).

Typical methods used to perform color image segmentation are described in Table 1 (Koschan and Abidi 2008) (Cheng et al., 2001).

## Morphology

### Size

Size is one of the most important features for accurate classification and sorting of fruit. It is in this context that the field of automatic inspection and machine vision comes to play an important role in quality control for agricultural products. Fruit size estimation is also helpful in planning packaging, transportation and marketing operations (Khojastehnazhand et al., 2010).

**Table 1. Color image segmentation methods and techniques.**

| *Fundamental* | *Methods* | *Description* |
|---|---|---|
| Pixel-Based Segmentation | Histogram Techniques | Use of histograms and threshold values in the segmentation of an image. |
| | Cluster Analysis in the Color Space | This technique is expanded to include vector-valued color images. |
| Area-Based Segmentation | Region-Growing Techniques | Viewing the pixel values, for example, in an 8-neighborhood around the given start values. |
| | Split-and-Merge Techniques | A region is split or merged depending on the decision criteria for a given uniformity condition. |
| Edge-Based Segmentation | Segmentation by Watershed Transformation | Processing steps:1. Calculation of the gradient image 2. Sorting of the gray-levels of the gradient image–.3. Calculation of the watersheds and reservoirs. |
| Physics-Based Segmentation | Dichromatic Reflection Model | The regions and their neighboring relationships are represented in a graph. The calculation of watersheds on graphs is carried out as analogous to the calculation of watersheds on the basis of pixels. |

## Shape

As with size, shape is another geometric measurement of food products. Furthermore, shape plays an important part in the consumer purchase decision process (Leemans and Destain 2004); and this establishes the significance of shape measurement in the applications of computer vision for food quality inspection. Shape-based analysis methods are appropriate strategies for recognition of fruit (Lak et al., 2010).

## Shape Identification

The shape of objects or regions of interest are important features used for content representation, and require good segmentation to detect objects or regions. Basically, shape characterization is of two types: boundary-based and region-based. Boundary-based shape features include rectilinear shapes, polygonal approximation, and finite element models, and Fourier-based shape descriptors. Region-based features include statistical moments and grid-based approaches. Methods employing shape features are suited for applications where objects or regions are readily available (Prasad 2004).

For example, Hough Transform is a technique to identify specific shapes in an image. It converts all the points in a curve into a single location in another parametric space by coordinate transformation. This method intends to map global features into local features. This concept can also be applied to detect circles, ellipses, or other geometric shapes (Shih 2010) (Haidekker 2010).

The Hough Transform for circles and ellipses can be applied to determine the shape of fruit and vegetables and measure their size to calculate or estimate the volume of the fruit. Figure 13 shows an example of the Hough Transform applied to determine the radius of an orange as well as the center on the image.

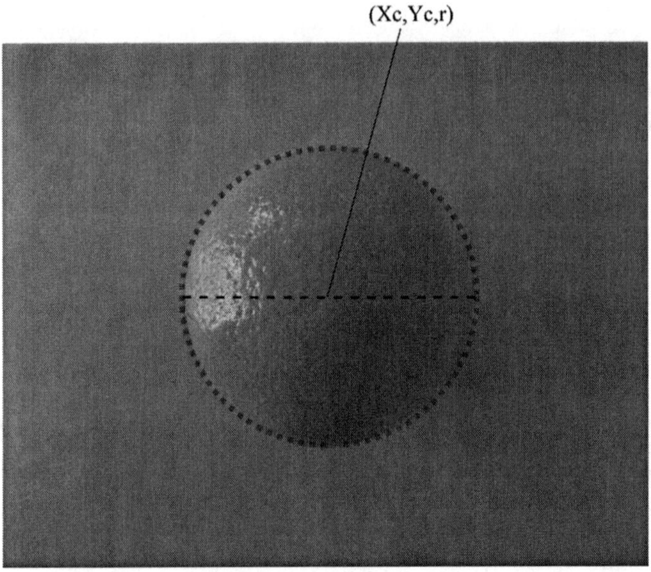


**Figure 13.** Application of the Hough Transform to measure the radius of an orange.

### Texture

Texture has been one of the most active topics in machine intelligence and pattern analysis since the 1950s. It tries to discriminate different patterns of images by extracting the dependence of intensity between pixels and their neighboring pixels or by obtaining the variance of intensity across pixels. Recently, different features of color and texture have been combined together for their application in the food industry (Arivazhagan et al., 2010).

### Evaluation of Surface Defects

The presence of skin defects is one of the most influential factors in the quality and price of fresh fruit, since most consumers associate quality with a good appearance and the total absence of external defects. The detection of defects during packing operations ensures that only good-quality fruit reaches the market. Moreover, the identification of the type of each defect will increase both the quality of the fruit and also the producer's profit. Most current computer vision systems used in the automatic quality inspection of food are limited to the visible region of the electromagnetic spectrum, as they tend to imitate the human eye. However, non-visible information, such as that provided by near-infrared or ultraviolet regions of the spectrum, can improve the inspection by detecting specific defects or allowing the detection of non-visible damage (Blasco et al., 2007).

## Object Classification Methods

As mentioned in previous paragraphs, a computer-based system that can mimic the human grading and sorting process may adequately expedite the process, in addition to sorting fruit into uniform and consistent quality groups. For this purpose, intensive research work is being conducted to design and build intelligent, reliable, flexible and effective systems that can quickly sort a variety of fruits and other agricultural produce. The feasibility and applicability of such a system are being explored in every agriculture-oriented country (Blasco et al., 2007) (Wang et al., 2011).

## Image Classification

Classification is a task that depends on the results from previous phases of the algorithm (see Figure 14). Sorting the sample images to the adequate class is an interesting challenge for the algorithm. This task requires a good performance from the training phases, due to the difficulty of identifying the appropriate parameters that distinguish one class from other (see Table 2). The trainer needs to be careful to train the algorithm with different sample images from a different class to the training one.

**Table 2. Techniques used for image classification.**

| *Method* | *Description* |
|---|---|
| Euclidian distance transform | Uses Euclidian distances to determine the difference between the parameters from the training samples (database) and the test sample (Sudha 2005). |
| Neuronal Networks | Are based on the training database of classes and compare the sample to that database and find the most similar class to which the sample belongs (Du and Sun 2006). |
| Statistical learning | Based on statistical inference (Bousquet et al., 2004). |
| Genetic Algorithm | A method that can represent complex, multivariate conditions straightforwardly, and also a method for approximating discrete-valued target functions (Abbasgholipour et al., 2011). |

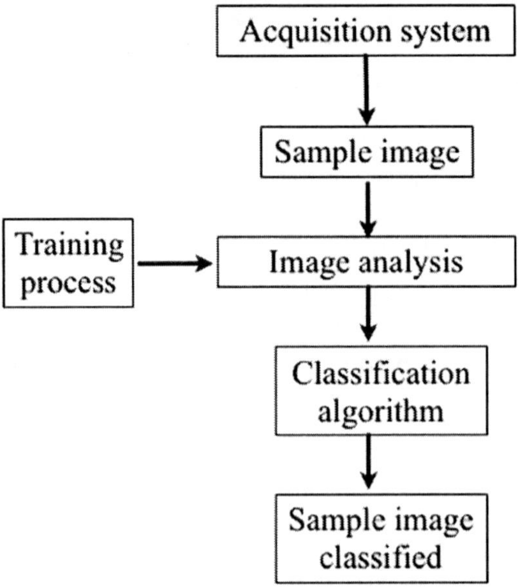


**Figure 14.** Automatic image analysis and classification process flow chart.

## Applications of Image Processing in Agriculture

### Aquaponic System

Aquaponic is a special form of recirculating aquaculture systems (RAS), namely polyculture consisting of fish tanks (aquaculture) and plants that are cultivated in the same water circle (hydroponic). The primary goal of aquaponic is to reuse the nutrients released by fish to grow crop plants. Most systems separate fish faeces as quickly as possible to reduce the Biochemical Oxygen Demand (BOD) load in the RAS, to enhance nitrification performance and to reduce clogging of plant roots, which could lead to a loss in crop productivity (Abbasgholipour et al., 2011)(Graber and Junge, 2009). Aquaponic systems are excellent candidates to use WSN (Wireless Sensor Networks) because they evaluate the quality of water in fish tanks as well and process plant images to determine the relationship between environmental parameters and plant growing.

The complexity of the biological micro-ecosystem involved in aquaponic cultures, due to aquatic organisms, bacteria and plants growing in the same recycled water (as shown in Figure 15), is a matter to be considered in the design of different instruments, control, image processing and WSN systems applied to this kind of agricultural crop. The health of aquatic organisms, bacteria and plants is an important issue

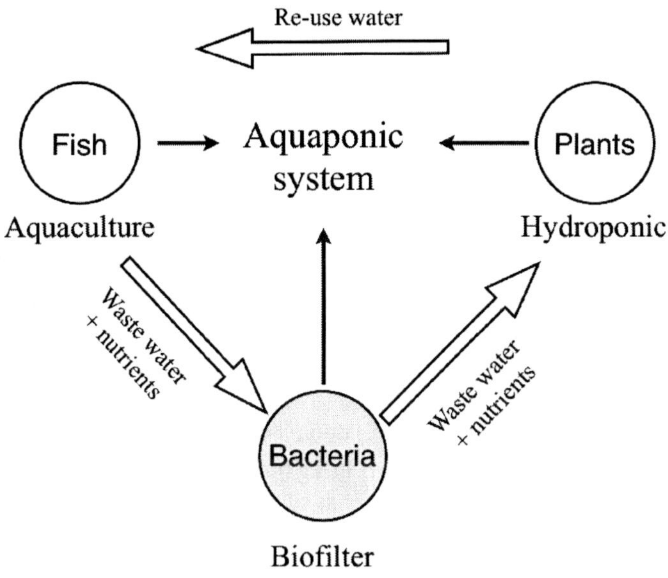


**Figure 15.** Schematic representation of an aquaponic system.

for growers who wish to keep their crops in optimal conditions. But each system requires specific attention; an example of this is pH control, because plants like more acidic solutions and aquatic organisms such as tilapia like alkaline solutions; herein began a trade-off about what is better for the system.

Some of the primary parameters used to determine the quality of fruit or agricultural products are color, firmness, and more recently, texture. Image processing has become a useful tool providing more information about what is happening in crops, especially color and the link it has with other basic measurements like dissolved oxygen, pH, and temperature in the aquaponic system (Wanitchang et al., 2010).

On the other hand, every system has the need to communicate the information acquired during the monitoring period and show it to the users on a friendly graphical user interface in order to take the appropriate decisions. The WSN is a framework used to collect the data from their sensor module and route them to a nearby base station, where the information is processed and presented to growers in a GUI.

### Vineyards

Current research efforts in precision viticulture shows a growing interest in remote sensing methods due to its potential for estimating vine biophysical variables such as shape, size and vigor, which are potential indicators of fruit quality and yield.

Progress on crop condition in vineyards has also been made at the leaf-level, studying absorptance in the visible region in the field, and detecting phenology and grape color at harvest to gather information about berry phenolics (Zhao et al., 2005).

Considering that digital images of individual grapes in the field can be easily acquired, with correct control of target distance and light conditions, by using computer vision to estimate the grape volume from its visible area, appropriate image processing algorithms could be developed in order to obtain detailed information such as berry color and size (see Figure 16) (Rabatel and Guizard 2007).

The development of image processing-based methods has advanced to automatically detect and count grapes to forecast yield with precision and accuracy, by taking conventional visible light cameras through a vineyard to image the vine, detect the crop and predict yield (Nuske et al., 2011). One of the most promising algorithms is divided in three distinct stages:

1. Detecting potential berry locations with a radial symmetry transform.
2. Identifying the potential locations that have similar appearance to grape berries.
3. Group neighboring berries into clusters.

**Figure 16.** Image segmentation of a bunch of grapes.

A critical issue to acquire images in vineyards is related to the camera assembly. Experimental tools have been developed (Rabatel and Guizard 2007) to easily acquire digital images of individual grapes in the field, with correct control of target distance and light conditions (see Figure 17). The main objective is to perform a non-destructive estimation of the bunch weight at an early stage by computer vision, using an elliptical model fitting to estimate the grape volume from its visible area. A grape image consists of a set of berries with nearly identical color properties, so that the information available for berry separation mainly relies on their contours (as seen in Figure 16 where a bunch of grapes is segmented from the background). Owing to the smooth 3D shape of the berries, these contours are accessible as luminance discontinuities in the image. However, the algorithm only works with grayscale images.

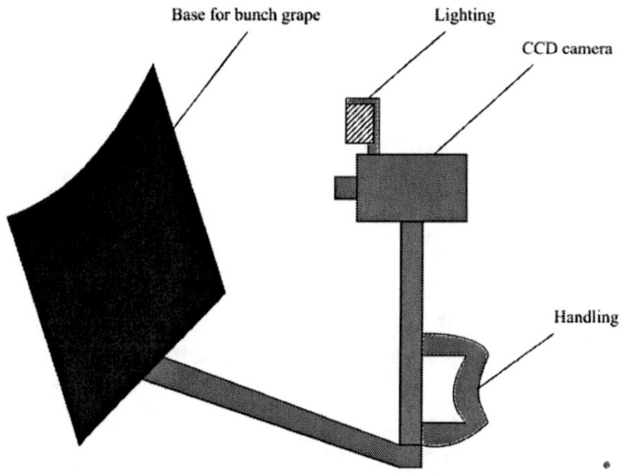


**Figure 17.** Fixture developed by Rabatel and Guizard.

## Image Processing Embedded Systems

Development of real time field monitoring systems is increasing in order to cover requirements in agriculture. On-site image acquiring and processing is important to measure different crop parameters. Since an embedded system is often adopted to reduce cost and size, it must run on a small computer with fewer resources compared to a general-purpose computer, such as a personal computer. As an example, a low-cost and robust field image processing system was developed to acquire real time field conditions from a remote site. It was also designed to work as a mobile gateway for Internet connection in a rural area. To reduce the total cost, inexpensive commercial USB devices such as USB cameras and 3G USB modem dongles were used. The total cost of the system was less than 300 USD (Yamaguchi and Ito 2009).

An embedded system is a dedicated computer-based system for a dedicated application or product (Kamal 2003). Embedded systems have an Operating System (OS) to control and manage the resources of the computer system. Embedded OS should be compact and efficient to achieve designated functions properly. Popular OS include ITRON, VxWorks, Nucleus PLUS, QNX, Windows CE, LynxOS and embedded Linux. This last OS is tuned for small computers with limited resources. However, Linux-experienced users find it very easy to use, because an embedded Linux is basically the same as a general Linux distribution. Moreover, some distributions are freely available on the Internet (Yamaguchi and Ito 2009).

## Conclusions

In this chapter, we have presented the basics of image processing and the fundamentals of computer vision technology including embedded systems for application in agriculture.

Color is a deciding factor in terms of the quality of agricultural products, in addition to their size and shape, but using human inspection to analyze these factors in high volume is inefficient. Herein lies the importance of having an automated system based on image processing for grading and sorting fruit and vegetables in order to speed up the process, as well as to maintain consistency uniformity.

The use of machine vision technology is becoming a routine operation for land-based and aerial-based remote sensing for natural resource assessments, precision farming, postharvest product quality and safety detection, classification and sorting, as well as process automation. In order to fully apply machine vision technology, vision systems for agricultural applications should take advantage of the differences in the characteristics of light absorption of vegetation, foods, and agricultural products. Many potential applications in the agricultural industry are found on the hyperspectral imaging technique, which combines the advantages of spectroscopy and imaging. Although the cost of acquiring hyperspectral images is typically high, for specific crops and in specific climates, hyperspectral remote sensing is becoming popular for monitoring the development and health of crops.

Advances in machine vision technology provide accurate, robust, and low-cost vision systems, where real-time operational requirements are being met with high-speed computers and frame grabbers. Moreover, machine vision systems incorporating artificial intelligence programming, neural networks, fuzzy logic and expert systems will improve their performance in agricultural applications.

It is also important to consider external factors like object illumination and background which are very important for the color image segmentation methods described.

There is a need for developing easy-to-use and low-cost commercial units so that growers can adopt them without too many difficulties. In the case of vineyards, sensing systems for soil properties and nutrients have been developed and some of them have been implemented for real-time variable rate nutrient applications. Such kinds of technological improvements could then be implemented in aquaponic systems with the appropriate sensors, thus both aquaponic systems and vineyards are excellent candidates to use image processing along with a WSN to sense desired parameters, process the data and send the result to a GUI.

## REFERENCES

Abbasgholipour, M., M. Omid, A. Keyhani and S.S. Mohtasebi. 2011. Color image segmentation with genetic algorithm in a raisin sorting system based on machine vision in variable conditions. Expert Systems with Applications. **38(4):**3671–3678. Elsevier Ltd. doi:10.1016/j.eswa.2010.09.023

Abdullah, M.Z., A.S. Fathinul-Syahir and B.M.N. Mohd-Azemi. 2005. Automated inspection system for colour and shape grading of starfruit (*Averrhoa carambola* L.) using machine vision sensor. Transactions of the Institute of Measurement and Control. **27(2):**65–87. doi:10.1191/0142331205tm 138oa

Acharya, T. and A.K. Ray. 2005. Image Processing. Hoboken, NJ, USA: John Wiley & Sons, Inc. doi:10.1002/0471745790

Al Ohali, Y. 2011. Computer vision based date fruit grading system: Design and implementation. *Journal of King Saud University—Computer and Information Sciences*. **23(1):**29–36. King Saud University. doi:10.1016/j.jksuci.2010.03.003

Alippi, C., G. Anastasi, M. Di Francesco and M. Roveri. 2009. Energy management in Wireless Sensor Networks with energy-hungry sensors. Instrumentation & Measurement Magazine, IEEE. **12(2):**16–23. IEEE. Retrieved from http://ieeexplore.ieee.org/xpls/abs_all.jsp?arnumber=4811133

Annadurai, S. and R. Shanmugalakshmi. 2007. Fundamentals of Digital Image Processing (440 p.). Pearson Education.

Arivazhagan, S., R.N. Shebiah, S.S. Nidhyanandhan and L. Ganesan. 2010. Fruit Recognition using Color and Texture Features. Information Sciences. **1(2):**90–94.

Bellis, S.J., K. Delaney, B.O. Flynn, J. Barton, K.M. Razeeb and C.O. Mathuna. 2005. Development of field programmable modular wireless sensor network nodes for ambient systems. Computer Communications. **28(13):**1531–1544. doi:10.1016/j.comcom.2004.12.045

Blasco, J., N. Aleixos, J. Gomez and E. Molto. 2007. Citrus sorting by identification of the most common defects using multispectral computer vision. *Journal of Food Engineering*, **83(3):**384–393. doi:10.1016/j.jfoodeng.2007.03.027

Blasco, J., N. Aleixos, J. Gómez-Sanchís and E. Moltó. 2009. Recognition and classification of external skin damage in citrus fruits using multispectral data and morphological features. *Biosystems Engineering*. **103(2):**137–145. doi:10.1016/j.biosystemseng.2009.03.009

Bousquet, O., S. Boucheron and G. Lugosi. 2004. Introduction to statistical learning theory. Advanced Lectures on Machine Learning (pp. 169–207). Springer.

Brosnan, T. 2002. Inspection and grading of agricultural and food products by computer vision systems—a review. Computers and Electronics in Agriculture, **36(2-3):**193–213. doi:10.1016/S0168-1699(02)00101-1

Chen, Y.R., K. Chao and M.S. Kim. 2002. Machine vision technology for agricultural applications. Computers and Electronics in Agriculture. **36(2-3):**173–191. Elsevier. doi:10.1016/S0168-1699(01)00177-6

Cheng, H.D., X.H. Jiang, Y. Sun and Jingli Wang. 2001. Color image segmentation: advances and prospects. Pattern Recognition, **34(12):**2259–2281. doi:10.1016/S0031-3203(00)00149-7

Chherawala, Y., R. Lepage and G. Doyon. 2006. Food Grading/Sorting Based on Color Appearance trough Machine Vision: the Case of Fresh Cranberries. 2006 2nd International Conference on Information & Communication Technologies. 1540-1545. IEEE. doi:10.1109/ICTTA.2006.1684612

Chmiel, M., M. S³owiñski and K. Dasiewicz. 2011. Lightness of the color measured by computer image analysis as a factor for assessing the quality of pork meat. *MESC*. **88(3):**566–570. Elsevier Ltd. doi:10.1016/j.meatsci.2011.02.014

Clouard, R., A. Renouf and M. Revenu. 2011. Human-computer interaction for the generation of image processing applications. *International Journal of Human-Computer Studies*. **69(4):**201–219. doi:10.1016/j.ijhcs.2010.12.002

Collewet, C. and E. Marchand. 2009. Colorimetry-based visual servoing. 2009 IEEE/RSJ International Conference on Intelligent Robots and Systems. IEEE. doi:10.1109/IROS.2009.5354416

Daley, W.D., R. Carey and C. Thompson. 1993. Poultry grading/inspection using color imaging. Proceedings of SPIE (pp. 124–132). SPIE. doi:10.1117/12.144816

Du, C. 2004. Recent developments in the applications of image processing techniques for food quality evaluation. *Trends in Food Science & Technology*. **15(5):**230–249. doi:10.1016/j.tifs.2003.10.006

Du, C.-J. and D.-wen Sun. 2006. Learning techniques used in computer vision for food quality evaluation/: a review. *Journal of Food Engineering*. **72:**39–55. doi:10.1016/j.jfoodeng.2004.11.017

Graber, A. and R. Junge. 2009. Aquaponic Systems: Nutrient recycling from fish wastewater by vegetable production. *Desalination*. **246(1–3):**147–156. doi:10.1016/j.desal.2008.03.048

Haidekker, M. (2010). Advanced biomedical image analysis (211 p.). Wiley Online Library. doi:10.1002/9780470872093.fmatter

Jain, R., R. Kasturi and B.G. Schunck. 1995. Machine vision (Vol. 5). New York: McGraw-Hill. Retrieved from http://www.cse.usf.edu/~r1k/MachineVisionBook/MachineVision.pdf

Jarimopas, B. and N. Jaisin. 2008. An experimental machine vision system for sorting sweet tamarind. *Journal of Food Engineering*. **89(3)**:291–297. doi:10.1016/j.jfoodeng.2008.05.007

Kalomiros, J. and J. Lygouras. 2008. Design and evaluation of a hardware/software FPGA-based system for fast image processing. Microprocessors and Microsystems. **32(2):**95–106. doi:10.1016/j.micpro.2007.09.001

Kamal, R. (2003). Embedded Systems, Architecture, Programming and Design (4 p.). Mc Graw Hill.

Khojastehnazhand, M., M. Omid and A. Tabatabaeefar. 2010. Development of a lemon sorting system based on color and size. *Plant Science*, 4(April), pp. 122–127.

Kocer, H.E. and K.K. Cevik. 2011. Artificial neural networks based vehicle license plate recognition. *Procedia Computer Science*. **3:**1033-1037. Elsevier. doi:10.1016/j.procs.2010.12.169

Koschan, A. and M. Abidi. 2008. Digital Color Image Processing (376 p.). Wiley-Interscience.

Lak, M.B., M, S., Amiriparian, J., and B. Beheshti. 2010. Apple Fruits Recognition Under Natural Luminance Using Machine Vision. *Journal of Food Science*. **2(6):**325–327.

Lee, P.G. 1995. A review of Automated Control Systems for Aquaculture and Design Criteria for their Implementation. *Aquacultural Engineering*. **14(3):**205–227. doi:10.1016/0144-8609(94)00002-I

Lee, W.S., V. Alchanatis, C. Yang, M. Hirafuji, D. Moshou and C. Li. 2010. Sensing technologies for precision specialty crop production. Computers and Electronics in Agriculture. **74(1):**2–33. doi:10.1016/j.compag.2010.08.005

Leemans, V. and M.-F. Destain. 2004. A real-time grading method of apples based on features extracted from defects. *Journal of Food Engineering*. **61(1):**83–89. doi:10.1016/S0260-8774(03)00189-4

Lunadei, L., P. Galleguillos, B. Diezma, L. Lleó and L. Ruiz-Garcia. 2011. A multispectral vision system to evaluate enzymatic browning in fresh-cut apple slices. *Postharvest Biology and Technology*. **60(3):**225–234. Elsevier B.V. doi:10.1016/j.postharvbio.2011.02.001

Mendoza, F., P. Dejmek and J. Aguilera. 2006. Calibrated color measurements of agricultural foods using image analysis. *Postharvest Biology and Technology*. **41(3):**285–295. doi:10.1016/j.postharvbio.2006.04.004

Miskam, M.A. 2009. Preliminary Design on the Development of Wireless Sensor Network for Paddy Rice Cropping Monitoring Application in Malaysia. *European Journal of Scientific Research*. **37(4):**649–657.

Nuske, S., S. Achar, T. Bates, S. Narasimhan and S. Singh. 2011. Yield Estimation in Vineyards by Visual Grape Detection (in process of publication). *2011 IEEE Conference on Intelligent Robots and Systems*.

Pedreschi, F., J. Leon, D. Mery and P. Moyano. 2006. Development of a computer vision system to measure the color of potato chips. *Food Research International*. **39(10):**1092–1098. doi:10.1016/j.foodres.2006.03.009

Petrou, M. and C. Petrou. 2010. Image Processing: The Fundamentals (2nd edn., 818 p.). John Wiley and Sons.

Prasad, B. 2004. Region-based image retrieval using integrated color, shape, and location index. *Computer Vision and Image Understanding*. **94(1–3):**193–233. doi:10.1016/j.cviu.2003.10.016

Pratt, W.K. 2007. Digital image processing: PIKS Scientific inside (4th edn., 782 p.). Wiley-Interscience.

Rabatel, G. and C. Guizard. 2007. Grape berry calibration by computer vision using elliptical model fitting. 6th european Conference on Precision Agriculture "ECPA 2007." Skiathos, Greece: ECPA 2007.

Sanz-Uribe, J.R., P.J. Ramos-Giraldo and C.E. Oliveros-Tascon. 2008. Algorithm to Identify Maturation Stages of Coffee Fruits. Advances in Electrical and Electronics Engineering—IAENG Special Edition of the World Congress on Engineering and Computer Science 2008 (pp. 167–174). IEEE. doi:10.1109/WCECS.2008.28

Shih, F.Y. 2010. Image Processing and Pattern Recognition: Fundamentals and Techniques (537 p.). Wiley-IEEE Press.

Sudha, N. 2005. A pipelined array architecture for Euclidean distance transformation and its FPGA implementation. Microprocessors and Microsystems, 29, pp. 405–410. doi:10.1016/j.micpro.2004.10.003

Sun, D.-W. 2007. Computer vision technology for food quality evaluation (583 p.). Academic Press.

Tao, Y., P.H. Heinemann, Z. Varghese, C.T. Morrow and H.J.I. Sommer. 1995. Machine Vision for Color Inspection of Potatoes and Apples. Transactions of the ASABE. **38(5):**1555–1561.

Umbaugh, S.E. (2010). Digital Imaging Processing and Analysis: Human and Computer Vision Applications with CVIPtools (2nd edn., 977 p.). CRC Press.

Valous, N. A., Fernando Mendoza and D.-W. Sun. 2010. Emerging non-contact imaging, spectroscopic and colorimetric technologies for quality evaluation and control of hams/: a review. *Trends in Food Science & Technology*. **21(1):**26–43. Elsevier Ltd. doi:10.1016/j.tifs.2009.09.003

Wang, J., K. Nakano, S. Ohashi, Y. Kubota, K. Takizawa and Y. Sasaki. 2011. Detection of external insect infestations in jujube fruit using hyperspectral reflectance imaging. *Biosystems Engineering*. **108(4):**345–351. IAgrE. doi:10.1016/j.biosystemseng.2011.01.006

Wanitchang, J., A. Terdwongworakul, P. Wanitchang and S. Noypitak. 2010. Maturity sorting index of dragon fruit: Hylocereus polyrhizus. *Journal of Food Engineering*. **100(3):**409–416. Elsevier Ltd. doi:10.1016/j.jfoodeng.2010.04.025

Yamaguchi, C. and R. Ito. 2009. A Field Image Monitoring System Based on Embedded Linux. *Proceedings of the 16th Tri-University International Joint Seminar and Symposium* (pp. 1–4). Retrieved from http://www.mie-u.ac.jp/kokusai/tri/TRIU09CN-01.pdf

Zhao, J., J. Tow and J. Katupitiya. 2005. On-tree fruit recognition using texture properties and color data. Intelligent Robots and Systems, 2005.(IROS 2005). 2005 IEEE/RSJ International Conference on (pp. 263–268). IEEE.

# 9

# Tag4M, An Ultra-Low Power Wi-Fi Embedded System for Measurements

*Silviu Folea, Mihai Hulea and Doru Ursuţiu*

**ABSTRACT**

The evolution in the field of wireless sensor technologies provides new opportunities for developing distributed sensor architectures. Wireless measurement devices that do not require connecting cables need to be small, reliable, very low power, and must have a very low production cost. One example of such a device is the Tag4M Wi-Fi tag which is not only a cable replacement measurement device, but also a very good solution for bringing measured data to web pages on the Internet. This material will briefly describe the tag's hardware architecture, its firmware and application software. A number of tag extension boards have been developed with the purpose of extending sensor support for measurements like temperature, humidity, pressure, light, etc. The applications employing them will be presented in this material. Wi-Fi tags connect sensors to web pages on the Internet and facilitate the introduction of new types of instruments which use web pages as their user interface. The concept of Cloud Instruments is presented in this chapter, along with some examples. Further on, the space of Cloud Instrumentation, which includes all of these cloud instruments, is defined. A multitude of independent vendors of application software on the Internet which offer their services for analysis and presentation of sensor digitized data are also mentioned.

## Introduction

A smart environment is a "physical world that is richly and invisibly interwoven with sensors, actuators, displays, and computational elements, embedded seamlessly in the everyday objects of our lives, and connected through a continuous network" (Cook 2005). In such a system, distributed and remote measurements are done using wireless modules or nodes. These modules may use Bluetooth, ZigBee, Wi-Fi, or serial proprietary communication protocols and firmware that supports hopping, meshing, TCP/IP, serial bit banging etc. The recent advances in wireless sensor networking technology have led to the development of low cost, low power, multifunctional sensor nodes. Sensor nodes enable environmental sensing together with data processing. Instrumented with a variety of sensors, such as temperature, humidity and volatile compound detection, that monitoring different environments. They are able to network with other sensor systems and exchange data with external users (Ruiz-Garcia et. al., 2009). One such devices is the Tag4M Wi-Fi tag which is an embedded system incorporating a sensor interface, a 32-bit CPU, memory, eCos real time operating system, complete Wi-Fi networking solution, network stack, crypto accelerator, power management system and real time clock (G2 Microsystems[1] 2008; G2 Microsystems[2] 2008). The Tag4M Wi-Fi tag has ultra low power Wi-Fi capability which makes it suitable for sensing applications where battery power management is critical. The requirement for battery power is represented by a current peak up to 0.5 A, with a very short pulse duration of about 1-2 ms (Folea and Ghercioiu 2008; Ghercioiu et al., 2007).

The current paper presents a series of sensor extension boards and applications developed to increase the sensing capabilities of the base Tag4M board. In order to demonstrate the utilization of the presented boards, some experimental results are also presented.

## Background

There are a number of wireless communication technologies which can be taken into account when designing a wireless sensor device such as Wi-Fi, ZigBee or Bluetooth. Some basic criteria for comparison between these wireless technologies are power consumption, bandwidth, application domains, typical range, network connectivity and network topology, as well as other key attributes (Labiod et al., 2007; Sidhu et al., 2007). The strength of Bluetooth lies in its ability to allow interoperability and replacement of cables. ZigBee networks can support a larger number of devices and in most cases, longer range between devices as compared to Bluetooth. ZigBee offers hopping and meshing, is cheaper and has lower power consumption, but the transfer rate is quite small if a greater amount of information has to be sent. Considering all the pros and cons of these technologies, Wi-Fi wireless communication technology has been used for developing the Tag4M device presented in this chapter.

A reduced number of companies produce Wi-Fi chips or System on Chip (SoC) solutions which incorporate a low power Wi-Fi, a sensor interface and a user-programmable CPU, all on the same SoC. One of these companies is RF Monolithics, which produces the WSN802G chip that integrates the GainSpan GS1010 System on Chip (SoC), the lowest power Wi-Fi chip solution on the market today. Redpine Signals is another company which produces modules that integrate the 802.11n MAC/BBP, RF, configuration flash, and other required components. Lantronix, with its PremierWave technology, provides a complete 802.11 a/b/g/n dual band radio, integrated solution with ARM9 Core 32-bit processor (440 MIPS at 400MHz) which offers extra on board memory for application development, up to 64 MB SDRAM and 256 MB NAND Flash, or up to 16 MB serial SPI Flash.

The Tag4M device has been developed starting from the G2M5477 chip, which is the second-generation of ultra low-power Wi-Fi SoCs from Roving Networks (G2 Microsystems[3], 2008; G2 Microsystems[4], 2009). The G2M5477 module offers the following functions: sensor interfaces, CPU and memory with user programmability and Wi-Fi communication, all of these in the same integrated chip. Because of this strong integration and miniaturization, a lower cost than for other solutions was possible. The accuracy of the sensor interface is very good and it is adequate for almost the entire range of sensors.

The usage of the Internet Protocol (IP) in a resource constrained device such as the Tag4M provides a high level of flexibility for the solution developers, making the development of the so called smart objects that form the Internet of Things (IoT) possible. As defined by CASAGRAS, the IoT is *"a global network infrastructure, linking physical and virtual objects through the exploitation of data capture and communication capabilities. This infrastructure includes existing and evolving Internet and network developments. It will offer specific object-identification, sensor and connection capability as the basis for the development of independent federated services and applications. These will be characterised by a high degree of autonomous data capture, transfer event network connectivity and interoperability"*. In (Hart and Kirk 2006), the concept of Environmental Sensor Networks (ESN) is defined. It comprises automatic sensor nodes and communication systems which actively communicate their data to a central server where they are integrated with other environmental data sets.

The deployment and adoption of high speed Internet networks along with the increase in computing power have also made significant advances in the field of distributed computer systems possible. The geographically distributed resources can be interconnected and managed in a unified way. A new computer paradigm emerged: cloud computing. Cloud computing can be described as a distributed computing and storage system in which services are provided to other computers and devices in the form of an utility over a network. Its main goal is to interconnect and to share computing resources in a transparent, reliable and scalable way. Cloud computing provides small and medium-sized enterprises with the ability to outsource IT services and applications. Another concept which evolved is represented by the grid computing paradigm. Grids are very large-scale virtualized, distributed systems which basically provide computing services for processing a single task for different users. This task is then divided into subtasks, and each subtask is sent to be processed in one of the available computing resources. The results are then aggregated and sent back to the user. In this way, the time to solve complex tasks is reduced. Grid computing technologies had a great success not only in the scientific world, but also in commerce. The use of grids in commercial applications has helped define the utility computing concept. This is a simple concept which refers to the externalisation of computing-intensive services to computing service providers. Payment is required only for the used hardware and software resources (Itu 2009).

This new concept inspired the work in (Ghercioiu[1] 2011) which introduced Cloud Instrumentation. Here, Internet-based instrumentation is defined as a concept opposed to PC-based instrumentation and the family of all cloud instruments which route digitized sensor data to the Internet is presented for computation, simulation, modelling and analysis.

The following diagram graphically illustrates this concept (Figure 1).

A sensor becomes a Cloud Instrument when:

- the sensor is connected to a wireless tag such as the Tag4M Wi-Fi tag;
- the tag digitizes the data to send it on to an Access Point, where the data is routed to the Internet and a server IP;
- a customized engine collects the data for feed into applications like metering, charting, control, analysis, modeling, data mining, etc. with display on web page Instruments and other web widgets.

In the PC-based data acquisition model, wireless measurement nodes send data to a PC running the application software. Another alternative is for the wireless modules to send data straight to the Internet via off-the-shelf access points. This very simple decision, to send digitized data straight into the Internet, if adopted, not only frees the wireless hardware from the confines of cabling, but also eliminates the link of the software to a specific box or PC. Instead, measurement front ends connect to the Internet and a Web page or a Web widget becomes the instrument, which users can access from anything that can surf the Web including mobile devices such as smart phones.

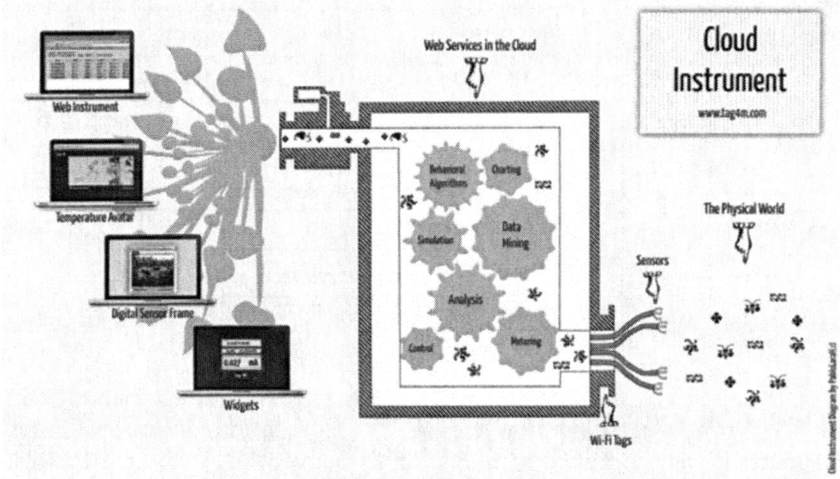


**Figure 1.** Generic Cloud Instrument (Ghercioiu[2] 2011).

A federation of sensors connected to Web Page Instruments running on a wide range of devices can provide any number of services: from a simple data presentation to an alarm service that sends a text message or tweet or even a correcting signal to the sensor tag; a metering service that calculates consumption and costs; a database service that stores data; machine health monitoring, home-based medical supervision and many others in what will become an enormous application space. What is even more important than the diversity of applications is that with processing taking place in the cloud, information from all these applications can be shared, data-mined and refined using behavioral algorithms, in order to detect patterns that can help us make predictions and define strategies for a better life together.

## Practical Applications of the Tag4M Technology

This section proposes a series of practical Internet applications employing the Tag4M device. These are intended to facilitate easy user access to measured data provided by Tag4M devices. In order to do this, a distributed multi tier solution is proposed in order to facilitate the integration of embedded equipment (Tag4M and others) within Internet applications. The implementation is based on Java technologies for developing server side components and on C++ for developing low level programming at Tag4M level. A series of hardware extension boards are also developed in order to extend the Tag4M sensing capabilities.

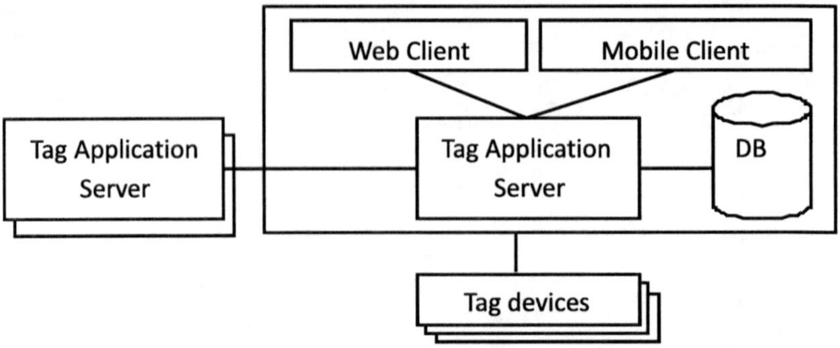


**Figure 2.** The block diagram of the system.

A multilevel, distributed architecture, Figure 2, which implements a set of services for the remote monitoring and supervision tasks in monitoring applications, is proposed. The system is composed of the following main components: Tag4M devices—Wi-Fi data acquisition modules; Tag Application Server (TAS)—a software application responsible for the implementation of the monitoring and supervision functions; remote clients (Web Application and Mobile Application) which provide user access to the system functions; DB—the database where the measured and processed data are stored.

The functions implemented at the Tag Application Server (TAS) level are:

- Handling communication with Tag4M devices; the communication protocol is based on the UDP transport protocol; the devices report data directly to the TAS.
- Storing of the received data in the database (DB).
- Processing and interpreting of the sensors data.
- Handling communication with clients; the application server implements a set of connectors by which it can communicate with remote clients using UDP and TCP protocols; these clients can be Web applications, desktop applications or mobile clients.
- Virtual Instruments Manager; implementing the functions for managing the life-cycle of the virtual instruments.
- Communicating with peer servers; the Tag Application Server also implements a communication mechanism through which tag data can be exchanged between application servers installed in different locations.

The Tag Application Server is implemented using the Java language, the J2SE and the J2EE technologies. The system functions are logically organized in the following levels:

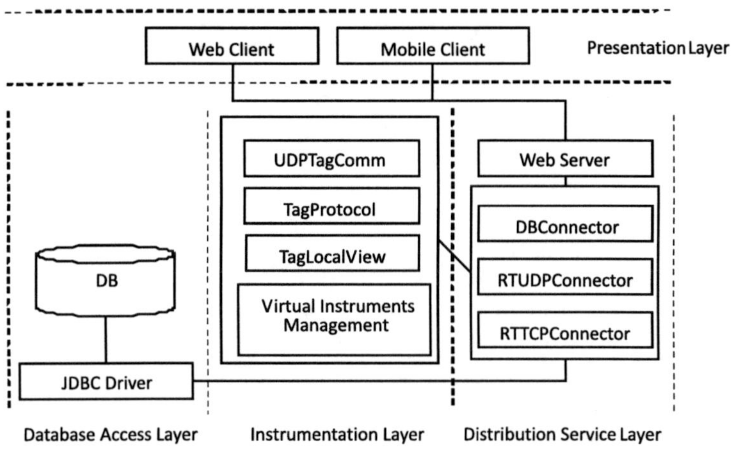


**Figure 3.** The tag application server diagram.

- Instrumentation Layer—implements the functions for managing virtual instruments; this level is responsible for data acquisition and for mapping the tag devices to software components.
- Database Access Layer—implements the functions for storing persistent data and providing access to these data.
- Distribution Services Layer—implements the interaction mechanism between the application server and remote clients. This level also implements security mechanism.
- Presentation Layer—implements the user interaction components (user interfaces and user interface processing components).

Figure 3 represents the block diagram of the Tag Server Application. The principal components are:

- UDPTagComm implements the low level UDP communication mechanism with tag devices; this component receives UDP packages from the tag devices;
- TagProtocol—implements the application level of the tag communication protocol; at this level, tag messages are interpreted and sent to the TagLocalView;
- TagLocalView—is a repository containing a local image of all active tag devices which transmit data; for each tag device, a separate component is responsible for storing its data; the last reported data are stored at this level;
- JDBC Driver—is used to access the local database where the tag data are stored; the database is also used for the tag data processing and the generation of reports;

- Web Server components are responsible for implementing server side components;
- VirtualInstrumentsManager—for real-time sensitive applications, which need to access as fast as possible the tag data, a local repository has been designed. In this repository real-time processing components are installed and managed. These components run as treads and are parts of the server application. These are the first components notified when new data arrive. At this level, the behavior of the virtual instruments is implemented. These components expose a set of interfaces which will be accessible through the server connectors and the Web Server components presented in this section.

If a remote application intends to access tag data or virtual instruments, it will have to connect to one of the following server connectors:

- RTTCPConnector—is a data feed connector used for sending tag data to remote applications; this component is implemented as a server and waits for connections from remote applications interested in tag data;
- RTUDPConnector—implements the same logic as RTTCPConnector, but uses the UDP protocol for communicating with clients;
- DBConnector—is a connector used by the client application which needs to access the data from the database.

This architecture is flexible enough to support a wide range of different applications. Based on this architecture, a series of application has been developed and is presented in the next part of this chapter.

**Tag4M Hardware Architecture**

The Tag4M Wi-Fi tag, shown in Figure 4, is a very good example of a measurement device which can send sensor measurements to the Internet and Web Pages which does not require connecting cables. The Tag4M Wi-Fi tag is a Wi-Fi RFID active tag with measurement capabilities that is intended to be used to build wireless measurement solutions for a wide range of applications. Sensors are connected to tag I/O terminal blocks. The tag is a complete Wi-Fi and networking solution, incorporating a 32-bit CPU, a memory unit, eCos real-time operating system, UDP and TCP/IP stack, an analog sensor interface, a power management unit, hardware cryptographic accelerator and a real time clock (G2 Microsystems[1] 2009). The tag has very small dimensions (4.7 cm × 7.0 cm), light weight, and it is battery powered.

**Figure 4.** Tag4M Wi-Fi tag data acquisition system.

The Tag4M platform has been designed from inception to maximize battery life and achieves ultra low-power consumption with a combination of proprietary technologies. The power management module includes logic to power-down functions not required and to control switching between sleep and active modes. When in sleep mode, the real-time clock and sensor interface remain active. The real-time clock can be programmed to wake up the Tag4M Wi-Fi tag at any required interval. Similarly, the Tag4M Wi-Fi tag can be programmed to wake up when a specified condition is detected. This "instant-on" feature of the Tag4M Wi-Fi tag can be utilized by mobile devices, allowing them to remain in the low-power standby mode until Wi-Fi is needed, and then powering on the Wi-Fi "instantly". The Tag4M Wi-Fi tag can wake up, and join a network using WPA2-PSK in under 35 msec. This allows it to stay asleep until required, but wake up and join a network without any noticeable user delay.

The tag device hardware architecture contains a Wi-Fi radio, conditioning circuitry for 5 analog inputs and 4 digital I/O lines, two I/O connector blocks, a JTAG connector and a battery holder as presented in Figure 5 (G2 Microsystems[3], 2008).

The tag device is based on 32-bit SPARC V8 processor, clocked at 40MHz, with 512 KB on-chip ROM. It contains an eCos operating system, a lwIP TCP/IP suite, security software and drivers. The lwIP is a lightweight implementation of the TCP/IP protocol stack and its role is to reduce RAM usage. The processor includes 128 KB of RAM for stack, user code and ROM code patches and 2 KB NVM (Non Volatile Memory), available when the AO Domain is powered (partial components from CPU, presented in Figure 5). The cryptographic accelerator can provide AES-128, RC4, MD5, SHA-1, CRC-32 and TKIP functions.

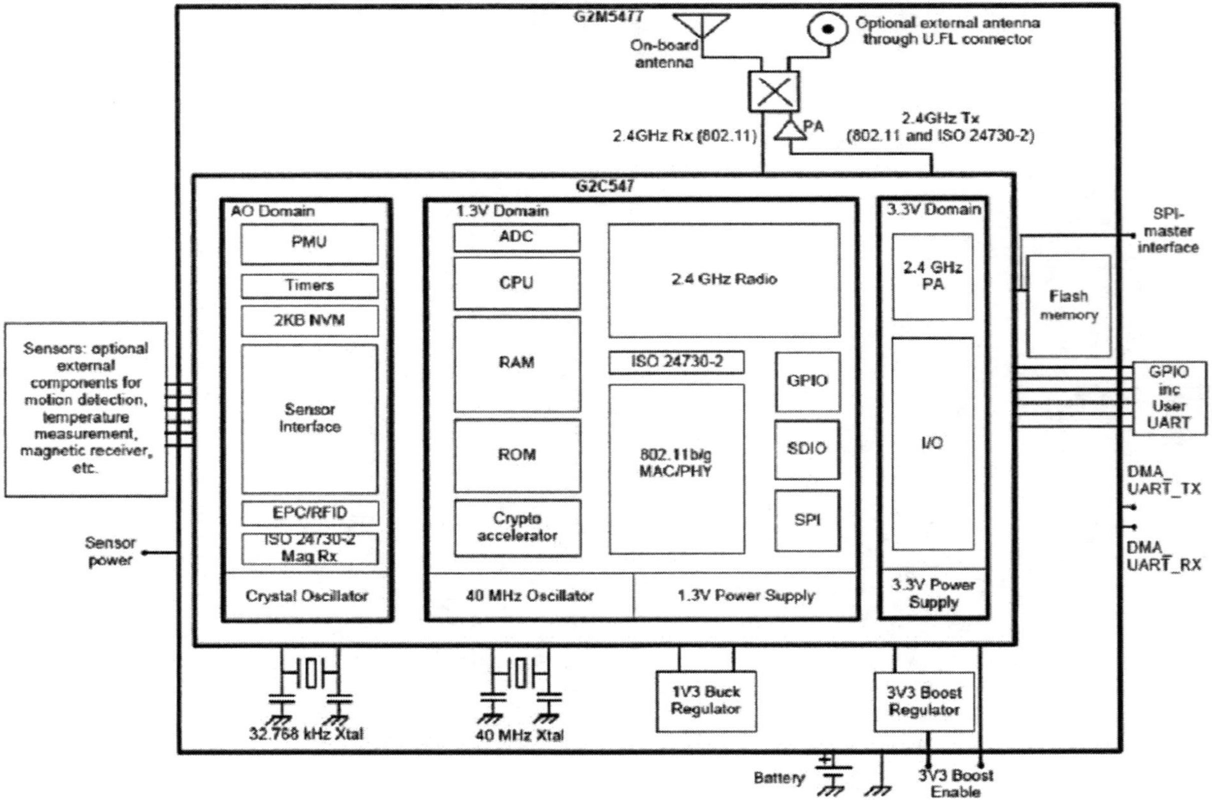


**Figure 5.** The hardware architecture of the Tag4M Wi-Fi tag data acquisition system.

The Wi-Fi transceiver with an on-chip synthesis of 2.4 GHz provides a complete solution for 802.11 b/g operation and supports a wide range of security suites (WEP, TKIP, WPA and WPA2) with a high throughput 4 Mbit/s sustained TCP/IP with WPA2. Transmit power works with the internal or an external power amplifier to provide consistent RF output power levels (up to +14 dBm) (G2 Microsystems[1] 2008).

The tag device firmware contains the eCos embedded operating system, a scaled down version of TCP/IP stack named LWIP, the start-up code, an application loader, power saving features, and a device driver. This functionality resides and it is run from a tag ROM location, so it does not need to be loaded at start-up. The device driver contains all of the tag opcode (operation code) definitions (Cores Electronic[1] 2011), which are numerical codes that describe tag settings and tag measurement functions, and are loaded from an EEPROM location at start-up (Folea and Ghercioiu 2010). By tag application software it is understood the application level code that interacts with the tag but runs in a computer. The applications can be developed in any high level programming languages like Java, C++ or LabVIEW.

**Distributed Environment Parameters Monitoring**

This section presents the extension boards developed for the Tag4M device. These extension boards have been developed to expand tag functionality and capabilities. Each tag extension board contains some dedicated sensor circuitry that covers a particular measurement application and connects to either the tag DIO or AIO lines. The following measurements and sensors populated on Tag4M Wi-Fi tag extension boards will be discussed in this section:

- Temperature, using the TC1046VNB sensor connected to tag line AI2 (Microchip 2004);
- Humidity, using the HIH-5030/5031 sensor connected to tag line AI0 (Honeywell 2010);

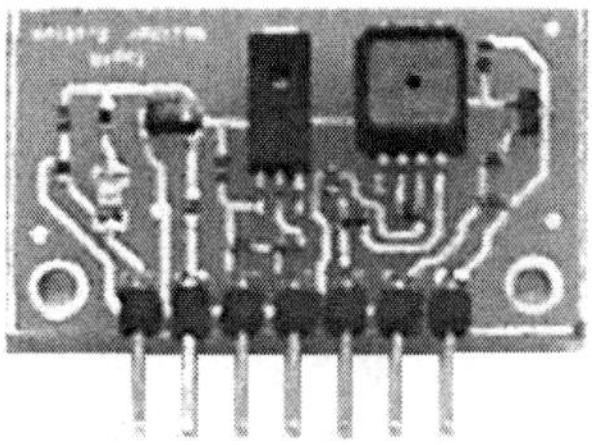

**Figure 6.** Weather station extension.

- Light intensity or solar radiation, using the LX1792 sensor connected to the AI0 line of the tag (Microsemi 2005);
- Atmospheric pressure, using the MP3H6115A6U sensor connected to tag line AI1 (Freescale Semiconductor 2009);
- Dust air density, using the GP2Y1010AU0F sensor connected to tag line AI2 (Sharp 2006).
- Wind speed and direction, using the WS-2310-15 sensor connected to the DIO0 and 1 lines of the tag (La Crosse Technology 2002).

One application that uses some of the sensors presented above is weather stations. A weather station extension (Figure 6) for the Tag4M Wi-Fi tag was built. It includes sensors for temperature, humidity, light and atmospheric pressure.

The weather station board schematic (Figure 7) includes a voltage reference component, REF3030, which is used to regulate power to sensors and improve measurement accuracy.

The weather station extension is powered from the tag 3V battery. The tag itself operates in a succession of sleep and wake up periods, which means that the extension will not have access to power during sleep time and also that sensor power up periods are very short, in the msec range.

The sensors used for building the board are: LX1972 light sensor, the HIH-5030 Humidity sensor, the TC1046VNB temperature sensor and the MP3H6115A atmospheric pressure sensor.

In order to test the weather station extension board, a LabVIEW application has been developed. This application is running on a PC connected to the Internet. The application receives data from the station and feed them in a Pachube stream (http://www.pachube.com/feeds/7277). One version of the weather station application has been developed using LabVIEW 2010. Pachube is a data infrastructure site for the Internet of Things and allows devices to feed data on demand or periodically at predefined (by the client) time intervals. The LabVIEW application panel is shown in Figure 8, followed by the block diagram in Figure 9.

The Virtual Instrument (VI) implements the following functions:

- Reads sensor data from the Tag4M tag;
- Packages the data into an EEML script format that is understood by Pachube;
- Opens a TCP/IP connection between the computer running this application and the Pachube server. Issues a TCP/IP Write in order to write the EEML script and read reply in 1024 byte blocks from Pachube. Close TCP/IP connection and start all over again by reading sensor data from the tag.

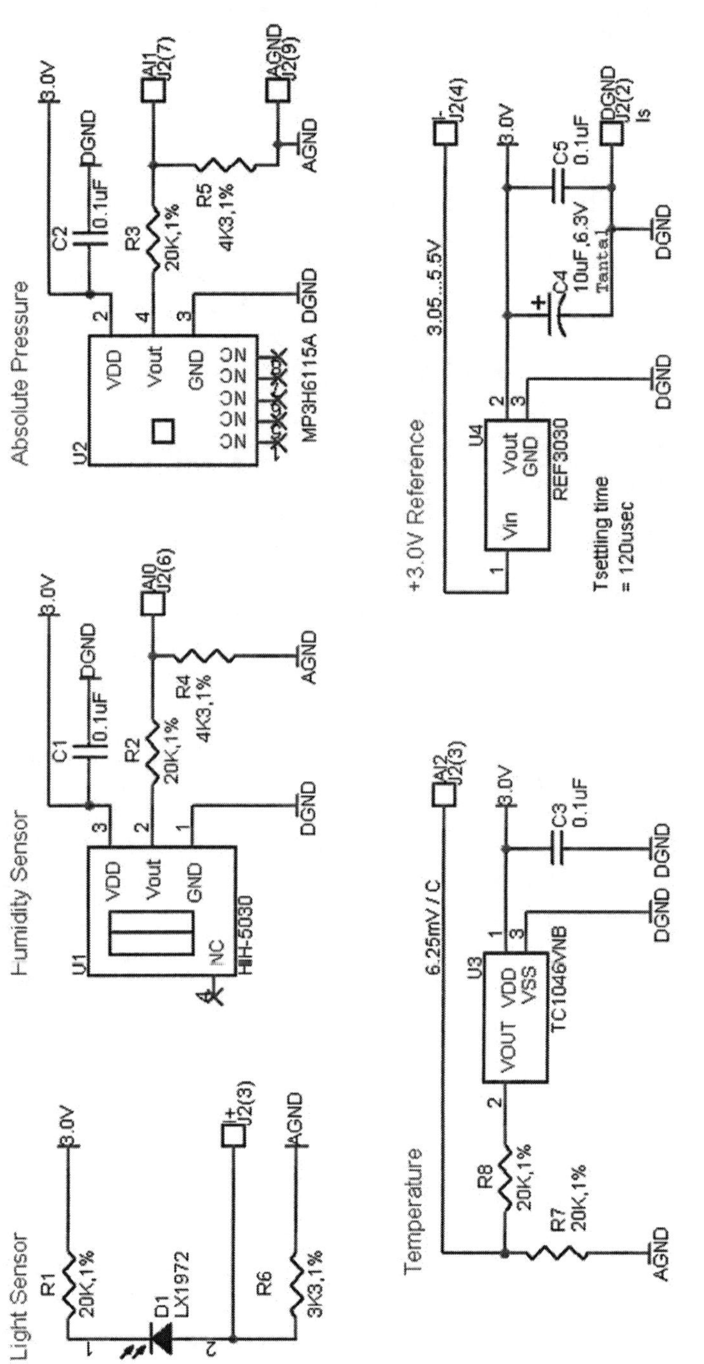


**Figure 7**. Scheme of the weather station extension.

*(Color image of this figure appears in the color plate section at the end of the book.)*

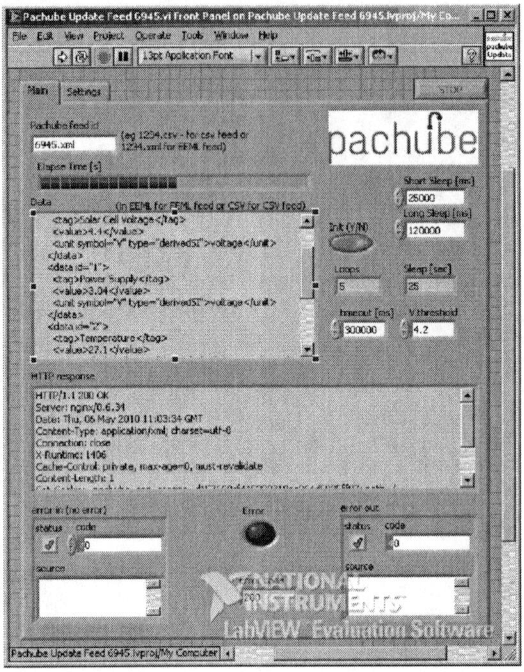


**Figure 8.** The Tag4M weather station application panel.

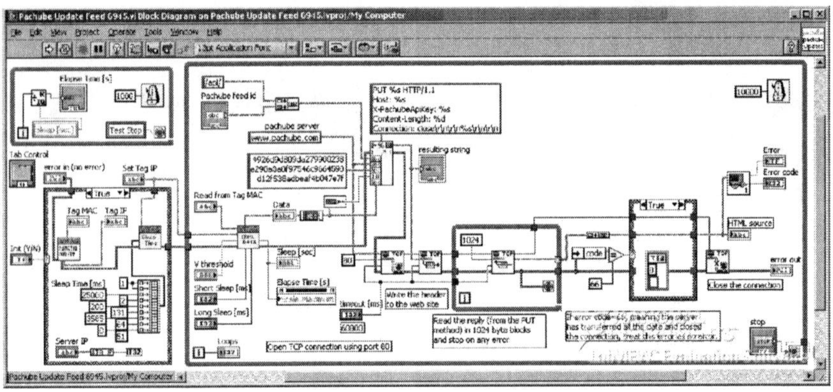


**Figure 9.** The Tag4M weather station diagram.

The data feeds can be accessed directly from the Pachube site. In Figure number 10, bellow, the data feeds sent from a Tag4M device connected with a weather station extension board are presented.

The weather station extension board is populated with the TC1046VNB temperature sensor which can accurately measure temperature between

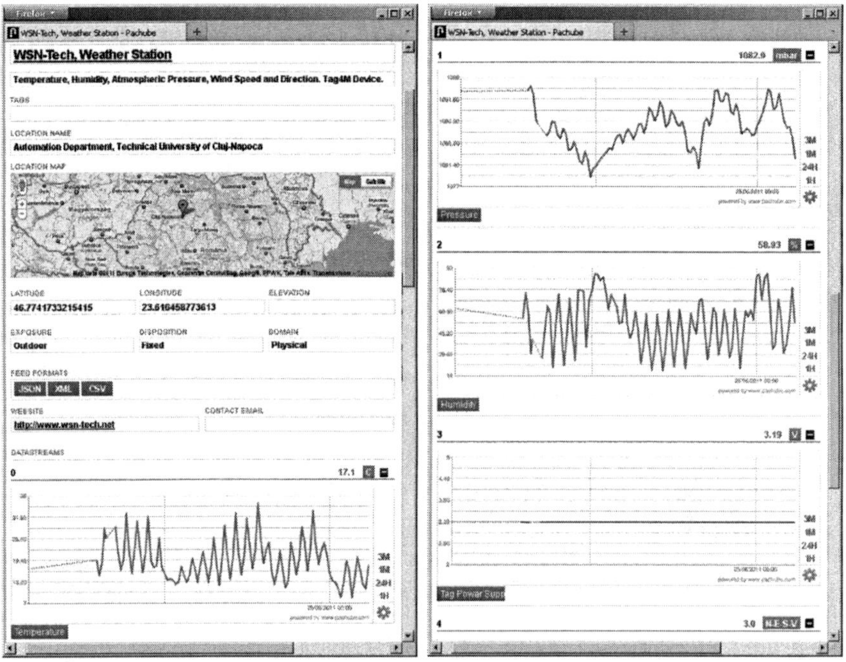


**Figure 10.** Weather station data posted by Pachube.

*(Color image of this figure appears in the color plate section at the end of the book.)*

-40°C and +125°C, generating a linear voltage output with a 6.25mV/°C slope (Microchip 2004). Figure 11 shows a sample of the sensor output signal corresponding to over seven days of continuous monitoring.

The weather station extension board is populated with the HIH-5030 humidity sensor (Honeywell 2010). The humidity output signal converted to units of humidity displayed as a graph is presented in Figure 12. The humidity output signal depends on the temperature values.

The weather station is populated with the LX1972 light sensor. The output signal from the light sensor extension is presented in Figure 13. The LX1972 is a low cost silicon light sensor which emulates the human eye, producing a spectral response. The response is produced at 520nm, with IR response less than 5% of the peak response, above 900nm. According to (Microsemi 2005) it is usable at ambient light from 1 to more than 5000 Lux.

The weather station is populated with the MP3H6115A6U pressure sensor (Freescale Semiconductor 2009). The output example shown in Figure 14 represents measured atmospheric pressure for a period of seven consecutive days, where the variations are small.

Another sensor developed extension is the one populated with a dust air density detector module, the GP2Y1010AU0F. This is a dust sensor

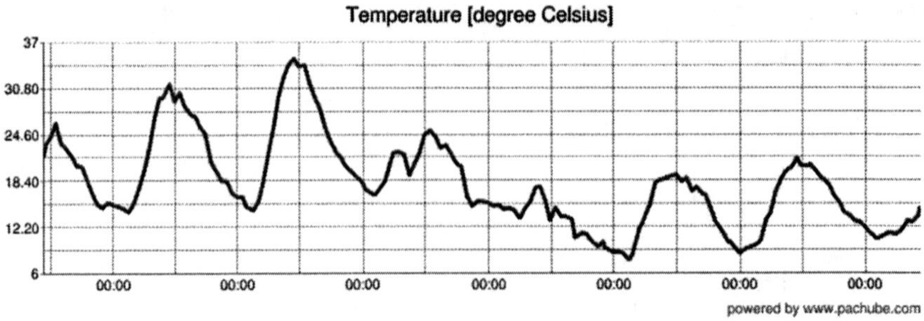


**Figure 11.** Temperature sensor, measurement data.

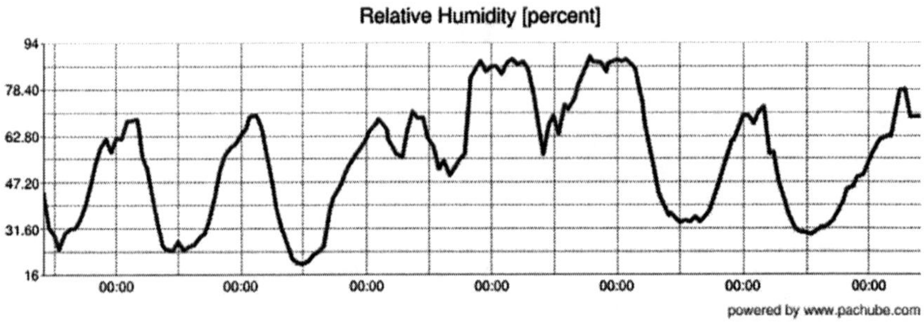


**Figure 12.** Humidity sensor, measurement data.

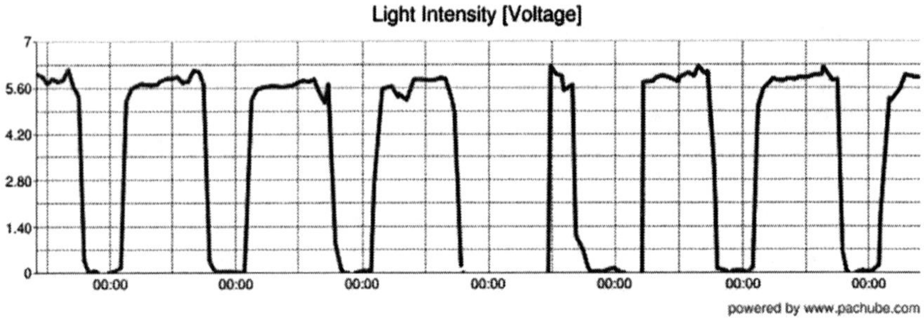


**Figure 13.** Light sensor, measurement data.

based on an optical sensing system. The basic principle on which the sensor detects dust is by quantifying reflected light of dust in the air using an infrared emitting diode (IRED) and a phototransistor which are diagonally positioned inside the device (Sharp 2006). The sensor has high sensitivity in detecting very small dust particles (like the ones contained in cigarette smoke) and also has the capability to distinguish house dust and smoke. Figure 15 presents the schematic of our dust air density detector module.

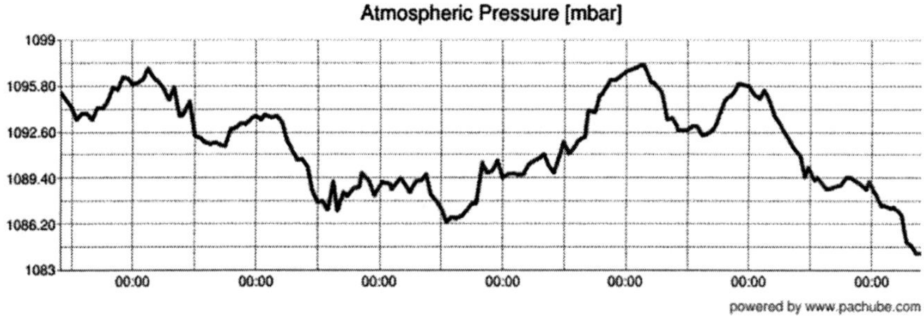


**Figure 14.** Pressure sensor, measurement data.

The sensor extension module containing the dust air density detector is presented in Figure 16.

A second phase in the development of the weather station extension board was to attach a sensor for monitoring wind speed and direction. With respect to this, a WS-2310-15 sensor was selected (La Crosse Technology 2002). This is a serial sensor which sends data in a compact serial package with the following format:

| Start | | | | | Direction | | | | Speed | | | | | | | | | | | | Direction | | | |
|---|---|---|---|---|---|---|---|---|---|---|---|---|---|---|---|---|---|---|---|---|---|---|---|---|
| 1 | 1 | 0 | 1 | 1 | 0 | 1 | 2 | 3 | 0 | 1 | 2 | 3 | 4 | 5 | 6 | 7 | 8 | 9 | 10 | 11 | 0 | 1 | 2 | 3 |

The wind sensor has a digital input command line to control its state. The Tag4M Wi-Fi tag turns the sensor ON/OFF in order to reduce its power consumption. Once the sensor is turned off, it takes about 2 to 2.5 seconds for the sensor to boot up and send the first output values. Wind speed values from this sensor are shown in Figure 17. The sample corresponds to a session of continuous samples for a seven day period.

Sensor readings of wind direction are presented in Figure 18. The numerical values represent wind direction in the following way: North (N) = 0, North-East (NE) = 2, East (E) = 4 etc.

Providing the possibility of powering the weather station from a solar cell represented a serious challenge. For this setup, a small solar cell AM-8801CAR, an accumulator and a charger were used. The purpose of this experiment was to create a solar cell circuitry with high efficiency that provides power to the tag under any condition of illumination—poor illumination, shadow, inside building, night etc.—and also covers the worst case scenario of retransmissions and associations between the tag and the Access Point.

The solar cell has a conditioning circuitry which uses a high efficiency, fixed frequency, Buck-Boost DC/DC converter (LTC3440) to provide high efficiency, low noise, burst power for the tag (Linear Technology, 2005).

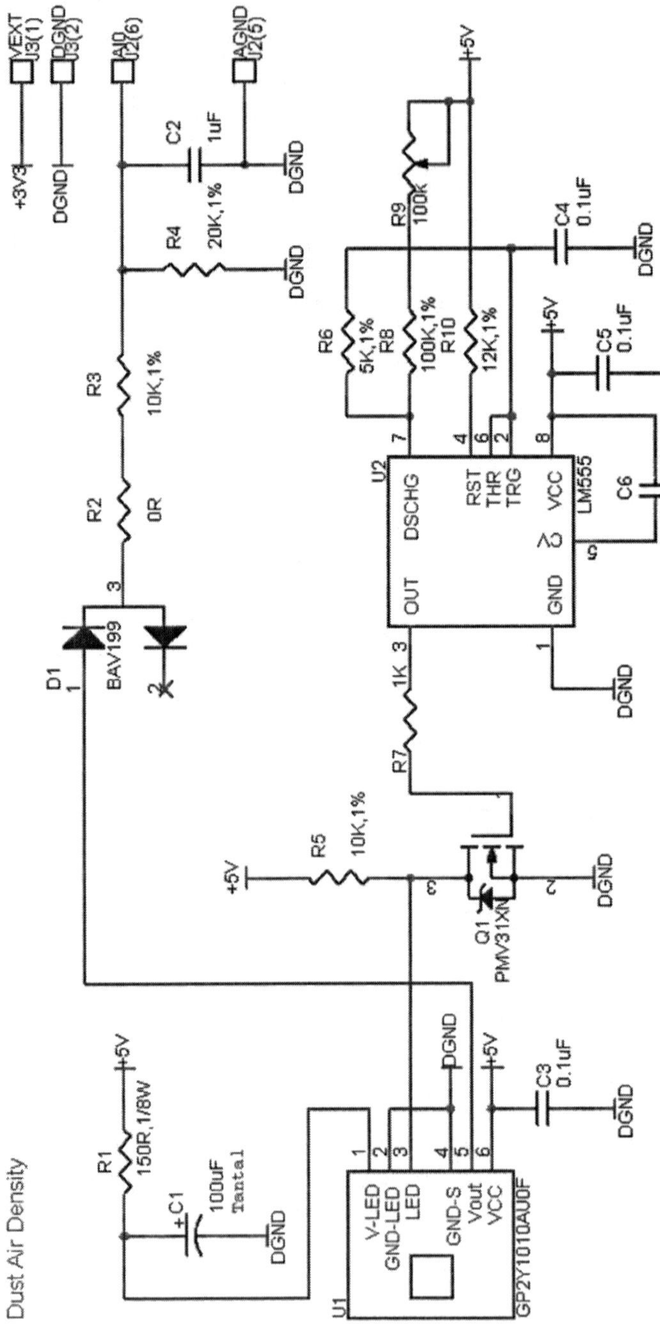


**Figure 15.** Dust air density sensor schematic.

*(Color image of this figure appears in the color plate section at the end of the book.)*

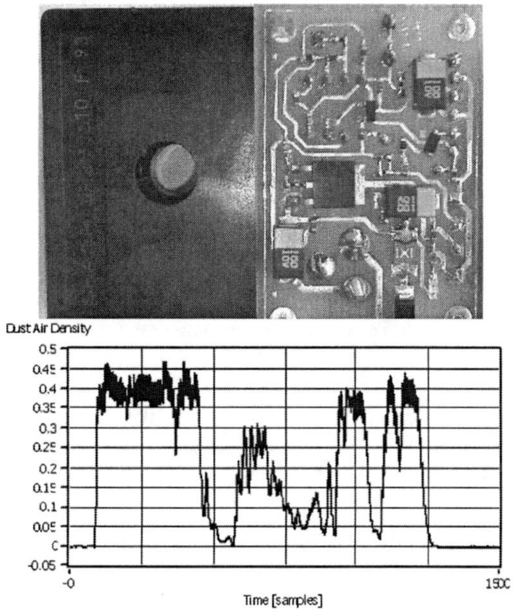


**Figure 16.** Dust air density sensor, PCB and measurement data.

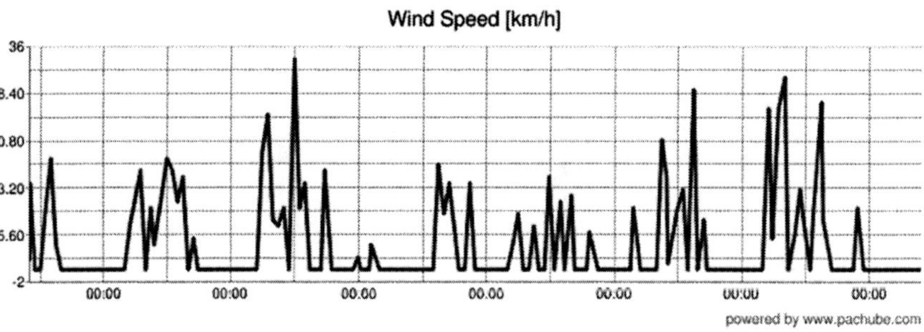


**Figure 17.** Wind speed, measurement data.

Burst power means almost no power during tag sleep time periods but highly efficient output voltage that is equal or greater than the tag input during tag Wi-Fi transmission sessions inside wake-up periods. The Buck-Boost DC/DC converter uses only 25µA and has an efficiency of 65 to 95% when generating the 3.3V necessary for tag operation from an input voltage as low as 1.8V. In comparison, a linear voltage regulator (LDO) provides a current consumption of 0.3 to 1mA which represents almost 10% of what the solar cell can generate. The schematic of the solar cell power supply extension is presented in Figure 19.

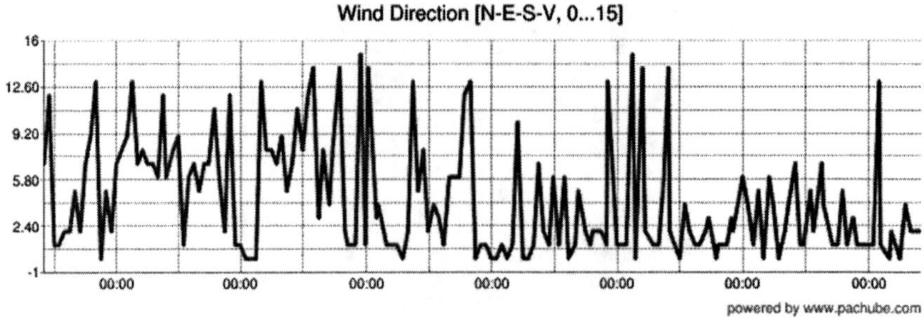


**Figure 18.** Wind direction, measurement data.

The solar cell conditioning circuitry has to insure power to the tag and the attached sensors under any conditions of illumination. The increased power consumption in the Wi-Fi association and transmission phases needs to be accommodated too. These problems have been solved by using an accumulator of type Li-Ion and a charger circuitry. The accumulator offers larger storage capability than what a capacitor array would. The solar cell charges the accumulator during the day while also powering the tag. If the accumulator is too small (for example 3.7V/200mAh) then the setup needs to have a capacitor array.

The autonomous solar cell has been tested for powering the weather station presented in a previous section. The resulting setup formed by the Tag4M, the weather station and the solar cell supply is presented in Figure 20.

Starting from the proposed architecture already presented, a web based weather monitoring application has been implemented. In Figure 21 the way in which the system components are interconnected is presented. This application collects data from multiple weather stations and stores them in a central database. A system where multiple solar cell powered weather station devices are gathering environmental data was designed. The data are stored in the Weather Station Database (WSD). A Web server application is used for accessing data stored in the database.

The following functionality has been implemented at the Tag Application Server level:

- Handling communication with weather station devices; the communication protocol is based on the UDP transport protocol; the SCWS devices report data directly to the Tag Application Server;
- Storage of the received data in the Weather Station Database (WSD);
- Processing and interpretation of the real-time data;
- Handling communication with clients; the application server implements a set of connectors by which it can communicate with

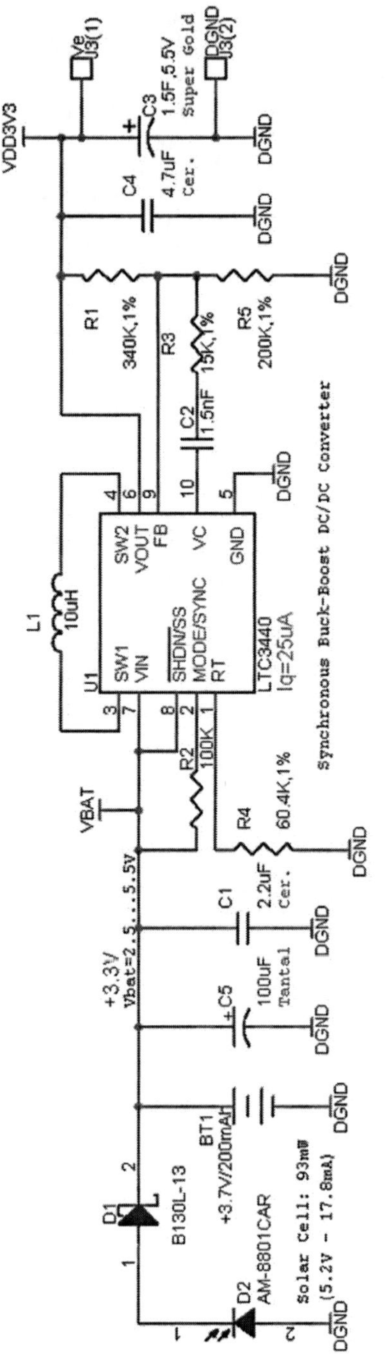


**Figure 19.** The solar cell power supply circuitry.

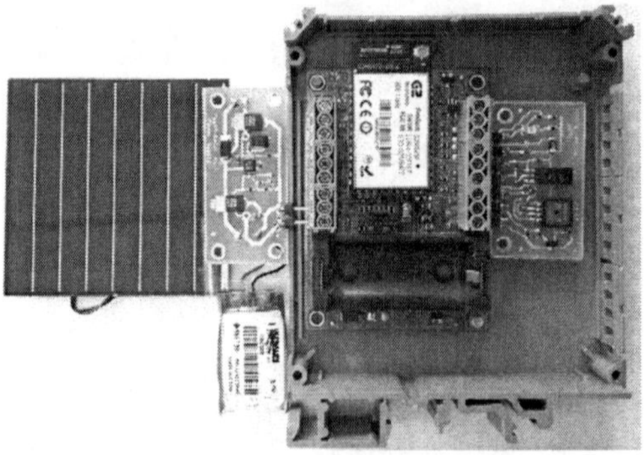

**Figure 20.** The solar cell powered autonomous weather station.

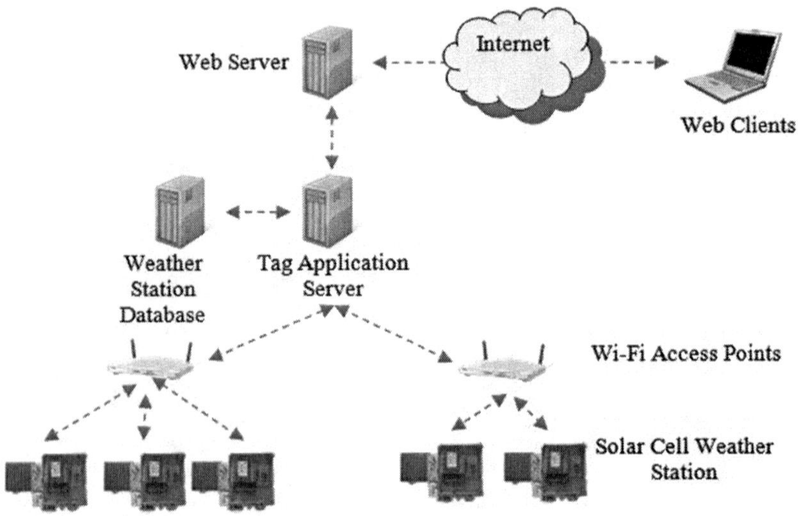


**Figure 21.** Distributed web based application using autonomous weather station devices.

remote clients using UDP and TCP protocols; these clients can be Web applications, desktop applications or mobile clients;

- Communications with peer servers; the Tag Application Server also implements a communication mechanism through which tag data can be exchanged between application servers installed in different locations.

In order to provide easy access to weather data, a Facebook application has been developed. To access this functionality, users need to install and grant access to the Weather Station Facebook application. A Facebook user can automatically receive a message when weather parameters (temperature, humidity or atmospheric pressure) are changing and some preset alarm conditions are reached. Figure 22 displays the Facebook page of the Environment Monitoring Demo application with the sensor location and environmental data.

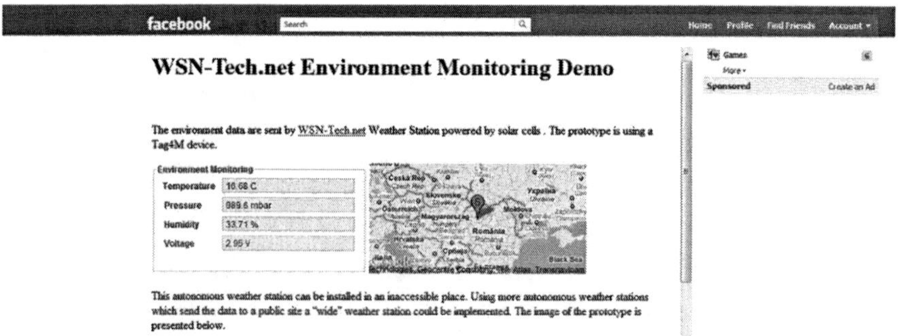


**Figure 22.** The Weather Station Facebook application.

As a conclusion regarding the weather station application, the following benefits of the proposed solution are enumerated:

1. *Lower cost of hardware*—Existing Wi-Fi infrastructure will be used for communication between weather station devices and the application server;
2. *Scalability*—The system is designed from ground-up to handle large number of sensor and weather station devices without imposing software restrictions. Server-based processing ensures minimal network communication;
3. *Better performance*—Core processing is done by the server processing power, which can handle large amounts of data, unlike microcontroller devices. This ensures better performance.

### Weather Sensor Application for Smartphone

Wi-Fi sensors need a router or access point in to communicate with applications located in the Internet space. One of the most important developments in Smartphone technology is the growing support for local Wi-Fi hotspots. The Smartphone becomes an access point which allows other devices, like Wi-Fi tags, for example, to connect to the Internet via Wi-Fi.

This feature is available for the Android-based Smartphone. The Android operating system is an open-source platform developed by Google along with major hardware and software developers, forming the Open Handset Alliance. The Android operating system is based on a Linux kernel and the software stack consists of Java applications running on a Java-based, object-oriented application framework on top of Java core libraries running on a Dalvik virtual machine featuring JIT compilation.

A prototype of a weather station application for the Android Smartphone was developed. This application is a mobile version of the Web Server-based alternative. Smartphone users can access nearby weather stations through a direct Wi-Fi link without the need to connect to the central server. Using this direct communication link, the user can visualize real-time sensor data and configure weather station parameters for measurements.

The Smartphone application has been developed in Java language for Android OS 2.2. The application acts as a server which listens for broadcast messages on the UDP port 50007 (which is the standard port also used by the application server). The Smartphone application has the structure presented in the figure below.

The UDPBroadcastServer layer is responsible for listening to messages coming from weather station devices on the UDP port. The server stores sensor data and makes this data available to the other application components. The UDPBroadcastServer runs as a background thread in the Android application.

The CommProtocol layer interprets data coming from tag devices and is also responsible for notifying the upper level components when new data arrives from tag devices. Upper level components need to register as listeners in order to be notified.

The ConfigurationActivity and DataViewActivity layers (Figure 23) are exclusive tasks which the Android user can choose to execute. The chosen task is notified by CommProtocol when new data arrives. ConfigurationActivity provides a user interface to a set of lower level configuration parameters of weather station device. DataViewActivity displays real-time sensor data received from the weather station device. This interface has been designed to display data from one weather station:

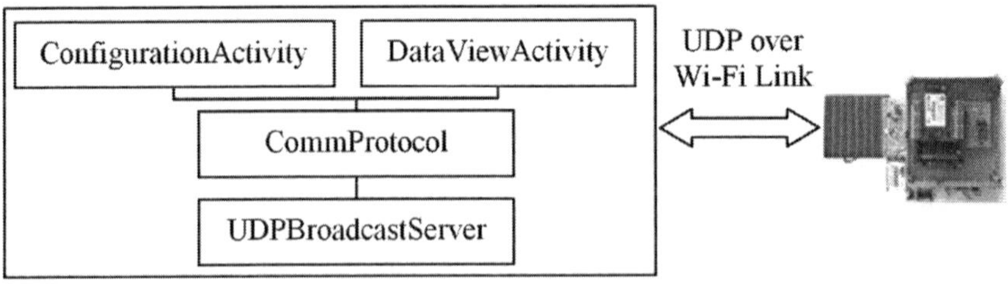


**Figure 23.** Android application components.

temperature, atmospheric pressure, humidity and light level recorded by the device. The Smartphone application displays data from one weather station at one time for an unlimited number of devices (Figure 24). The user can choose which device to display from the dropdown menu button located at the top of the phone screen. Weather stations are uniquely identified by the Tag4M MAC address which is displayed on the dropdown menu button.

Figure 25 shows the Android application running on a Samsung Galaxy S i9000 phone. Data received by the Smartphone can also be forwarded to the web based Tag Application Server to be stored in the database.

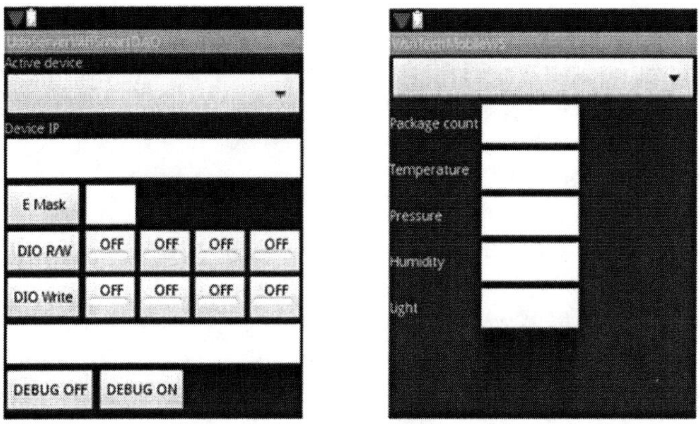


**Figure 24.** Android application user interfaces.

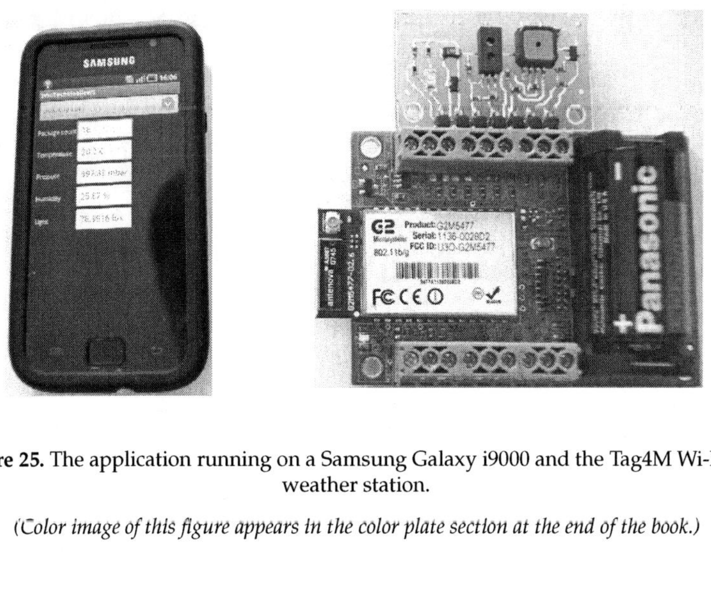


**Figure 25.** The application running on a Samsung Galaxy i9000 and the Tag4M Wi-Fi tag weather station.

*(Color image of this figure appears in the color plate section at the end of the book.)*

## Future Research Directions

There are a number of directions in which the work presented in this chapter can continue. More sensor attachments for the Tag4M tag can be developed in order to implement new applications. Some of the future applications consist of measuring radiation and elements of control using solid state relays. The control function will be more difficult to implement because relays are connected to the tag digital lines which are reset before each sleep time. Additionally, the tag digital lines are active only during the wake up time, a very short interval, so the relays need to be very responsive and they should not be altered by the digital line reset.

The idea of associating art and sensors data has been defined by Marius Ghercioiu and the artist Arlissa Vaughn. They defined the concept of Multi-Sensory Moving Art in which a digital frame is used to display artistic images which are dynamically modified in correlation with data received from sensors.

An application employing concept is the usage of the wind sensor presented in the previous section to display digital images which are correlated with the wind speed and direction.

Another direction in which the presented sensors can be used is related to the concept of Cloud Instrumentation (Ghercioiu 2011). A sensor becomes a cloud instrument when it is connected to a wireless device such as a Wi-Fi device.

The cloud instrumentation idea will be developed further and more tag web-based applications will be built along with specialized widgets that incorporate data processing. A community of tag users who all contribute to widget instruments needs to be built. There is a direct connection between this community of Cloud Instrumentation and the Internet of Things movement. As the Internet of Things concept becomes reality, and definitions are set for how objects connect to the Internet, how objects are identified based on objectives and functionalities, how security is implemented, etc, cloud instruments will have to incorporate these standards, so much work will have to be done in this direction too. Then, there is the art movement: Multi-Sensory Moving Art which also bounds with the Internet of Things because it enables art objects to participate and have an identity on the Internet.

## Conclusion

The first step in the research which is the subject of this paper is represented by the development of the Tag4M Wi-Fi tag. This was a challenging task because of the need for reduced device dimensions, reduced production cost and low power consumption. The sensor extensions developed by the authors can be used for solving real-life problems. The measuring instrument that results from the association of Wi-Fi capabilities to sensors becomes a Cloud Instrument. A society formed by different types of such devices can efficiently provide information for metering and charting applications, data logging, etc. The weather station presented in this paper is the strongest proof that supports this assumption. This application can be further developed in such a way that the weather station becomes a sensory extension of the building to which it is associated. The data obtained in this way can also be used in a Multi-Sensory Moving Art application.

The work presented in this paper emphasizes the advantages of using the Tag4M Wi-Fi tag in wireless measurement applications.

## REFERENCES

Cores Electronic[1] (2011). Technical Report from http://tag4m.com/.

Cores Electronic[2] (2011). Tag4M—Web Instrument from http://demo.tag4m.com/.

Cook, D. and S. Das. 2005. Smart Environments: Technology, Protocols and Applications. Wiley-Interscience, ISBN 0-471-54448-5.

Corning. 2011. From site: http://www.youtube.com/watch_popup?v=6Cf7IL_eZ38&vq=medium.

Freescale Semiconductor. 2009. High Temperature Accuracy Integrated Silicon Pressure Sensor for Measuring Absolute Pressure, On-Chip Signal Conditioned, Temperature Compensated and Calibrated. *Technical Report*, 2009, from www.freescale.com.

Folea, S. and M. Ghercioiu. 2008. Ultra-Low Power Wi-Fi Tag for Wireless Sensing, 2008 IEEE International Conference on Automation, Quality and Testing Robotics, May 22–25, Cluj-Napoca, Romania.

Folea, S., M. Ghercioiu and D. Ursuţiu. 2010. Cloud Instrument Powered by Solar Cell Sends Data to Pachube. *i-JOE, International Journal of Online Engineering*. **6(4)**:20–25.

Folea, S. and M. Ghercioiu. 2010. Tag4M, a Wi-Fi RFID Active Tag Optimized for Sensor Measurements. C. Turcu [ed.], Radio Frequency Identification Fundamentals and Applications Design Methods and Solutions (pp. 287–310). Austria, PA: InTech Education and Publishing.

Ghercioiu, M., S. Folea and I. Monoses. 2007. The WiTAG—a WiFi Sensor TAG, The 2007 International Conference on Wireless Networks (ICWN'07: June 25–28), Las Vegas, Nevada, USA, 376 p.

Ghercioiu, M.[1] 2011. Cloud Instrumentation: Data without Infrastructure, 2011 Control Engineering, March 15, Chicago, USA.

Ghercioiu, M.[2] 2011. Cloud Instrumentation, the Instrument is "in the Cloud", 2011 Ree Revue De L'electricite Et De L'electronique, Nr. 2–2011, France.

G2 Microsystems[1]. 2008. G2C547 SoC Data Sheet, *Technical Report*, 2008, from www.g2microsystems.com.

G2 Microsystems[2]. 2008. G2C547 Software, Example Application. *Technical Report*, 2008, from www.g2microsystems.com.

G2 Microsystems[3]. 2008. G2M5477 Wi-Fi Module Data Sheet. *Technical Report*, 2008, from www.g2microsystems.com.

G2 Microsystems[4]. 2009. Data Sheet G2C547 and G2M5477, *Technical Report*. G2 Microsystems, Inc., Campbell, CA, USA.

Hart, J. and M. Kirk. 2006. Environmental sensor networks: A revolution in the earth system science?, Earth-Science Reviews. 78. pp. 177–191.

Honeywell. 2010. HIH 5030/5031 Series, Low Voltage Humidity Sensors. *Technical Report*, from http://sensing.honeywell.com.

Integrating Social Networks and Sensor Networks. 2009. W3C Workshop on the Future of Social Networking, 15–16 January 2009, Barcelona , John G. Breslin, Stefan Decker, Manfred Hauswirth, Gearoid Hynes, Danh Le Phuoc, Alexandre Passant, Axel Polleres, Cornelius Rabsch, Vinny Reynolds.

International Telecommunication Union. 2009. Distributed Computing: Utilities, Grids and Clouds, ITU-T Technology Watch Report 9.

Labiod, D.H., H. Afifi and C. De Santis. 2007. Wi-Fi, Bluetooth, ZigBee and WiMAX, Springer Verlag Publisher. 2007, XVI, 316 p. Hardcover ISBN: 978-1-4020-5396-2.

Linear Technology. 2005. Micropower Synchronous Buck-Boost DC/DC Converter, Datasheet, LT 0507 REV B, USA.

Microchip. 2004. High Precision Temperature-to-Voltage Converter, Technical Report, from http://ww1.microchip.com.

Microsemi. 2005. Ambient Light Detector, LX1972. Technical Report, from www.microsemi.com.

Pachube. 2011.WSN-Tech, Weather Station from site: http://www.pachube.com/feeds/6671

Ruiz-Garcia, L., L. Lunadei, P. Barreiro and J.I. Robla. 2009. A Review of Wireless Sensor Technologies and Applications in Agriculture and Food Industry: State of the Art and Current Trends, *Sensors 2009*, ISSN 1424-8220.

SensedIn. 2011. This site is an example of a LinkedIn environment for sensors and devices that are connected to the Internet, from site: http://sensedin.blogspot.com/.

Sharp. 2006. GP2Y1010AU0F Compact Optical Dust Sensor. *Technical Report*, from www.sharp-world.com.

Sidhu, B., H. Singh and A. Chhabra. 2007. Emerging Wireless Standards—WiFi, ZigBee and WiMAX, World Academy of Science, *Engineering and Technology* Vol. 25.

Vaughn, A.[1]. 2011. Multi-Sensory Moving Art Explanation, from site: http://www.youtube.com/watch?v=FKdQCygws8A

Vaughn, A.[2]. 2011. From site: http://arlissawind.temperatureavatar.com/

WSN-Tech. 2011. Technical Report from http://www.wsn-tech.net/home

## ADDITIONAL READING SECTION

Hojoon, Y., H.S. Kwak, J.H. Kim and C.J. Lee. 2010. Interference analysis between ZigBee transceivers at 2.4 GHz frequency bands in a multiple integrated laser engagement system, in Microwave and Optical Technology Letters. Vol. **52(3):**662–665.

Hlavacs, H., R. Weidlich, K.A. Hummel, A. Houyou, A. Berl and Meer de Hermann. 2008. Distributed energy efficiency in future home environments, *Annals of Telecommunications, Home networking: performance and architecture challenges.* **63:**9-10. 473–485.

Hussain, S., S.Z. Erdogen and J.H. Park. 2009. Monitoring user activities in smart home environments, Kluwer Academic Publishers Hingham, MA, USA. pp. 1–12.

Jamil, M., M.F.M. Zain, V. Krishnamurthy and E. Sazonov. 2009. Scientific Data Acquisition for Structural Health Monitoring by Using Wireless Intelligent Sensor, *European Journal of Scientific Research.* ISSN 1450-216X, Vol. 33, No. 4, pp. 585–593, Euro Journals Publishing.

Jun-jie Peng, Hui, He, Ping-an Zhu, Yan-ping Liu, Xue-Jun Zhang and Yi Jin. 2010. Zigbee-based New Approach to Smart Home, *Journal of Shanghai University,* Volume 14, Number 1, pp. 12–16.

Lu, G. and B. Krishnamachari. 2007. Minimum latency joint scheduling and routing in wireless sensor networks, *Science Direct, Ad Hoc* Networks, Vol. 5, pp. 832–843.

Neumann, P. 2007. Wireless Asset Management, Studies in Computational Intelligence (SCI) 42, SpringerLink, pp. 123–128.

Stankovic, J.A. 2008. When Sensor and Actuator Networks Cover the World, *ETRI Journal,* Volume 30, Number 5.

Stojmenovic, I. 2005. Handbook of Sensor Networks, Algorithms and Architectures", *Wiley Interscience.*

Varchola, M. 2007. Zigbee Based Home Automation Wireless Sensor Network, *Acta Electrotechnica et Informatica* No. 4, Vol. 7.

## KEY TERMS AND DEFINITIONS

**Tag4M Wi-Fi tag:** Wi-Fi tag manufactured by Tag4M.

**Web Page Instrument:** The User Interface of the Tag4M tag engine. It displays live Tag4M tags on the Internet.

**Cloud Instrument:** A sensor becomes a Cloud Instrument when it is connected to a wireless tag such as a Wi-Fi tag. The tag digitizes the data to send it on to an access point, where the data is routed to the Internet and a server IP. Here a customized engine is collecting data to feed into applications like metering, charting, control, analysis, modeling, data mining, and so forth, with display on Web page and Web widget instruments.

**Widget Instrument:** This is a small measurement instrument that can be installed and executed within a web page. The widget gets data from a Cloud Instrument and either displays the value as it is or it applies some processing on the data before display.

**Multi-Sensory Moving Art:** This is Art that uses a special frame type named Digital Sensor Frame. The digital sensor frame is connected to sensors via wireless Wi-Fi tags to digitally record natural phenomena like temperature (or other environmental effects) and turn that data into useful information that can be manipulated to create art.

- The development of an online application to manage the customer orders and the production of a rapid prototyping/rapid manufacturing centre (involving mostly software).

The applications presented in this chapter are related to the first objective. More than a year has been dedicated to it and several results have been obtained, including the knowledge and applications presented below.

In order to have an overview of the application context (i.e., teleengineering/telemanufacturing), it is important to note that there are many teleengineering application fields. Examples include: education—from virtual laboratories employing e-learning (Kostal et al., 2010) to telelaboratories part of the iLabs project like the one presented in (Orishaba et al., 2011) or the Sytron virtual room for collaborative learning (Chellali et al., 2008) -, medicine—Assisted Living Laboratory (Ras et al., 2007), TeleMedSys (Langmann et al., 2011) -, robotics—humanoid robots that recognize objects, with remote processing (Saidi et al., 2007) -, remote maintenance in various fields etc.

Regarding the telemanufacturing field, one has to mention the "Digital Factory" Hungarian project which includes, amongst others, process monitoring (with recording for future investigation regarding the rejected parts) and telepresence (Kovacs et al., 2005). A project which complements and evolves this concept is the "Factory-of-things" (Zuhlke 2010) implemented at the German Research Center for Artificial Intelligence from Kaiserslautern. At the Wurzburg telematics centre, there is ongoing research for the remote control of industrial robots, along with mobile robot control (Zeiger et al., 2009). An important achievement was the remote monitoring of the internal parameters and partial control, through the Internet, of a CNC lathe (Alvares and Ferreira 2006). The application has been included into an online CAPP (Computer-Aided Process Planning) application for revolution solids (Alvares and Ferreira 2008). More references, related to the respective topics, can be found in the following parts of this chapter.

Most of the developed applications presented in what follows are intended to be used for telemanufacturing systems. However, they have been developed with consideration to generally applicable principles and can be easily adapted for various other application fields.

Because the programming of the MC9S12C32 microcontroller—which is found on the CSM12C32 development module and is included by default—may seem hard at the first glance, the author decided to include helpful hints on how to proceed, for various scenarios.

# 10

# Applications Developed with the Microcontroller Student Learning Kit for the Teleengineering Field

*Bogdan-Alexandru Deáky*

## Introduction

The author works as a postdoctoral researcher in the teleengineering field (especially oriented towards telemanufacturing) and found the Microcontroller Student Learning Kit (MCU SLK) from Freescale Semiconductor to be of great assistance, not only for research work, but for educational purposes as well. This chapter presents a few new applications developed with the MCU SLK and focuses, in particular, on information regarding the embedded part.

The research is being conducted in the AMTS (Advanced Manufacturing Technologies and System)—D05—research department and its applications target, among others, the remote monitoring and control of modern manufacturing machines acquired within the PLATEDINO (Platform for Innovative Technological Development) interdisciplinary platform (Udroiu and Ivan 2010).

Currently, the author's main research objectives are:

- Developing an external remote monitoring solution for fabrication machines (involving both hardware and software and achieved by using sensors),
- Developing an internal remote monitoring solution for a fabrication machine (achieved by talking to the machine),

An additional type of module (the MC13192U application module for the Freescale MC13192 2.4GHz RF transceiver), not to be found by default in the MCU SLK, has been used to achieve wireless communication.

This chapter is split into different parts, according to the topics presented, as follows:

- Information about the MCU SLK (kit content, presentation of included modules, about the MC9S12C32 microcontroller, CodeWarrior etc.).
- A data acquisition application for monitoring.
- Using the data acquisition application to monitor the output of a renewable energy source.
- Wireless communication applications (using the MC13192U module).
    - o General considerations for wireless application development with the MCU SLK.
    - o Wireless transmission and reception of the keyboard keys pressed between 2 PCs.
    - o Wireless control of remote DC motor.
    - o Wireless communication of data acquired by one or more data acquisition boards.
- Conclusions (including other existing practical usage examples for the MCU SLK).

## About the MCU SLK

The MCU SLK enables students to learn microcontroller design and programming and is compatible with the National Instruments Educational Laboratory Virtual Instrumentation Suite (NI ELVIS). The MCU SLK contains the MCU Project Board, the CSM-12C32 module (development module for the MC9S12C32 microcontroller), a serial cable, an AC/DC adapter, a jumper wires kit, the appropriate CodeWarrior development environment, a DIP (dual in-line package) sample HC(S)08 family microcontroller, documentation and others.

### The MCU Project Board

The Freescale Microcontroller Student Learning Kit or MCU SLK includes the main plug-in board (called MCU Project Board) which allows students to rapidly design and prototype electronic systems around a variety of 8-, 16- and 32-bit computing platforms from Freescale Semiconductor (FS

2011a). The MCU Project Board itself can also be used as a full-featured circuit prototyping board, for designing and testing electronic circuits. It is applicable to both non-microcontroller circuits and circuits with microcontroller.

Figure 1 shows the version of the MCU Project Board that has been used by the author (with the CSM-12C32 module attached) and figure 2 (source: Freescale Semiconductor) shows the latest version of the MCU Project Board.

Both versions come with many useful peripherals: LEDs (Light Emitting Diode), push button switches, DIP switches, Banana connectors, BNC connector, serial port, potentiometer, buzzer etc. Maybe the most important difference between the two versions is the presence of a LCD (Liquid Crystal Display) on the latest version. The older, used version has only an LCD module interface.

**Figure 1.** MCU Project Board - 2 (used version).

**Figure 2.** Latest version of the MCU Project Board.

Its functionality can be leveraged by connecting it to a NI ELVIS II platform. The ELVIS platforms are tightly integrated with LabVIEW and make it easier to develop complex embedded applications (NI 2011/a). On the MCU Project Board, a PCI style connector is the interface for connecting it to NI-ELVIS platforms. The applications developed in this chapter were developed directly in C, using the CodeWarrior IDE, and only involve hardware and software resources from Freescale Semiconductor.

To help students get accustomed with Freescale microcontrollers, there are a number of MCU development modules that include a Freescale microcontroller and different peripherals, which can be connected to the MCU Project Board through one of the 60 pin connectors, called MCU Ports. To allow the programming, erasing and debugging of the microcontrollers located on Freescale modules, an integrated USB BDM port is provided. There are development modules that plug in directly into the MCU Project Board for CS(S)12(X), HC(S)08, DSP and Coldfire microcontrollers.

**The CSM-12C32 Module**

The CSM-12C32 educational/development module for the Freescale MC9S12C32 microcontroller (fig. 3) has been used for data acquisition, to control the wireless communication process and to have access to the peripherals (LEDs, potentiometer, serial port, motor etc.).

The module can be used in a stand-alone configuration but, by connecting it to the MCU Project Board, it gets also direct access to a part of the board's peripherals (and can be connected to all of them by means of wires).

**Figure 3.** The CSM-12C32 module.

For our application scenarios, the important features of the CSM12C32 module (depending also on the characteristics of the MC9S12C32 microcontroller which lies at its core) are:

- MC9S12C32 MCU, 48 QFP form, with:
  - 32K Byte Flash EEPROM program memory
  - 2K Bytes RAM
  - 8 MHz Internal Bus Operation Default
  - 25 MHz Bus Operation using internal PLL
  - +3.3VDC to +5VDC operation
- 40 pin connector provides access to most MCU I/O signals
- On-board, regulated +5V power supply but also optional power input from and output to the MCU Project board
- 16 MHz Ceramic Resonator
- RS-232 Serial Port w/ DB9 Connector
- 8-Ch, 10-bit, Analog Comparator with full rail-to-rail operation and external trigger capability
- 8-Channel, 16-bit Timer with Input Capture, Output Compare, and PWM capabilities

It has to be mentioned that the CSM12C32 educational module, on which the MC9S12C32 microcontroller is placed, and the MCU Project Board are prebuilt to allow the microcontroller to access some components (4 LEDs, 4 push buttons, the potentiometer, the buzzer etc.) of the MCU Project Board without additional wiring. A few jumper connectors on the hardware part allow enabling and disabling access to these peripherals.

**The Codewarrior Integrated Development Environment**

CodeWarrior Development Studio is a complete integrated Development Environment (IDE) that provides a highly visual and automated framework to accelerate the development of complex embedded applications. It includes editor, debugger, software upload interfaces etc.

Figure 4 presents a screenshot of the CodeWarrior IDE v5.1, for the HC(S)12 family, taken during the coding of one of the wireless communication applications.

Attention is advised when choosing the version of Codewarrior, because there are separate versions for different families of microcontrollers In order to develop software for the MC9S12C32 microcontroller, the author used CodeWarrior v5.1 for the HC(S)12 family. The MCU SLK contains the HC(S)08/ColdFire v1 and HC(S)12 versions of CodeWarrior.

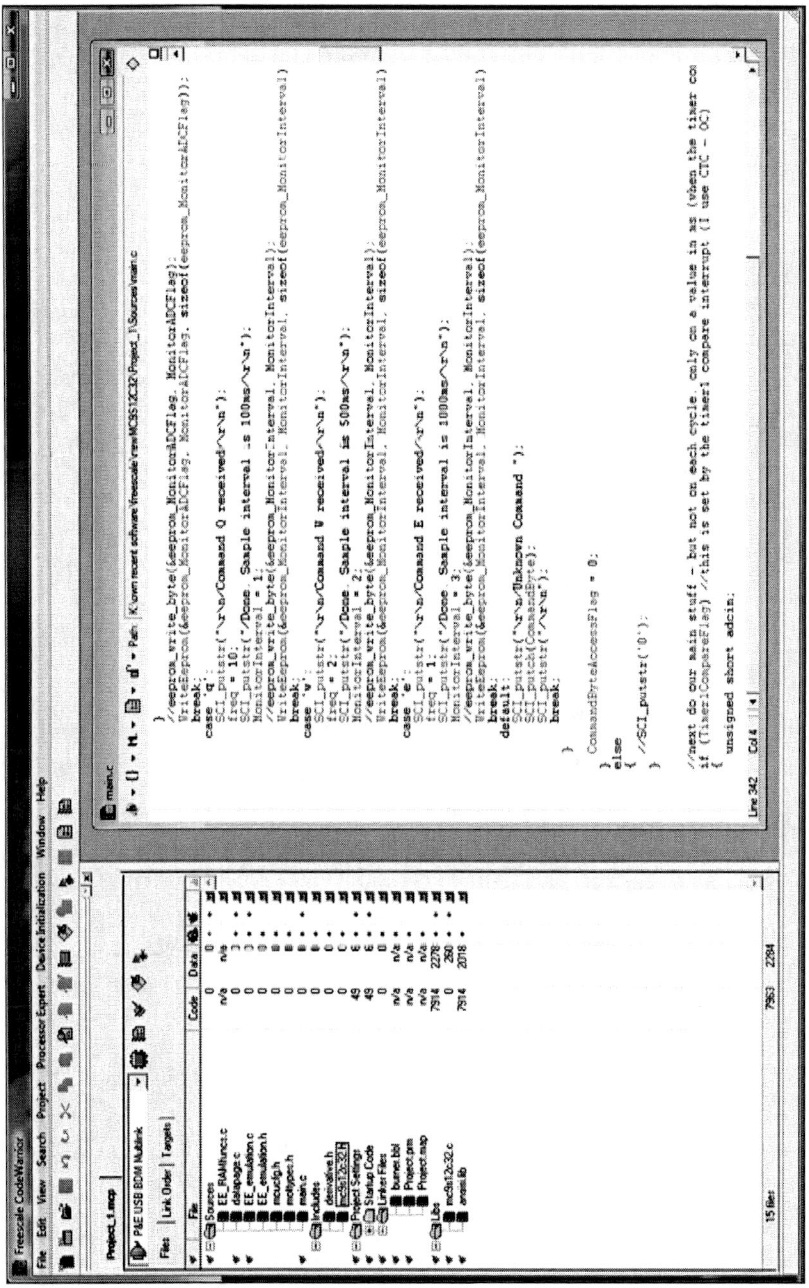

**Figure 4.** CodeWarrior v5.1 (for the HC(S)12 family).

At the time this chapter is being written, Freescale's CodeWarrior Development Studio for Microcontrollers v10.1 just appeared and the good news is that it integrates the development tools for the RS08, HCS08, ColdFire, ColdFire+, Kinetis and MPC56xx architectures into a single product based on the Eclipse open development platform.

### The MC13192U Transceiver Module

Although it is not included by default in the MCU SLK, the MC13192U module (fig. 5) is presented here as well, due to the fact that it has been used in all presented wireless communication applications.

The MC13192U transceiver module is a low cost demonstration and application module for the Freescale MC13192 2.4GHz RF, Zigbee capable or IEEE802.15 transceiver device. Module operation is controlled by any host microcontroller with a Serial Peripheral Interface (SPI) port. A minimum of two of these modules were used to achieve wireless communication.

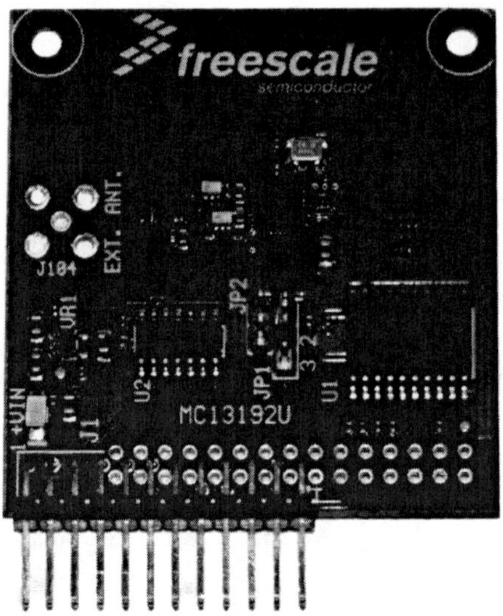


**Figure 5.** The MC13192U module

## Application 1: Data Acquisition for Monitoring

### Background and Motivation

In regard to developing remote monitoring solutions, one of the problems that arose right at the beginning was to achieve the sensor data acquisition and the transmission of this data to a PC. It is an essential part of any monitoring problem, no matter whether one needs to achieve local monitoring or to have a real-time situation report regarding a factory on a mobile phone.

Of course, there are available state-of-the-art solutions. Programmable logic controllers are available from many manufacturers. The reader might be already aware of the high-quality hardware platforms offered by National Instruments, like Compact RIO, and their versatility. NI is following the new trends and now their tools can be used even for developing remote monitoring applications (there is a special browser plugin that helps display and use a LabVIEW application over the Internet and, for some time now, outputting applications to mobile platforms has also been made possible). An application of this technology is presented in (Le Roux et al., 2010) and concerns cooperative remote learning in chemical engineering.

The new myDAQ from National Instruments is a low cost portable data acquisition (DAQ) device that uses LabVIEW based software instruments, allowing students to measure and analyze real-world signals. Combined with LabVIEW on the PC, students can analyze and process acquired signals and control simple processes anytime, anywhere (NI 2011/b).

Tag4M is also worth mentioning: a new device which digitizes the data to send it on to Access Points, where the data is routed to the Internet and a Server IP (CE 2011), thus creating an instrumentation cloud. (Ursutiu et al., 2010) discusses its use in predictive maintenance and machine control. (Folea et al., 2010) present a solar powered implementation to submit data to Pachube, along with the cloud instrumentation concept.

Small manufacturing centers and researchers form a niche that requires a smaller number of these features and might not afford to pay the prices involved with the full set. Because of the reasons enumerated above, and to gain better understanding of the field, the author decided to develop a couple of solutions of his own (including both hardware and software).

The author discovered several products and technologies that promise to be of valuable assistance in developing own solutions. These can be developed either by designing hardware and software from scratch or by adapting and complementing existing solutions to achieve the objectives.

Designing and manufacturing new hardware along with developing new software falls in the first category and the researcher already manufactured a PCB (Printed Circuit Board) with an AVR ATMega8 microcontroller at its core. The embedded software for achieving data acquisition and communication with a PC, as well as the PC software for receiving the data, processing and displaying it in an intuitive form, has been developed as well. It is a simple and cheap solution that functions well.

In regard to adapting existing solutions, using a CSM-12C32 module (from the MCU SLK), a Tag4M or a MyDAQ device would certainly fall in this category. The new Tower System from Freescale Semiconductor is also worth a good look because of its modular approach and the possibility to use the MQX RTOS (Real Time Operating System). The MQX is royalty free and offers also high level functions for the developers. The author has already developed an embedded software application for the Tower System, similar to the one that will be presented in this chapter. Details will be published in the future.

Fig. 6 shows the self-developed PCB, with an ATMega8 core. Fig. 7 presents the TWR-MCF5225X-KIT component of the Tower System, connected for data transmission to the PC (the on-board accelerometer and potentiometer were used for data sampling).

The existing adaptation, for data acquisition, of the CSM-12C32 module (part of the MCU SLK) and the embedded software application for it are presented bellow. A short presentation of the data receiving desktop application is included as well.

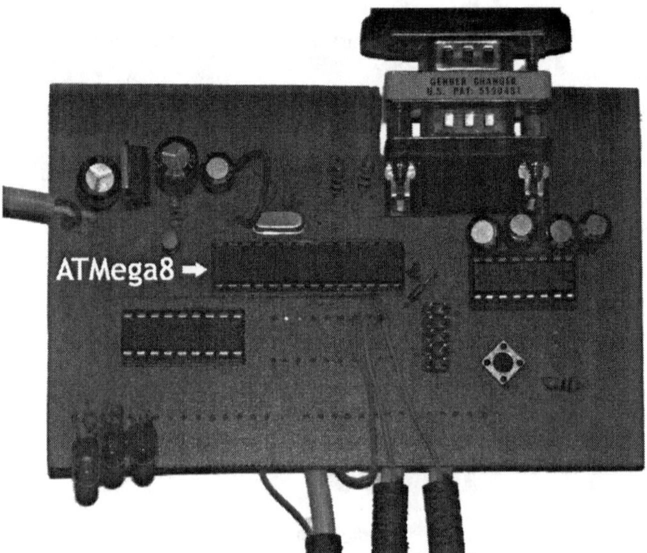


**Figure 6.** Self-developed PCB (ATMega8 MCU).

**Figure 7.** TWR-MCF5225X-KIT.

## The External Remote Monitoring System

To put this application in the right context, fig. 8 presents the whole external (based on sensor information) remote monitoring system that is being developed. The application focuses only on that part of the system where the MCU SLK is involved (sensor data acquisition and communication with the local PC).

## Hardware Setup

In order to avoid soldering and to make an easy testable environment, the MCU Project Board has been used as well. The hardware setup consists of a few steps presented below:

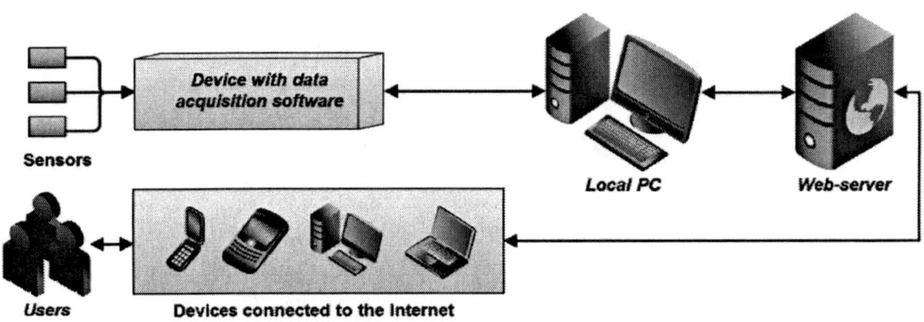


**Figure 8.** External remote monitoring system using the MCU SLK.

- The CSM-12C32 module (with the MC9S12C32 microcontroller) was inserted in the left-most MCU Port.
- Three variable resistors were mounted on the bread-board component of the MCU Project Board in order to obtain a variable signal and emulate sensors. They were wired to the regulated 5V power supply and GND (ground). Using the variable resistors, one obtains an environment that is easy to setup and test.
- The output signals of the three variable resistors were connected to the MCU Port. Wires were inserted into the 18, 20 and 29 holes of the second left-side MCU Port, which are internally connected to the PAD00/AN00, PAD01/AN01 and PAD02/AN02 pins of the MC9S12C32 microcontroller.
- The jumpers at the JP10 selector were checked to be on the connectors, to make sure that the MC9S12C32 microcontroller has direct access to the peripherals mentioned in part 2 (four LEDs, four push buttons, potentiometer and buzzer). For this application, only the connection to the four LEDs is needed, therefore the existence of the jumper at the LED connector of the JP10 selector would have been and is sufficient. To make sure that the on-board potentiometer does not interfere, the POT jumper was removed from the connector (although access would have to be enabled from the software as well).
- A RS-232 to USB converter cable was used to make the connection to the PC. The serial port connector (DB-9 male) end of the cable was attached to the serial connector (DB-9 female) of the CSM-12C32 module. The USB end of the cable was inserted into a USB port of the PC.

Figure 9 shows the MCU Project Board with the CSM12C32 module installed after all these steps have been done.

Other unmentioned steps, such as connecting the power supply to the MCU Project Board and the USB cable from the PC to the USB BDM Port of the MCU Project Board—to enable programming—are general steps required for most development setups of the MCU SLK.

## Software Development

### *Embedded Software for the MC9S12C32 Microcontroller*

Several software application versions have been developed for data acquisition, implementing more and more functionality. (Deaky et al., 2011a) includes a brief description of the functionality included at that time

**Figure 9.** Hardware setup for Application.[1]

and a comparison with similar software developed for the solution based on the ATMega8 microcontroller.

The current version used to program the MC9S12C32 achieves the following:

- Microcontroller port initialization (inputs and outputs definition).
- Initialization of the serial port communications (over SCI).
- Initialization of the timers.
- Initialization of the ADC (Analog to Digital Converter).
- Reading and making the analog to digital conversions of sensor data (10-bit resolution) at certain intervals.
- Bidirectional serial port communication (the MC9S12C32 sends the sampled data at predefined intervals; the user can specify which sensors to query and the sampling interval, by pressing PC keyboard keys)
- Turning 3 LEDs on to communicate that the sensor input is greater than a pre-programmed value (each LED for one variable resistor).

The software system engineering process had the goal to obtain a simple and reliable, permanently functioning system, with decent response time (immediate response is not needed in this case; there is no problem if data is being sampled or if a command is processed several milliseconds later). Portability was also an important requirement and this functionality has already been implemented on the ATMega8—very similar software architecture—and the Tower System—multi-tasking software architecture by using the MQX RTOS (Real-Time Operating System).

Input commands were restricted to one byte (enough to contain one ASCII character—one key press in a terminal on the host PC) to eliminate the need for command processing and allow fast porting of the code. The disadvantage is, of course, that the maximum number of sensors and the possible data sampling frequencies are hard-coded. These were predetermined according to the author's necessities. The code can be easily changed to fit new options.

The data sampling, analog to digital conversions and outgoing serial port communications are done in an infinite loop.

In order to help the system react faster, all serial port input polling and elapsed time checks done in the loop have been removed and interrupts were used instead. The interrupts are used on incoming serial port communication (on receiving data/commands) and for the timer output comparison with a certain value (for sampling data at certain intervals). The interrupts include only minimal code and set flags that are checked in the infinite loop. The infinite loop processes the commands and samples data from the sensors. It acquires and transmits data to the PC only when the timer output compare flag is set.

Figures 10 and 11 present a simplified flowchart of the main function.

The flowcharts in figs. 10 and 11 focus on the most important functions and exclude simple features such as the enabling and stopping of LEDs based on the values of sampled data.

Figure 12 shows the flowchart for the timer output compare interrupt and fig. 13 presents the flowchart for the serial port communication (over SCI) interrupt.

As mentioned previously, the interrupts include only minimal code and leave most of the work for the infinite loop. They only set flags and, the case of the serial port interrupt, put aside incoming data for later processing. This is done to ensure that the task that was executed when the interrupt was triggered is resumed as fast as possible.

### *Desktop Software for the Local PC*

This information is included to introduce the reader to the big picture of the data acquisition system.

The desktop application, currently called Industrial Sensor Monitor, has been presented in detail in (Deaky et al., 2011b). It is being developed to:

- Receive data from the data acquisition board(s) over the PC's serial port.

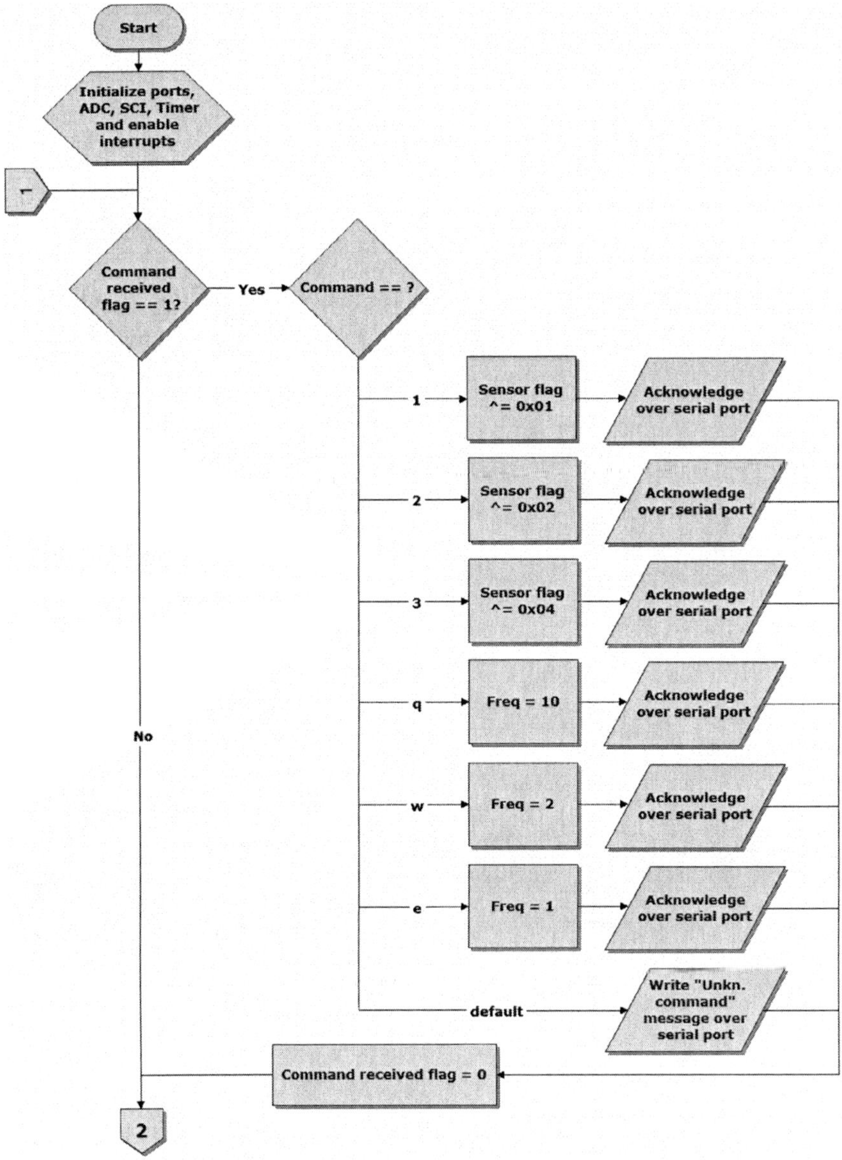


**Figure 10.** MC9S12C32 main function flowchart (part 1).

- Process the received messages.
- Display the sensor readings to the user in a more intuitive way.
- Enable the user to send commands to the data acquisition board (regarding which sensors to sample data from and the sampling frequency).

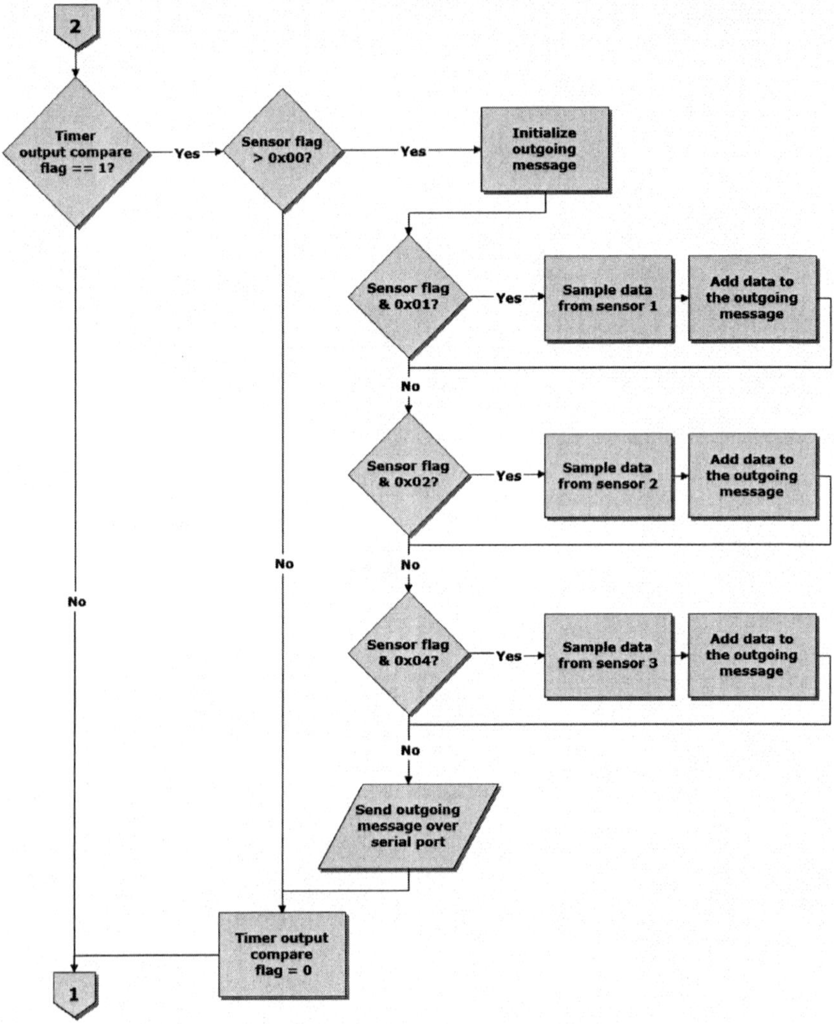


**Figure 11.** MC9S12C32 main function flowchart (part 2).

- Save the data locally for logging and charting purposes (recent addition).
- Forward the data to the web-server and accept commands from it, for remote monitoring and control (implementation is pending).

Figure 14 presents the GUI (Graphic User Interface) of the latest version.

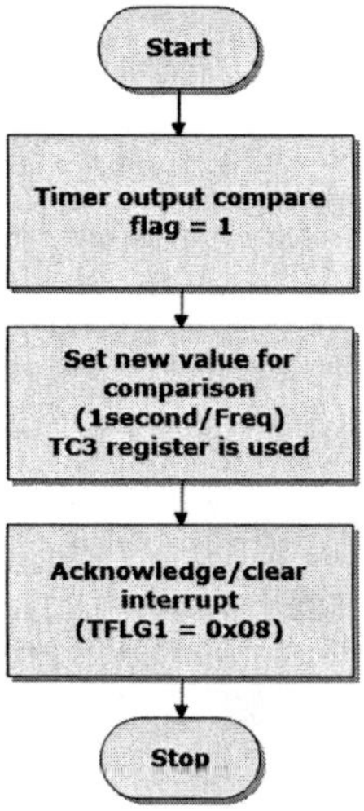


**Figure 12.** Timer output compare interrupt flowchart.

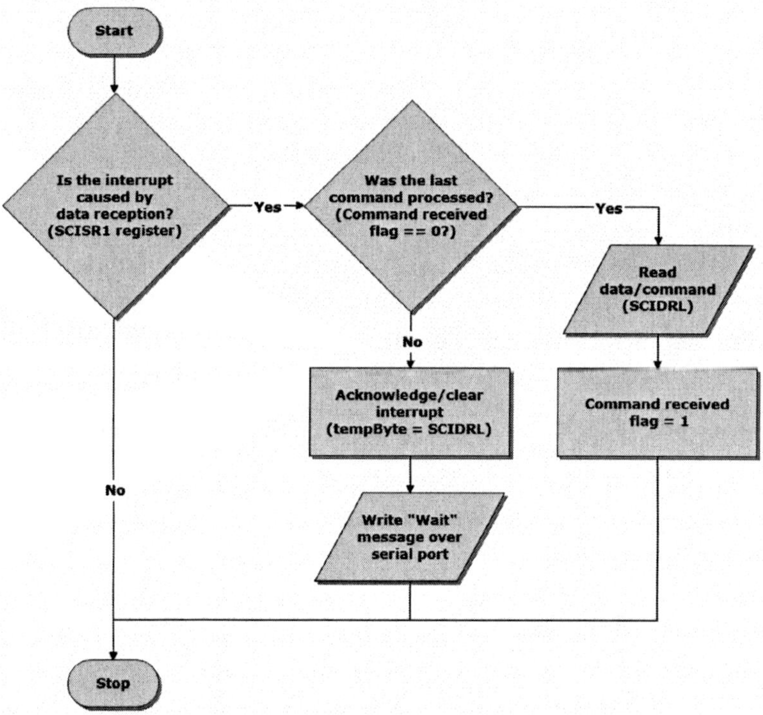


**Figure 13.** Serial port communication interrupt flowchart.

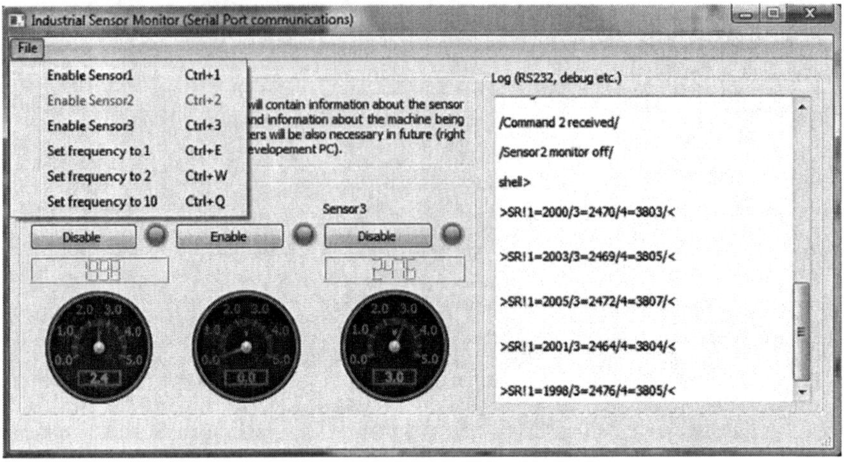


**Figure 14.** Industrial Sensor Monitor desktop application.

### Other Considerations and Information

As the reader may notice by looking at fig. 14, the encoding protocol for transmitting the sensor data (also used to form the outgoing message composed by the MC9S12C32 microcontroller in fig. 11) is:

>SR!sensor_number_1=sensor_reading_1/
sensor_number_2=sensor_reading_2/.../< (1)

In practice, readings may look like

>SR!1=0/3=1023/4=501/< (2)

The Industrial Sensor Monitor application interprets these messages to show the values read acquired from the sensors in a more meaningful way (by means digital display and gauge widgets).

Of course, the interpretation of the readings depends on the ADC reference voltage, the ADC resolution, existing signal amplification and others.

## Application 2: Using the Data Acquisition System to Monitor the Output of a Renewable Energy Source

### Background and Motivation

In the new data acquisition solution development process, some of the most important factors were the cost-awareness—(Juca et al., 2011) also takes into account this aspect when presenting a data acquisition solution for decentralized renewable energy plants—and the usage of state-of-the-art technologies. Recently, needed attention was directed towards the energy consumption as well. This issue is being addressed in both hardware and software design and, by using renewable energy sources, additional advantages will be gained (reliable remote monitoring for isolated locations, independence of existing power socket location—with a portable energy source, reducing the non-renewable energy consumption of the monitoring grid etc.).

The work that did lead to the solution presented in this part of the chapter actually started as an experiment done to use a cheap solar energy source for powering the existing data acquisition systems. However, it became rapidly clear that the cheap solar energy source was not suitable. Because of this, there was a turnover and, from powering a data acquisition board with a solar energy source, the author ended up monitoring the solar energy source with the board that it was supposed to power.

The MCU SLK looked like the most convenient tool to make a fast implementation and was used to develop the solution. The data acquisition software presented in the previous part was adapted to fit the new scenario.

The presented solution will be improved and used to monitor and estimate the output of photovoltaic or other renewable energy sources before committing to their usage. It was also a convenient scenario to test the data acquisition solution developed with the MCU SLK for longer periods. More details and actual results of the monitoring process will be published in a future paper.

**The Cheap Solar Energy Source**

The author considered a combination of solar (photovoltaic) cells and (rechargeable) batteries to be needed, in order to power a data acquisition board both at daytime and nighttime and in order to be able to supply an approximately constant voltage and current.

To make a cheap solution fast and because such products already include a solar cell—battery system, the author decided to try to use parts from solar powered garden lamps. A professional (and more expensive) solution, which ensures daylight functioning of a Tag4M board, achieved by using better photovoltaic cells and capacitors, is presented in (Folea et al., 2010).

In our case, which is a common one for garden lights, the included battery's nominal voltage is supposed to be around 1,2 V. Because approx. 5 V are needed to drive the data acquisition boards (if we bypass voltage regulator circuitry), four cell/battery packs had to be connected in series (see fig. 15) thus summing the battery voltages.

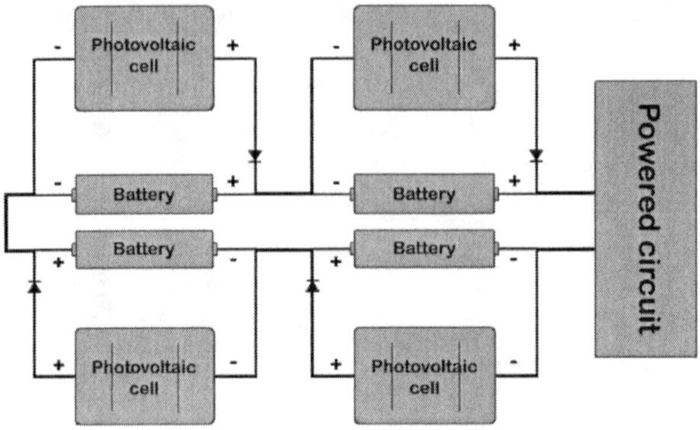


**Figure 15.** Four-pack solar energy source for powering the data acquisition boards.

However, when one of the data acquisition boards has been connected to this energy source, it did only work for a few minutes. This seemed to point out that the energy source was not powerful enough but the author needed to be sure. This clearly showed that a solution to monitor a sample source, before committing to using it, would be beneficial and the development of such a solution, for energy source output estimation, has been started.

## Hardware Setup

Wires were attached to the single photovoltaic cell—battery circuit and connected to ADC (Analog-to-Digital Converter) inputs of the MC9S12C32 microcontroller. Wires were either directly connected to the MCU Port on the MCU project Board or to the bread-board prototyping component in order to allow for additional circuitry (ex: a voltage divider). Figure 16 shows where the wires were attached in the photovoltaic cell—battery circuit.

As one can see, voltage generated by the Photovoltaic cell is measured on channel 0 of the ADC (pin PAD00/AN00, accessed through MCU Port–18) and voltage at the battery's "+" end is measured on channel 1 of the ADC (pin PAD01/AN01, accessed through MCU Port–20). The connection to the ground wires of the board is mandatory in order to have common ground for all circuits and to be able to make the measurements. Later on, circuitry to measure the electrical current was added to the same single photovoltaic cell—battery circuit. A resistor of a known value (a 10 Ω metallic resistor, with a 1% value precision) was installed up-stream (near the cell, before any powered circuit) and the voltage that dropped on it was monitored. In order to do this, two wires were soldered, one before the resistor and one after it. The wires were connected to channel 3 (pin PAD03/AN03, accessed through MCU Port–31), respectively 4 (pin PAD04/AN04, accessed through MCU Port–33), of the ADC (fig. 16). The current that flows through the circuit is computed easily, with the equation:

$$I = (U_{ADC3} - U_{ADC4})/\mathrm{R} = (U_{ADC3} - U_{ADC4})/10 \qquad (3)$$

As the reader might suspect, the wire connected to ADC3, soldered closer to the resistor than the wire connected to ADC0, was added only to make sure that the voltage drop measurements are as precise as possible. The later measurements showed that the difference between the two readings is indeed undetectable at the available resolution, thus the wire to ADC3 could be removed.

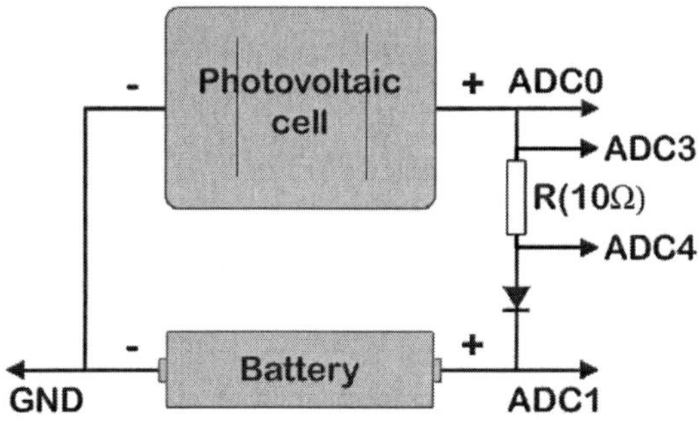


**Figure 16.** Single pack wirings for current measurements

When connecting the four-pack solar energy source, no extra soldering was needed, since the "+" wire was directly connected to channel 2 of the ADC (pin PAD02/AN02, accessed through MCU Port–29)and the "-" wire to the common ground (GND). However, because the actual voltage of the 4-pack source became greater than 5 V and the analog-to-digital convertor's reference voltage (and maximum measurable voltage) is 5 V by default, a voltage divider circuit was made, on the MCU Project Board, and used to halve the measured voltage.

Of course, the RS-232 to USB cable was used again, to help transmit data to the PC over the serial port.

## Software Development

### *Embedded Software for the MC9S12C32 Microcontroller*

The existing embedded data acquisition software application (on the MC9S12C32 microcontroller) was modified to sample data over the five ADC channels at every ten seconds. It was also modified in order to completely shut down the ADC circuitry between taking samples to avoid draining too much power from the battery and make sure that the monitoring process is correct.

Regarding the minimal precision of the measurements, affected by the 10-bit resolution of the ADC (which means that measured analog values are converted into numbers contained in the 0–1023 interval) and the reference voltage of 5 V (considered also as the maximum possible reading), the following can be stated:

- The voltage measurement precision is $5\ \text{V}/1024 = 0.00488\ \text{V} = 4{,}88\ \text{mV}$.
- The current measurement precision (considering the $10\ \Omega$ resistor) is $4{,}88\ \text{mV}/10\ \Omega = 0{,}488\ \text{mA}$.

### *Desktop Software for the Local PC*

The Industrial Sensor Monitor was modified in order to add the data logging capabilities using the CSV (Comma Separated Values) format. The logged data has been further processed and diagrams were created with MS Excel. Future development of the Industrial Sensor Monitor might include real-time charting.

The monitoring system was used to draw diagrams containing the variation of the measured values during different parts of the day. Fig. 17 shows a file containing measurements (in CSV format). Fig. 18 shows a chart of the temporal variation of the electric current that flows from the photovoltaic cell into the battery. Microsoft Excel was used to import the data, process it and generate charts.

## Other Considerations and Information

After monitoring the photovoltaic energy source for a few days, it was confirmed that it could not possibly deliver enough power to the data acquisition boards (the highest constantly measured current was 12mA). This solution will be improved and applied in the future to test any photovoltaic cells (or other renewable energy sources with similar output) before committing to using them as an energy source to power the data acquisition boards. The lessons learned in regard to making a photovoltaic power source to power data acquisition boards will be published in the

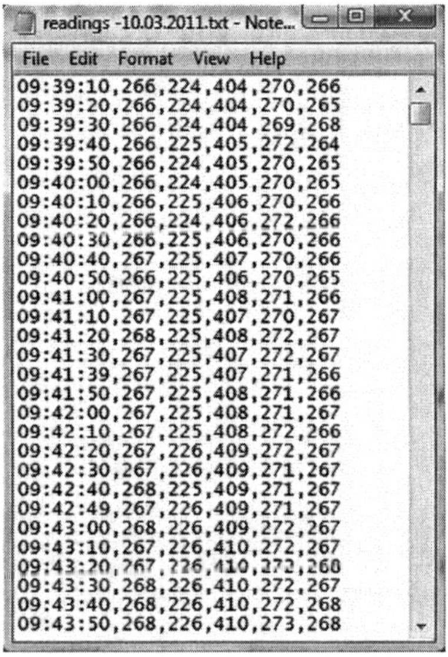

readings -10.03.2011.txt - Note...

File Edit Format View Help

```
09:39:10,266,224,404,270,266
09:39:20,266,224,404,270,265
09:39:30,266,224,404,269,268
09:39:40,266,225,405,272,264
09:39:50,266,224,405,270,265
09:40:00,266,224,405,270,265
09:40:10,266,225,406,270,266
09:40:20,266,224,406,272,266
09:40:30,266,225,406,270,266
09:40:40,267,225,407,270,266
09:40:50,266,225,406,270,265
09:41:00,267,225,408,271,266
09:41:10,267,225,407,270,267
09:41:20,268,225,408,272,267
09:41:30,267,225,407,272,267
09:41:39,267,225,407,271,266
09:41:50,267,225,408,271,266
09:42:00,267,225,408,271,267
09:42:10,267,225,408,272,266
09:42:20,267,226,409,272,267
09:42:30,267,226,409,271,267
09:42:40,268,225,409,271,267
09:42:49,267,226,409,271,267
09:43:00,268,226,409,272,267
09:43:10,267,226,410,272,267
09:43:20,267,226,410,272,268
09:43:30,268,226,410,272,267
09:43:40,268,226,410,272,268
09:43:50,268,226,410,273,268
```

**Figure 17.** Measurements saved in CSV format

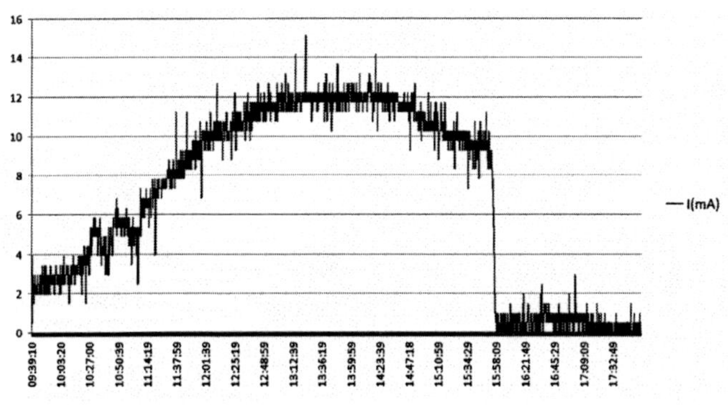


**Figure 18.** The monitored temporal variation of the electric current.

future. This part of the chapter only focused on briefly presenting how the MCU SLK was used to develop a solution to monitor and estimate the output of a photovoltaic energy source.

## General Considerations for Wireless Application Development with the MCU SLK

### General Background and Motivation

While the initial telemanufacturing research objectives concern the monitoring of only one machine, the system will obviously have to be extended to several machines. For this scenario, it is a good idea to design a solution involving more than one data acquisition boards (each monitoring at least one machine) and also a local centralization point where data can be displayed for a user and routed to the Internet. At least in this case, it should prove very beneficial to employ wireless transmission of the data acquired from sensors.

From the examined solutions, Tag4M does the wireless transmission job well from the start, because it was designed to be a part of an instrumentation cloud. It uses a wireless connection to a local wireless access point to send acquired sensor data to a web-server. In turn, the web-server generates the HTTP page with the received readings. However, this strategy is not meant to display data offline and it is preferred to also have local access to centralized data. The author admits that a local web server and a more complex web application might do the job.

Another attractive solution consists of the ZSTAR development tool from Freescale Semiconductor. A ZSTAR3 system can accommodate up to 16 accelerometer transmitter boards, connected through RF ZigBee 2.4GHz communication to a single USB node connected to a PC (Freescale 2011b). While it is a great solution, and very similar to what is desired, it only includes accelerometer information. The final design is similar to the concept behind ZSTAR3, but each transmission will include information sampled from more sensors connected to independent data acquisition boards.

Electromagnetic fields might be measured as well and, as stated in (Severdaks et al., 2011), choosing wireless communication should minimize the interference with the field that is measured.

Because of the reasons enumerated above, experiments with wireless communication have been started. The author had been using the MCU SLK previously with very good results (see the previous parts of this chapter) and, given the fact that also the necessary MC13192U ZigBee wireless communication modules were already available, it looked like the right platform to use in order to learn about and experiment with wireless communication.

This chapter includes information about initial experiments and the wireless applications that have been developed with the MCU SLK up to this date. It organizes some of the previously published information in (Deaky et al., 2011c) and brings new information and insight.

In the future, if the data acquisition network will grow and spread so much that relays will become necessary, it might prove very useful to use new relay placement algorithms like the Efficient Relay Sensor Placing Algorithm (ERSPA) presented in (Chang 2011). The author's experience in multicriterial optimization as well as Banica's genetic optimization algorithms mentioned in (Chira et al., 2009) will also prove helpful. Network design should also take into account the energy consumption. For now, this is addressed only primitively by using technology based on the ZigBee specification, but a good source of information for this purpose are papers like (Lloret et al., 2010) which address network and protocol design in a more complex manner, for energy saving.

**SMAC (Simple Media Access Controller)**

The SMAC, version 4.1a, was used to achieve wireless communication with the used version of the MCU SLK. The Freescale SMAC is a simple ANSI C based code stack available as sample source code. The SMAC is used for developing proprietary RF transceiver applications using a Freescale 802.15.4 transceiver. The SMAC was built to work with an HCS08 based MCU with an SPI, but it can easily be adapted to almost any processor core.

The SMAC is incorporated into the Freescale BeeKit Wireless Connectivity Toolkit. As stated by Freescale Semiconductor, the incorporation of SMAC into BeeKit makes it easier for users to employ and customize SMAC. However, please be aware that the BeeKit does not support the CSM12C32 module.

The first wireless communication experiments consisted on building sample source code and programming the two microcontrollers involved.

Attention: They all failed and the debugging process lead to the cause: the included SMAC version was not the needed one. The author obtained the proper version of SMAC after contacting directly Freescale Semiconductor.

**General Hardware Setup**

In order to avoid soldering and to make an easy to change environment, the MCU Project Board has been used here as well. The general hardware setup consists of a few steps presented below:

- The CSM-12C32 module (with the MC9S12C32 microcontroller) was inserted in the left-most MCU Port of the MCU Project Board (fig. 19).
- The MC13192U transceiver module (fig. 5) was inserted into the right-most MCU Port of the MCU Project Board (fig. 20).
- A RS-232 to USB converter cable was used to make the connection to the PC. The serial port connector (DB-9 male) end of the cable was attached to the serial connector (DB-9 female) of the CSM-12C32 module. The USB end of the cable was inserted into a USB port of the PC. If not needed for the application itself, the cable is still

**Figure 19.** General hardware setup—CSM-12C32 module.

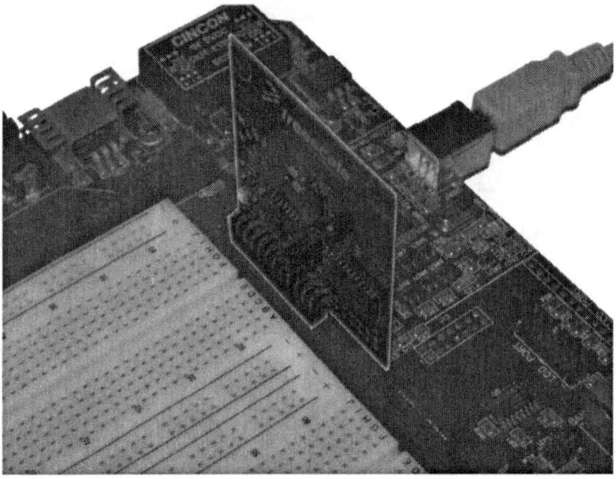

**Figure 20.** General hardware setup—MC13192U module.

always useful for displaying debug messages and testing the applications.

Again, the other unmentioned steps, such as connecting the power supply to the MCU Project Board and the USB cable from the PC to the USB BDM Port of the MCU Project Board—to enable programming—are general steps required for most setups of the MCU SLK.

The above general hardware setup is common to all presented applications and also to the transmitter and the receiver boards.

## Application 3: Wireless Transmission and Reception of the Keyboard Keys Pressed Between Two PCs

### Application Specific Background and Motivation

This first application's purpose was to test hardware functionality. It was developed by modifying sample source code obtained from Freescale Semiconductor.

Hardware testing was achieved by implementing sufficient functionality in order to use wireless transmission and reception of keyboard strokes between two personal computers.

Figure 21 presents the system structure for this application. All connections are bidirectional. In the author's case, the wired connections are materialized by RS-232 to USB converter cables. The wireless connection is made by using the MC13192U transceiver modules.

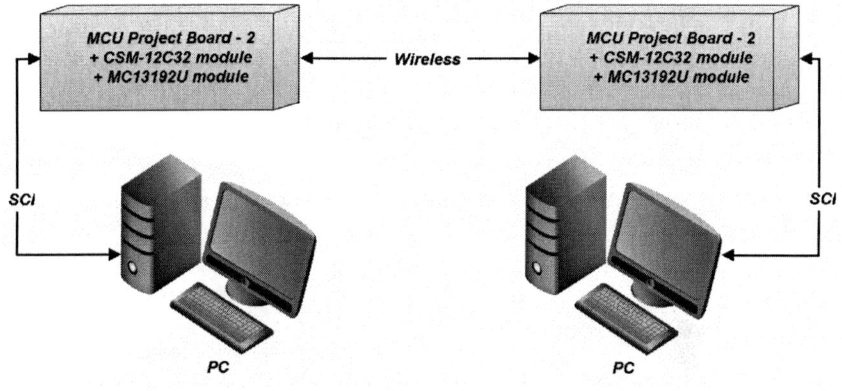


**Figure 21.** System diagram for the wireless transmission of the keyboard keys pressed between 2 PCs.

### Application Specific Hardware Setup

Both boards use the general hardware setup described in the previous part.

The only additional step taken was to make sure that the proper jumpers at the JP10 selector are on the connectors, to make sure that the MC9S12C32 microcontroller has direct access to the default connected MCU Project Board peripherals, as mentioned in part 2. For this application, only the connections to the LEDs and the push buttons are needed.

## Software Development

### *Embedded Software for the MC9S12C32 Microcontroller*

The same software application is used for both boards. To achieve the goals, the developed application does the following:

- Microcontroller port initialization (inputs and outputs definition).
- Initialization of the MC13192 transceiver (needs packet initialization, power level setting, clock source settings etc.).
- Initialization of the serial port communications (over SCI).
- Detection of 1-byte long serial port (RS-232) transmissions from the computer (when the user inputs a character in the serial port terminal by pressing a key) and reading the byte.
- Using the transceiver module to wirelessly transmit the byte read from the serial port (timeout and acknowledgment exist).
- Detecting the 1-byte wireless transmissions from the remote board (using the MC13192U transceiver) and reading the received byte.
- Transmitting the byte received by the transceiver, to the computer, by means of the serial port.
- Transmitting the acknowledgment.

The latest version of this application added new functionality, to help test the wireless communication even without any PC. The initial version was enhanced by adding peripheral access to the MCU Project Board's resources. This allowed the user to light LEDs on the remote board with different actions. In the end, each of the boards has the same application on it, to achieve the following:

- Turning on and off LED's on the MCU Project Board when certain bytes are received (LED 1 when receiving the '1' character and LED2 when receiving character '2').
- Transmitting a certain character wirelessly when a certain push button on the MCU Project Board is pressed (it transmits the '1' character when PB1 is pressed and the '2' character when PB2 is pressed).

Obviously, the result is that the user can turn on or off LED1 or LED2 on the remote board by pressing PB1 or PB2 on the local board.

### *Desktop Software for the Local PC*

Only embedded software for the MC9S12C32 microcontroller was developed. Any serial port terminal will do the job on the PC.

### Other Considerations and Information

Although the software seems to implement only 1-byte message transmissions, a 3-byte acknowledgement message is used internally to verify that the remote board received the messages. One can read more about it in (Deaky et al., 2011c).

## Application 4: Wireless Control of a Remote DC Motor

### Application Specific Background and Motivation

This application's purpose was to exercise remote control.

The first attempts to develop it involved modifying sample source code obtained from Freescale Semiconductor. These attempts were unsuccessful due to the improper version of the included SMAC. To make it work, its functionality was added on top of the first successfully developed application presented in the previous part. It was considered that it is easier to build on top of a tested and working SMAC implementation than to replace the SMAC code and recode the version which did not work.

Figure 22 presents the system structure for this application. All connections are unidirectional. The local board uses the ADC to read values from the potentiometer and transmits the values wirelessly to the remote board which, in turn, uses PWM (Pulse Width Modulation) to act on the connected direct current motor to make it spin with variable speed (which depends on the value sampled from the potentiometer on the local board).

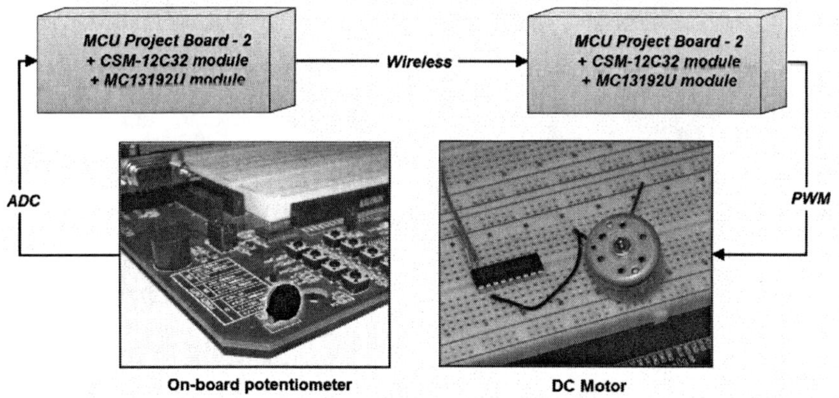


**Figure 22.** System diagram for the wireless control of a remote DC motor.

## Application Specific Hardware Setup

As the reader can see, the two boards have clear separate functionalities. The local board is used for data acquisition and transmission, while the remote board is used for data reception and DC motor control.

Both boards use and extend the general hardware setup described in the previous part. The serial port connections were kept only for debugging purposes.

An additional step taken for the local MCU Project Board, used for data acquisition, was to make sure that the POT and LED jumpers at the JP10 selector are on the connectors, to make sure that the MC9S12C32 microcontroller has direct access to the potentiometer and LEDs of the MCU Project Board.

The potentiometer is only needed on the local/data acquisition board, to be able to control the DC motor connected to the remote board. Access to the LEDs is necessary on both boards.

The MC9S12C32 microcontroller pins that were used to acquire data and drive the motor are:

- For the local data acquisition board, the potentiometer is directly connected to the PAD01/AN01 pin (ADC channel 1, through MCU Port—20) (it works only when the jumper is on the POT connector of selector JP10). It can be, however, connected with wires to any other ADC input available through the MCU Ports.
- For the remote DC motor control board, pin PT2/PW2/IOC2 (PWM channel 2, through MCU Port—30) was used to emit the signal that controls the motor.

While some motors could be directly driven with the PWM signal, the author decided to use a ULN2803 sink driver IC (Integrated Circuit) to drive the motor and to act as an interface between the low logic level circuitry and eventual higher voltage or current needs of the DC motor. This allows to use DC motors which need higher voltage than the 5 V that the MC9S12C32 can provide and also, very important, protects the microcontroller (ULN2803A outputs use internal clamp diodes for transient voltage suppression to protect low-power microcontroller circuitry).

## Software Development

### *Embedded Software for the MC9S12C32 Microcontroller*

This time, different functionality is needed for the two boards and, therefore, each board needs its own different software application. However, following the way it was done in the initial sample code received from Freescale, the author managed to use the same source code files for both applications. To differentiate between the two applications, a constant, named TRANSMITTER, was used.

Any of the two applications can be compiled from the same source code by settings the constant to the appropriate value (1 for the data acquisition and transmitter board and 0 for the receiving and motor driving board). For a better understanding, figure 23 is included. The flowchart also illustrates how the same source code is used to generate the two different applications.

The functionality of the two software applications is described in what follows.

The software on the data acquisition and transmitter board performs:

- Microcontroller port initialization (inputs and outputs definition) and MC13192 transceiver initialization (needs packet initialization,

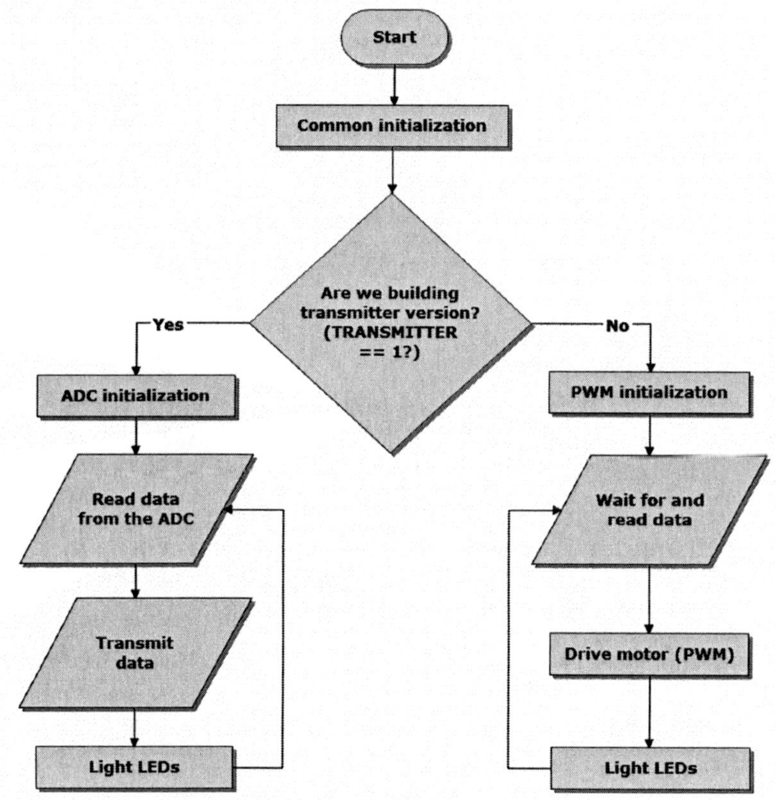


**Figure 23.** Simplified source code flowchart for Application 4.

power level setting, clock source settings etc.—just like in the previously presented application).

- Initialization of the serial port communications (over SCI).
- Initialization of the ADC. In order to have 1-byte message communications, the author initialized the ADC with a resolution of only 8-bit. 10-bit resolution can be used, but it forces the message size to increase to 2 bytes and requires processing on the receiving board.
- An endless loop where:
  - o It samples data from the ADCs channel 1 (to which the on-board potentiometer is connected).
  - o It transmits the data over the serial port, for testing and debugging purposes.
  - o It transmits the data wirelessly, using the module.
  - o It uses LEDs to signal whether the transmissions are successful and to give a visual clue about the transmitted value (turns a LED on if the value is greater than 127).

The software on the receiving and motor driving board performs:

- Microcontroller port initialization (inputs and outputs definition) and MC13192 transceiver initialization.
- Initialization of the serial port communications (over SCI).
- It initializes channel 2 of the PWM (Pulse Width Modulation) module.
- An endless loop where:
  - o It waits for wireless data.
  - o It detects incoming wireless transmission and reads the 1-byte message (using global variables that signal and store the message).
  - o It transmits the received data over the serial port, for testing and debugging purposes.
  - o It writes the received value to PWMDTY2 (PWN channel duty register 2–8-bit value) and thus drives the motor at the speed dictated by the received value.
  - o It flashes LEDs to signal when data is received.

Please note that the actual incoming wireless transmission detection and message processing is done in a function called MCPSDataIndication that is internally called when a message has been received. The function assigns the content of the message to a global variable used in the endless loop.

#### *Desktop Software for the Local PC*

Only embedded software for the MC9S12C32 microcontroller was developed. Desktop software is necessary only for testing the application and any serial port terminal will do the job.

### Other Considerations and Information

In order to speed up data transmission times for a first implementation, the acknowledgment protocol was not used for this application. The effect of turning the potentiometer on the local board can easily be observed on the remote board (visual and audible feedback from the motor).

For a future, more complex and robust, application, the author considers adding at least:

- Fast acknowledgment protocol.
- Automatic stopping of the motor if no message is received for a predetermined period.

## Application 5: Wireless Communication of Data Acquired By One or More Data Acquisition Boards

### Application Specific Background and Motivation

Although they can be useful on their own, the two applications introduced in the previous parts were especially developed to gain a fair understanding of the hardware and software components and of wireless communication in general.

After they have been completed, the development of a new application, better related to the telemanufacturing/remote monitoring objectives, could finally be started. The aim of this application is to achieve wireless transmission of the data acquired from sensors. The first tests have already been concluded and included only two boards.

Figure 24 presents the system structure for the wireless transmission of data acquired from sensors using two boards. One of the boards uses the ADC to acquire data from the sensors and transmits this data wirelessly. The other board receives the data and forwards it to a PC (over the serial port). The white ended arrows symbolize the intention to add the ability to send commands to the data acquisition and transmission board in the future.

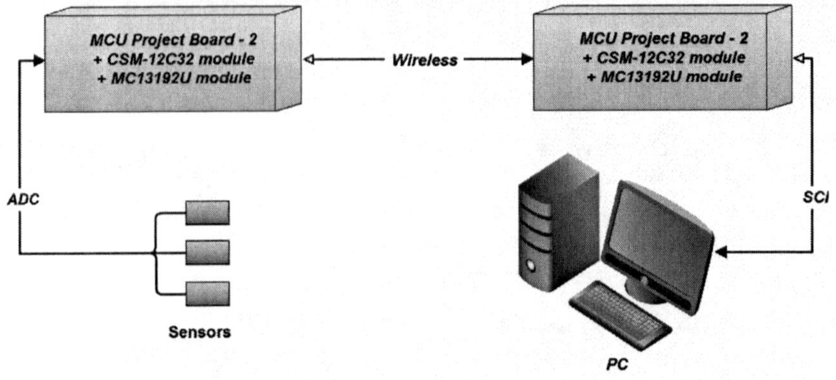


**Figure 24.** System diagram for the wireless transmission of sensor data.

## Application Specific Hardware Setup

The two boards have clear separate functionalities.

Both boards use and extend the general hardware setup described in the previous part. On the data acquisition and transmission board, the serial port connection was kept only for debugging purposes.

An additional step taken for both transmitter and receiver MCU Project Boards was to make sure that the LED jumper at the JP10 selector is on the connector, to make sure that the MC9S12C32 microcontroller has direct access to the LEDs on the MCU Project Board.

The data acquisition and transmission board makes use of the same variable resistors as in Application 1, to emulate analog sensors. However, because of the presence of the MC13192U module (which is inserted into the upper 28 locations of the MCU Port), they had to be connected lower on the MCU Port. For this application the MC9S12C32 microcontroller pins that were used to acquire data are:

- PAD02/AN02 pin (ADC channel 2, through MCU Port–29)
- PAD03/AN03 pin (ADC channel 3, through MCU Port–31)
- PAD04/AN04 pin (ADC channel 4, through MCU Port–33)

No additional setup is needed for the board that receives the wireless messages.

## Software Development

### *Embedded Software for the MC9S12C32 Microcontroller*

Again, different functionality is needed for the two boards and, therefore, each board needs its own different software application. The same source code structure as in the previous application has been used, to generate the two different embedded software versions. The source code flowchart is presented in fig. 25 and resembles fig. 23, except for the PWM functionality which is replaced by functionality used to forward the received data to the PC over the serial port.

The functionality of the two software applications is described below.

The software on the data acquisition and transmitter board is very similar to the software presented in Application 1 (the data acquisition application). It has the same functionality, except for the facts that it now transmits the data wirelessly instead of using the serial port (actually, serial port transmissions have been kept in place but only for debugging purposes) and that it cannot receive commands. The data acquisition and message composition functionality was added on top of a working SMAC implementation. Also, the data acquisition application now features

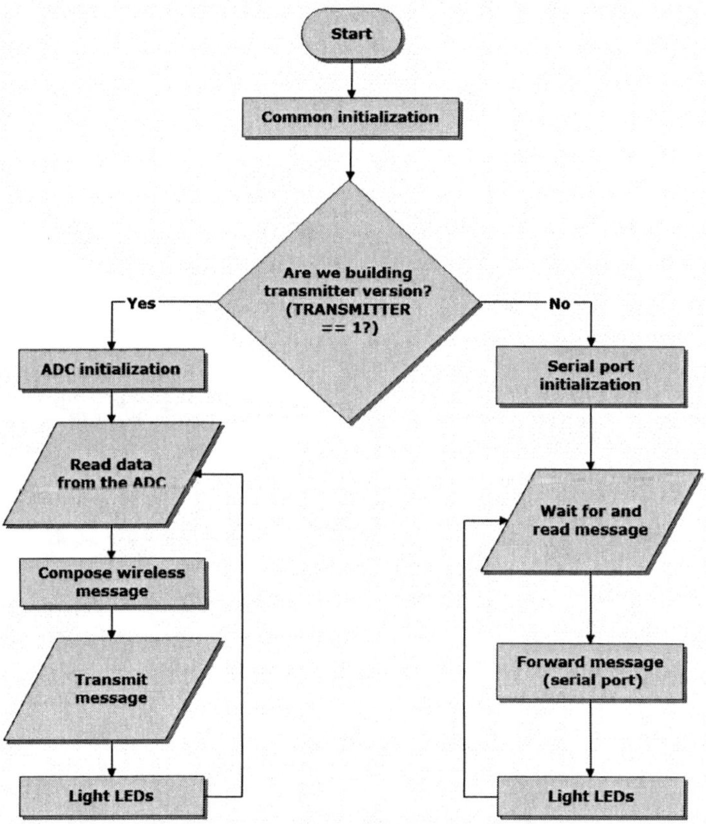


**Figure 25.** Simplified source code flowchart for Application 5.

multiple byte long wireless message transmissions to send all sensor readings at once.

The software on the data receiving and forwarding board performs:

- Microcontroller port initialization (inputs and outputs definition) and MC13192 transceiver initialization.
- Initialization of the serial port communications (over SCI).
- An endless loop where:
    - o It waits for wireless data.
    - o It detects incoming wireless transmission and reads the message (using the global variable mentioned below).
    - o It transmits the received data over the serial port, to the PC.
    - o It flashes LEDs to signal when data is received.

Also here, please note that the actual incoming wireless transmission detection and message processing is done in a function called MCPSDataIndication that is internally called by SMAC when a message has been received. The function assigns the content of the message to a global variable accessed in the endless loop.

### *Desktop Software for the Local PC*

The Industrial Sensor Monitor desktop application for the local PC works well in the two board scenario and can be adapted for multiple data acquisition and transmission boards.

## Other Considerations and Information

At this time, the connection is unidirectional (the data acquisition board reads the sensor data, forwards it to the receiver board which sends it to the PC). However, future development will allow the user to send commands to the data acquisition boards (using wireless communication) to enable/disable sensors, set sampling intervals etc. (this is already possible on the existing developed solutions that use wired connections).

The final objective is, of course, to monitor more than one manufacturing machine. Therefore, more data acquisition boards will exist and only one receiving board which centralizes the information and forwards it to the supervisor PC (see fig. 26).

The message encoding protocol will have to accommodate transmitter board identification and, in future, the messages may look like

>SR!B=3!1=0/3=1023/4=501/<, (4)

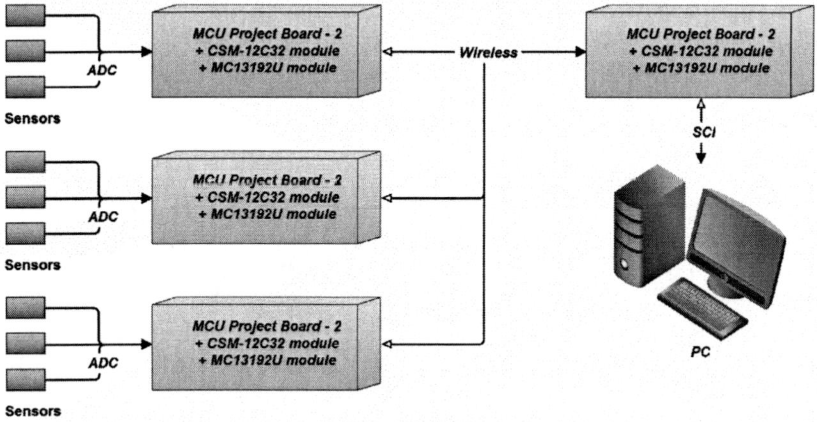


**Figure 26.** System diagram for the wireless transmission of sensor data acquired by multiple boards.

where B=3 identifies the transmitter board. Different messages and their encoding protocols are planned to be developed in the future.

In the end, the PC will also forward the data to a web-server, to achieve the remote monitoring of several machines (fig. 27).

## Conclusions

The embedded system applications, developed with the MCU SLK and presented in this chapter, handle important aspects of a practical monitoring paradigm, from data acquisition to wireless data transmission.

Learning to program the MC9S12C32 microcontroller might seem hard at first (there is a considerable amount of documentation to be studied) but the process leads to a good understanding of microcontrollers and embedded architectures. To help the reader and to signal eventual time-consuming mistakes, this chapter included hardware setup details, software development hints and flowcharts.

The author also used the MCU SLK to prototype different circuits (voltage regulator, LED controllers, voltage dividers etc.) and to test and get accustomed to several kinds of sensors (potentiometers, photoresistors, temperature sensors etc.). An extended educational usage of the MCU SLK, to help in learning and teaching other microcontroller architectures, was elaborated and published in (Deaky et al., 2011d). It involves a specific manner of using the circuit prototyping area to place different kinds of microcontrollers and common usage of peripherals. Microcontrollers with other packaging types than DIP (Dual In-line Package) can be used as

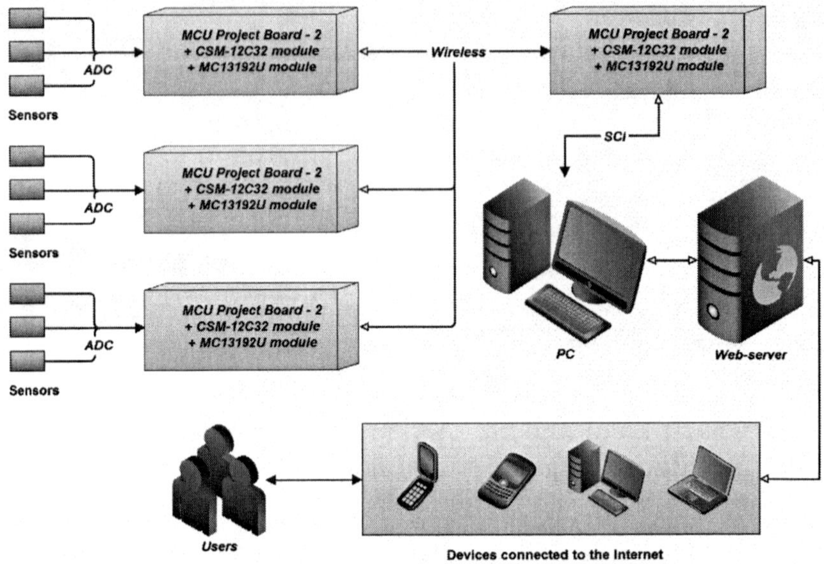


**Figure 27.** System diagram for a telemonitoring system that employs the wireless transmission of sensor data acquired by multiple boards.

well, by means of DIP adapters. This educational method helps reducing experiment setup times and costs and increases the ability to get familiar with different hardware implementations and architectures.

Some modules that extend the MCU SLK will be used as electronic parts of a new product and casing will be prototyped using the research department's existing expertise, exhibited by papers like (Udroiu and Mihail 2009), and the rapid prototyping machines that exist at the department.

Besides the educational value it brings, the aid that the MCU SLK offers in research and development is not to be neglected, as shown by the applications presented in this chapter. Some of these applications are being prepared to be on permanent display at the Creativity Center of the Transilvania University of Brasov.

This chapter also introduced the reader to a part of the author's work in the teleengineering field (as the context of the embedded system applications). The field has tough interdisciplinary requirements (but can be very rewarding) and the MCU SLK proved very useful to learn about and develop needed embedded solutions. The future R&D directions, for this field, involve application development for other embedded system platforms (including embedded HTTP servers) and implementations of the remote monitoring and the remote control of manufacturing systems.

## Acknowledgments

This paper is supported by the Sectoral Operational Programme Human Resources Development (SOP HRD), financed from the European Social Fund and by the Romanian Government under the project number POSDRU/89/1.5/S/59323.

The author would like to thank the following:

- Freescale Semiconductor for hardware, software and information support.
- The Creativity Laboratory—Center for Valorization and Transfer of Competence, at Transilvania University of Brasov, which continues to enable the author to gain access to state-of-the-art solutions (such as the MCU SLK).
- Assoc. Prof. Dr. Eng. Razvan Udroiu and Drd. Eng. Bogdan Neagoe for reviews of the pre-final draft of this chapter.

# REFERENCES

Alvares, A.J. and J.C.E. Ferreira. 2006. WebTurning: Teleoperation of a CNC turning center through the Internet. *Journal of Materials Processing Technology*. **179.1-3:**251–259.

Alvares, A.J. and J.C.E. Ferreira. 2008. A system for the design and manufacture of feature-based parts through the Internet. *International Journal of Advanced Manufacturing Technology*. **35:**646–664.

Chang, J.H. 2011. An Efficient Relay Sensor Placing Algorithm for Connectivity in Wireless Sensor Networks. *Journal of Information Science and Engineering*. **27.1:**381–392.

Chellali, R., N. Mollet, C. Dumas and G. Subileau. 2008. SyTroN: Virtual desk for collaborative, tele-operated and tele-learning system with real devices, 153–161. In: Proceedings of the 3rd International Conference on e-Learning and Games (Edutainment 2008), Nanjing, Peoples Rep. China.

Chira, F., M. Banica and M. Lobontiu. Optimising Involutes Asymmetrical Teeth Gears Software, pp. 363–376. In: B. Katalinic [ed.]. 2009. DAAAM International Scientific Book 2009, Vol. 8. DAAAM International Vienna, Vienna, Austria.

Deaky, B. and M.V. Dragoi. 2011b. Self-developed Software Applications for Remote Monitoring (the Desktop and Web Applications), 105–108. In: The

Proceedings of the 5th International Conference on Manufacturing Science And Education—MSE 2011, Vol. I, Sibiu, Romania.

Deaky, B., N.B. Lupulescu and D. Ursutiu. 2011a. Data Acquisition with the MC9S12C32 Microcontroller (by means of MCU SLK) For Monitoring Applications, 305–308. In: The Proceedings of Modern Technologies, Quality and Innovation (ModTech) International Conference—New face of TMCR 2011, Vol. 1, Vadul lui Voda—Chisinau, Rep. of Moldova.

Deaky, B., N.B. Lupulescu and D. Ursutiu. 2011c. Wireless Communication with the Microcontroller Student Learning Kit (MCU SLK), 180–185. In: The Proceedings of REV 2011 international conference (Remote Engineering and Virtual Instrumentation), Brasov, Romania.

Deaky, B., N.B. Lupulescu and D. Ursutiu. 2011d. Extended Educational Use of the Microcontroller Student Learning Kit (MCU SLK), 913–916. In: The Proceedings of IEEE EDUCON 2011, Amman, Jordan.

Folea, S., M. Ghercioiu and D. Ursutiu. 2010. Cloud Instrument Powered by Solar Cell Sends Data to Pachube. *International Journal of Online Engineering (iJOE).* **6.4:**20–25.

Juca, S.C.S., P.C.M. Carvalho and F.T. Brito. 2011. A Low Cost Concept for Data Acquisition Systems Applied to Decentralized Renewable Energy Plants. *Sensors.* **11.1:**743–756.

Kostal, P., A. Mudriková and D. Cagáňová. 2010 The Virtual Laboratory of Program Control, 1–9. In: The Proceedings of ASME 2010 10th Biennial Conference on Engineering Systems Design and Analysis (ESDA2010), Istanbul, Turkey.

Kovacs, G.L., G.Z. Haidegger and S. Drozdil. 2005. Some interactive multimedia applications in production engineering, 161–167. In: The proceedings of IEEE 3rd International Conference on Computational Cybernetics, ICCC 2005, Mauritius.

Langmann, R., H. Haehnel, S.D.C. Ortiz, Z.G. Videuax and A.B. Gonzalez. 2011. TeleMedSys—Web-based Telemedical System for Rural Areas, 32–35. In: The Proceedings of REV 2011 international conference (Remote Engineering and Virtual Instrumentation), Brasov, Romania.

Le Roux, G.A.C., G.B. Reis, C.D.F. de Jesus, R.C. Giordano, A.J.G. Cruz, P.F. Moreira Jr., C.A.O. Nascimento and L.V. Loureiro. 2010. Cooperative Weblab: A Tool for Cooperative Learning in Chemical Engineering in a Global Environment. *Chemical Engineering Education.* **44.1:**9–12.

Lloret, J., S. Sendra, H. Coll and M. Garcia. 2010. Saving Energy in Wireless Local Area Sensor Networks. *Computer Journal.* **53.10:**1658–1673.

Orishaba, D., R. Besiga, A.T. Asiimwe, S. Mwanje, P.I. Musasiziand and S.S. Tickodri-Togboa. 2011. Design of a Bi-directional Fiber Optic Link and Wavelength Division Multiplexing iLAB, 238–243. In: The Proceedings of REV 2011 international conference (Remote Engineering and Virtual Instrumentation), Brasov, Romania.

Ras, E., M. Becker and J. Koch. 2007. Engineering tele-health solutions in the ambient assisted living lab, 804–809. In: The Proceedings of the 21st International Conference on Advanced Networking and Applications, Vol. 2, Niagara Falls, Canada.

Saidi F., O. Stasse, K. Yokoi and F. Kanehiro. 2007. Online Object Search with a Humanoid Robot, 1677–1682. In: The Proceedings of the 2007 IEEE/RSJ International Conference on Intelligent Robots and Systems, San Diego, California, USA.

Severdaks, A., G. Supols, M. Greitans and L. Selavo. 2011. Wireless Sensor Network for Distributed Measurement of Electrical Field. *Elektronika Ir Elektrotechnika*. **107:**7–10.

Udroiu, R. and L.A. Mihail. 2009. Experimental determination of surface roughness of parts obtained by rapid prototyping, 283–286. In: The Proceedings of the 8th WSEAS Int. Conf. on Circuits, Systems, Electronics, Control & Signal Processing, CSECS '09, Puerto de la Cruz Tenerife, Canary Islands, Spain.

Udroiu R. and N.V. Ivan. 2010. Rapid prototyping and rapid manufacturing applications at Transilvania University of Brasov, Bulletin of the Transilvania University of Brasov, Series I: Engineering Sciences, Vol. 3. **52:**145–152.

Ursutiu, D., M. Ghercioiu, R. Oros and C. Samoila. 2010. Tag4M in Predictive Maintenance and Machine Control. *International Journal of Online Engineering (iJOE)*. **6.1:**28–30.

Zeiger, F., M. Schmidt and K. Schilling. 2009. Remote Experiments with Mobile Robot Hardware via Internet at limited Link Capacity. IEEE Transactions on Industrial Electronics. **56.12:**4798–4805.

Zuhlke, D. 2010. SmartFactory—Towards a factory-of-things. *Annual Reviews in Control*. **34.1:**129–138.

## URL REFERENCES

[CE] Cores Electronic LLC. 2011. Tag4M Cloud Instrument, Available from: http://tag4m.com/, Accessed: 15/05/2011.

[FS] Freescale Semiconductor. 2011a. PBMCUSLK Product Summary Page. Available from: http://www.freescale.com/webapp/sps/site/prod_summary.jsp?code=PBMCUSLK, Accessed: 20/05/2011.

[FS] Freescale Semiconductor. 2011b. RD3172MMA7456L Product Summary Page. Available from: http://www.freescale.com/webapp/sps/site/prod_summary.jsp?code=RD3172MMA7456L, Accessed: 20/05/2011.

[NI] National Instruments. 2011a. NI ELVIS: Educational Design and Prototyping Platform. Available at: http://www.ni.com/nielvis/, Accessed: 2010-06-23.

[NI] National Instruments. 2011b. NI MyDAQ user guide and specifications.

# 11

# IEEE 1451 Smart Plug and Play Sensors with Application in Wireless Sensor Networks

*Ismael Hernández-Capuchin, Manuel Moisés Miranda Velasco, Miguel Enrique Martínez-Rosas, Carlos Gómez Agis and Horacio Luis Martínez Reyes*

**ABSTRACT**

The purpose of this chapter is to help readers become familiar with the IEEE1451 standard (Smart Transducers Interface for Sensors and Actuators) and how to use it with Wireless Sensor Networks (WSN) in practical applications for various sectors, particularly crops. Due to the recent release of the IEEE1451 standard, users of this standard require some support about its essential components, and get a clear idea about the improvements made to implement this standard along with WSN, such as better data management and improved detection of critical events in the system under measurement.

The combination of the above technologies has the potential to become a powerful tool for developers of new applications and research projects with embedded systems.

## Introduction

With the advances in electronic and information technologies, various sensing systems have been developed to measure and know the value of variables which affect different production systems, such as determining accurate information concerning the spatial variability within crop fields for precision farming. This variability is affected by different factors, some of which are: soil properties and crop nutrients, crop canopy volume and biomass, water content, and pest conditions (disease, weeds, insects, among others). These factors can be measured using diverse types of sensors and instruments such as field-based electronic sensors, spectroradiometers, automatic vision systems, airborne multi spectral and hyper spectral remote sensing, satellite and thermal imaging, among others (Lee et al., 2010).

Actually, in agriculture, some of the measured variables include: relative humidity, air temperature, solar irradiance and rainfall, wind speed and direction because these have a major impact on the final product. Because the average farmer cannot be expected to be an expert in computer science or electrical engineering, just as they are not expert in agricultural pathology or nutrition (Lee 1995), sensors and instruments need to be as simple as possible, without sacrificing operations, to avoid rejection by farmers or abandonment after a short time.

This does not mean that farming or precision farming is the only application that can be found in the literature but, it is a good example of how technology can be used for improves many aspects that are affecting or hinder obtaining a final product with quality.

The recent development of hardware platforms and software frameworks has led to many significant advances in technology. This means that the instruments required by farmers, among others, have been made available by means of these new systems.

In this chapter, we lead the reader through one of these new developments, the IEEE 1451 standard, known also as "smart transducer interface for sensors and actuators". This chapter provides the reader with some basic concepts related to Wireless Sensor Networks (WSN), their versatility, along with some examples and proposed use of new tools and systems.

Because the IEEE 1451 standard is relatively new, this section provides some basic principles and explanations about the different components of this standard. On the other hand, because WSN are quite popular, this section presents only basics concepts, assuming that the reader has some knowledge in this area.

For this reason, more than a cursory review of the IEEE 1451 standard is provided and a more in-depth review of standard specifications are provided.

At the end of the chapter the reader will find some examples using both IEEE 1451 and WSN. These are not the only examples in the literature but the reader must know that because these topics are relatively new, commercial applications are not yet common. Most of the examples provided are research applications to improve or demonstrate the use of these standards.

This is a good opportunity for new research and design work based on these two tools, as both are available and ready to be applied in many areas, including farming.

## The Basics of Wireless Sensor Networks and IEEE 1451

A complete analysis of the IEEE 1451 standard and its relationship with WSN provides its characteristics, advantages, benefits and applications. However, to understand how these tools work one first needs to know the basics of its operation.

### Wireless Sensor Networks

A WSN is a system comprised of radio frequency (RF) transceivers, sensors, microcontrollers and power sources, which allow measurement of different values simultaneously (Wang et al., 2006). Due to the functionalities of a WSN, it has the ability to monitor and react to events and phenomena in a specific environment (Baronti et al., 2007). The environment can be a physical process, mechanical system, biological system or information in a technological framework. Common applications include telemetry, data collection and monitoring (Sohraby et al., 2007).

An area that has recently developed applications is agriculture. Here, WSN can be used to monitor climatic conditions of different zones of a large cultivated area and calculate different water or chemicals needs of the soil (Bencini et al., 2009). Precision agriculture can be defined as the art and science of utilizing technology to improve crop production. WSN has found a field of applications in precision agriculture, where WSN can play an important role in handling and managing crop information in order to better understand changes and take advantage of timely and valuable knowledge pertaining to crop conditions. For example, the data obtained can provide the precise time for harvesting a crop, as well as estimate the fertilizer requirements and water resources for irrigation (López-Riquelme et al., 2009).

The advantages related to the WSN are the ability to handle data correlation, the occurrence of events, system status and proper use of information. The wireless nodes perform these tasks. To achieve scalability, interoperability and reduced cost, wireless nodes generally follow a standard of communication. The IEEE 802.15.4 standard is the most popular; it was developed for wireless communications with low data rates, simple connectivity and applications with low power consumption. Because the nodes have low power consumption, it is possible to use limited energy sources such as batteries or charging systems like solar chargers. This means improved implementation of the systems because it is not necessary to have electricity nearby, allowing persons to install nodes in remote or isolated locations.

An example of this standard is the ZigBee protocol, whose main objective is to provide an easy and secure communication between the components of the network with a low data rate, with maximum energy savings. Figure 1 shows a simple network with all the component of a ZigBee protocol.

Other important elements in WSN are the sensors and actuators. These elements represent the interface that interacts with the system of production and the proper management of these elements is fundamental. The IEEE 1451 standard is one of the newest ways to manage sensors and actuators.

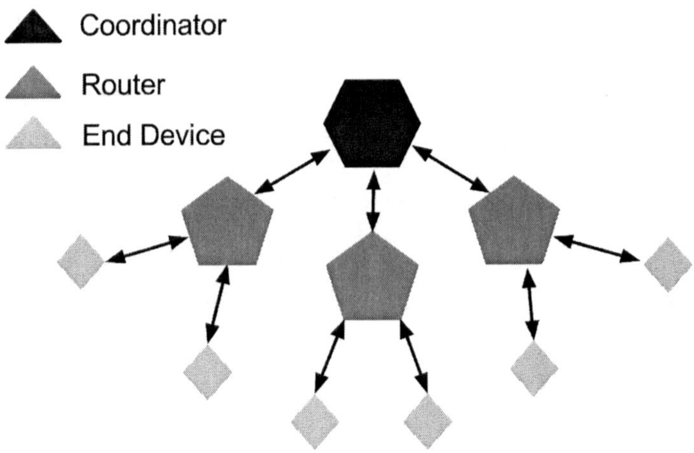


**Figure 1.** Example of ZigBee network.

**Transducer**

A transducer is a device that converts one type of energy into another that is easier to measure. Transducers can be found as part of sensors or actuators. A sensor consists of two parts: the measuring element and the transducer, which converts the measured signal into a signal representing this measurement. The sensor responds to physical variables with electrical representations as an actuator receives an electrical signal from outside, and acts showing physical reactions (Sammarco-JJ and Fries-EF 2007).

In order to analyze data from the sensors or submit information to actuators, tools (such as microcontrollers) can be configured to interpret information. A microcontroller normally combines the functions of a computer, but in a small package. They contain a central processing unit, memory and input/output ports. A microcontroller is a device that can monitor, manage and control electronic equipment, communication systems and systems for measuring physical variables.

In the memory of the microcontroller, there is only one program that has been made for a specific application, its input and output ports have been configured to support the sensors and actuators that allow for the assignment. Since it has been programmed and configured, the microcontroller only works making its assignment. It therefore develops its maximum potential, adding the fastest possible ways to connect and communicate with other peripheral components.

Because they can be programmed, microcontrollers can be adapted to the environment in which they are working and can act according to what happens around them and, if necessary, provide the correct response required to provide a specialized service to end users.

**Smart Transducer**

A smart sensor or smart actuator is obtained by combining a sensor (or an actuator), a transducer, an analog interface, a converter, the communication bus and a driver in a single housing. In a system that uses smart transducers, sensor (actuator) functions and interfaces are combined into a complete design. These functions include: measuring, signal conditioning, analog to digital conversion (ADC), the bus interface and data processing (Meijer and Meijer 2008). The IEEE 1451 standard describes a device that has one or more connections with transducers, which automatically identify, set and calibrate themselves as smart.

Smart sensor operation starts at the time a measurement is taken, when the analog output of the sensor is conditioned and converted to a digital signal using an analog to digital converter. The digitalized signal of the sensor can be easily processed by a microcontroller using a program designed by the user for this purpose. Measurements made with the device are sent to an administrator in a network; the management and delivery of the information is done through a device responsible for coordinating with the network, so protocols are used in the network to manage the information (Song and Lee 2008). In a contrary outcome, there is an actuator. The smart actuator receives the information sent over the network from an administrator. First, the network driver, which is the link to the network, receives a request. Then, it interacts directly with the transducer indicating which action to take. Finally, the transducer responds to the action requested.

### The Difference Between a Traditional Transducer and a Smart Transducer

A transducer it is an element that transforms one physical signal into another one. A smart transducer within a technological standard involves many other aspects. Typically, a smart transducer obtains its functionality using a device that incorporates firmware for signal conditioning, calibration, diagnosis and communication technologies, to carry out specific tasks.

Figure 2 shows the difference between a transducer and smart transducer, using the internal components. In the figure, the measuring device is called a sensor and is considered part of the transducer, just for understanding and graphical purposes.

For appropriate use and development of smart sensors (smart actuators), it is necessary to regulate a standard to address them in a specific way; IEEE 1451 was created to meet this need.

### IEEE 1451 Standard for a Smart Transducer Interface for Sensors and Actuators

In 1994 the National Institute for Standards and Testing (NIST) proposed a protocol "plug & play" for smart transducers that were associated

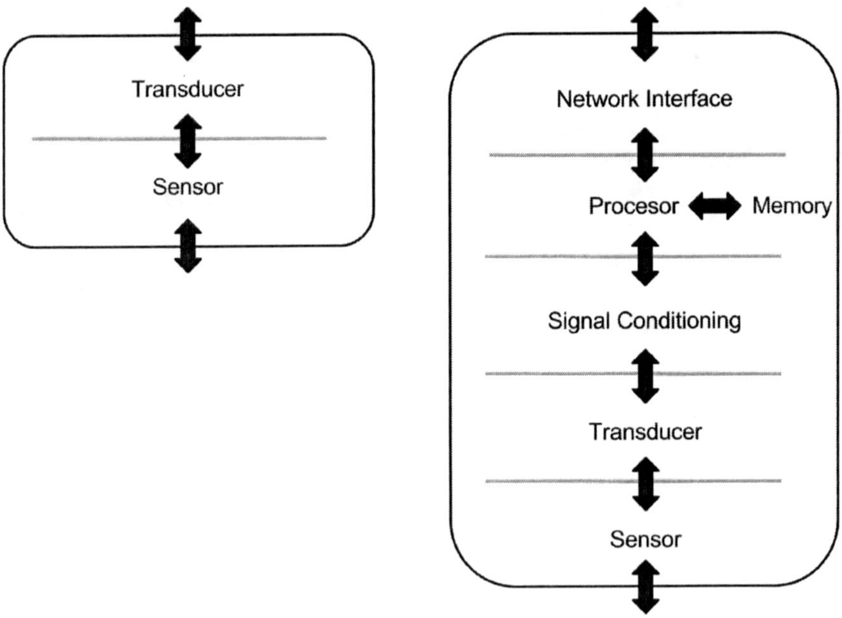


**Figure 2.** Transducer and Smart Transducer.

with network control; this protocol was formally adopted as the IEEE 1451 standard (Kher et al., 2006). The IEEE 1451 governs the hardware, software and communication protocols associated with smart transducers. The aim is quite simply to develop a family of standards to interconnect smart sensors; obviously, the resulting standards have been developed to facilitate the implementation of smart devices, which bring specific attributes such as networking, systems and tools (Rivera 2009). It is very important to note that once smart transducers are installed and running on any network within the area of measurement, it is easier to add new devices due to the versatility that the standard provides.

This standard is endorsed by the Institute of Electrical and Electronic Engineers (IEEE) and the Measurement Society's Technical Committee on Sensor Technology, who are still working on a suite of different standards to help interface smart transducers with different networks for a better understanding or application. Integration is the most important contribution offered by this standard, allowing the connection of devices (working under different communication standards) to larger networks in a standardized way.

The IEEE 1451 standard for smart transducers provides exceptional functionality by helping to get appropriate data collection and accurate interpretation of measured variables. This means that sensors based on IEEE 1451 have the ability to identify, describe, diagnose and calibrate by themselves, as well as recognizing their location, time, data processing, reasoning, data merging, report alerts, format and management of information and communication protocols (Song and Lee 2008).

Due to their ability to start measurements just at the moment when they are connected as mentioned in previous paragraphs, smart transducers provide enough information to the final user in order to assess whether the system is working according to the desired settings.

The standard makes definitions of the communication interface for connecting the transducer or transducers to the network through a linking device. A smart transducer must be able to interact as a sensor that is isolated. It must be able to interact with other transducers that are part of an array of transducers and, in turn, must be able to communicate to a network. The connection or interface between the transducer and the linking device to the network is specified in the standard and can be wired or wireless.

### Composition of a Smart Transducer

The first module of a smart transducer is the Transducer Interface Module (TIM), which as its name suggests, involves the instrumentation required for the operation of the transducer. The TIM includes the Transducer

Electronic Data Sheet (TEDS) with information for operation, calibration, and its unique identity. This module is attached to the Network Capable Applications Processor (NCAP), which is the second module that manages communication between the TIM module or modules connected to the NCAP and the network to which users wish to communicate. Finally, communication between TIM and NCAP modules is done by means of a data connection interface, which can be wired or wireless and is controlled by one of the family versions of the standard.

Figure 3 shows the basic architecture of the IEEE 1451 standard. Beginning from the left side of the figure, there is the module with the transducer interface for the external device, which records the measurements, and includes the electronic datasheet with the information related to the smart transducer. The next module is the interface between the previous module (TIM) and the NCAP module that enables communication with an external network. This interface is ruled by a family member of the standard, selected to do this task. The external network can be any possible network that can manage the data generated by the smart transducers.

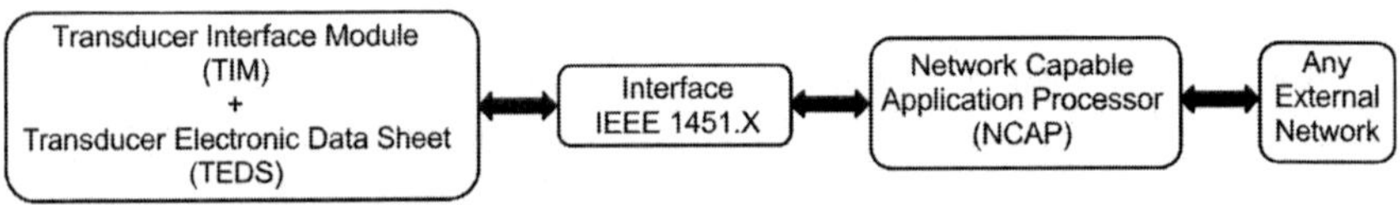


**Figure 3.** IEEE 1451 Standard Architecture.

### *Transducer Interface Module (TIM)*

When the TIM module is in operation, it is ready to act and respond to requests through the NCAP. The TIM module communicates directly to the NCAP module to check and maintain its status through one of the options that permits the IEEE 1451 standard.

The standard does not specify the instrumentation or the conversion rules that apply to the sensor or actuator; such rules are beyond the control of the standard. Consequently, it is the responsibility of the user to properly configure the instruments to handle these signals, making them more manageable for the system. However, this matter relies directly on the device manufacturer that uses the standard. Obviously, if users need to configure the sensors, they will be deployed to work with applications specifically defined by them.

When performing an instrumentation to work with a specific transducer, the accompanying electronic components usually become part of the hardware owned by the TIM. This means that the more complex the sensor or actuator used, the greater the number of components that form the TIM module.

Once the sensor signal has been conditioned to work with the standard, the TEDS are in operation, as these contain the specifications to allocate a wide range of data in digital format (Wiczer and Lee 2005).

The software that makes up the TIM can include a series of communications, data conversion and signal processing algorithms, or, in its most basic form, the TIM contains the software needed to respond to the requests and initiate certain services requested by the NCAP.

The TIM may also include software that supports various forms of communication with the NCAP or other TIM that are network members.

### *Transducer Electronic Data Sheet (TEDS)*

The purpose of the TEDS is to provide identification and detailed specification of the sensor or actuator that is attached to the TIM module. TEDS are normally within the TIM, however, a copy of the datasheet is transferred to make a connection with an NCAP device.

The information that composes the TEDS is configured by the manufacturer according to the standard and is usually stored in an external memory (Wobschall 2007).

The TEDS are tables containing specific information about the drivers that work with the TIM and also include global information about deploying IEEE 1451. The NCAP selects the TEDS information in order to identify the TIMs that are attached to the NCAP and also to identify which transducers are attached to each TIM. The TEDS supply information about each transducer, operating mode, trigger mode, calibration coefficients, functionality, and a wide-range of other use-specific information.

### *Network Capable Application Processor (NCAP)*

The NCAP is a vital gateway between the external network and the elements of the TIM module. The IEEE 1451 standard is not restricted to a particular external network; this means that different NCAPs may be developed. The NCAP functions like a gate with two links: one used to support a standard item and the other used to drive the data network bus selected by the user. Each NCAP is designed to interact with a specific network and a specific part of the protocol from IEEE 1451. This is a motivation for manufacturers and users, in general, as it can be developed and built a NCAP for almost any network communication standard that is compatible with market requirements, which allows a better access between devices (Wiczer and Lee 2005).

## Family Members of the IEEE 1451 Standard

Before working with the family of the IEEE 1451 standard, the developer needs to understand and choose the standard specifications to configure and implement the device. The interface data connection between TIM and NCAP is the path by which data and commands are exchanged between the TIM and the NCAP. In this section the reader will find a better explanation of every member of the IEEE 1451 standard that can be used to create an interface.

At present, the IEEE 1451 standard can be divided into eight different specifications. Figure 4 shows the family structure of the standard. The specifications of the standard can be split in two groups, one defines the functions of NCAP, and the other defines the functions of TIM.

Because the extension of this standard, the next subsections present a small explanation about the different specifications of the standard, which defines different interfaces between the TIM and NCAP modules and the configuration of the NCAP module.

### *IEEE 1451.0—Common functions, Communication protocols, and Transducer Electronic Data Sheet (TEDS) formats*

This standard defines a common set of commands, which enable interoperability between other standards within the same family (Ramos 2008).

Through a set of commands, any sensor or actuator can be accessed, regardless of being within a wired or wireless network based on IEEE 1451. This means that this part of the standard is independent of the medium used for communications between the transducer and the NCAP, thus simplifying the creation of new standards for different physical layers, while maintaining interoperability between different family members. This part of the standard is used in both TIM and NCAP.

### *IEEE 1451.1—Networked Smart Transducer Model*

This standard defines a common software interface connection between the NCAP and control networks, providing classes that represent control models for smart transducer networks. These classes (common network control information) are transducer blocks, function blocks of networks and their complementary structure. The information provided in this standard, if it is common with other parts of the standard, is just complementary. It is in no way competitive to some other model described in other parts of the standard.

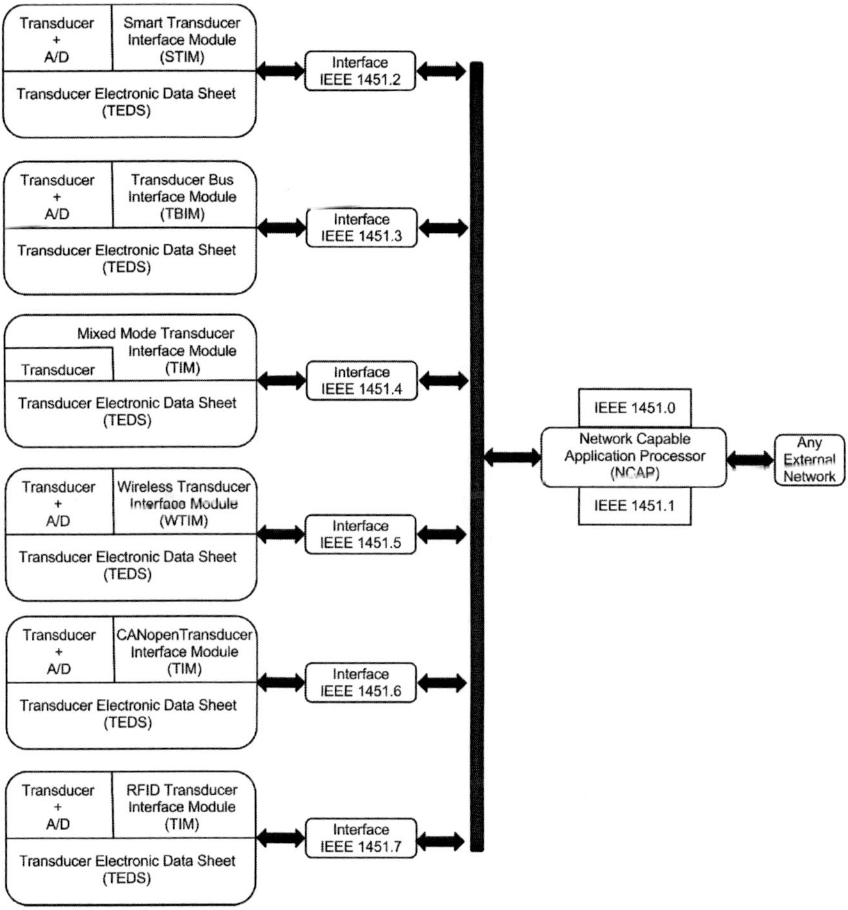


**Figure 4.** IEEE 1451 Standard Family.

It defines a data model for the smart transducer, which is supported by the same network. This part of the standard is used inside the NCAP module.

This specification, models the behavior of smart transducers as a "software card-cage", where the modules can be connected with a simple handshake and can interact with each other (NIST 2011).

The models of smart transducer networks encapsulate the hardware implementation details of the transducer with a simple programming model. This makes programming the hardware interface of the transducer less complex. The network service interfaces encapsulate the details of different network protocol implementations behind a small set of communication methods (Artaud et al., 2004).

### *IEEE 1451.2—Transducer to Microprocessor Communication Interface*

This standard defines the interface model for a smart transducer (STIM). They also specify components and the format of the TEDS to describe the full capabilities of the sensor (Ramos 2008).

The standard specifies the operation of the device and its calibration, the physical interface between the transducer and the NCAP, the use of standardized components to build smart sensors, the digital interface including the location of the connector terminals and communication protocols between the transducer and microprocessor (microcontroller).

Every smart transducer has its own TEDS, which details the features of the transducers and thus enables its "plug & play" ability. The TEDS contains information such as: the manufacturer, model and number of transducer, serial number, measurement range, accuracy and calibration (Artaud et al., 2004).

Currently, the standard is complete and published, but there is a review in progress to include other communication interfaces to reduce costs and complexity when implementing it.

### *IEEE 1451.3—Distributed Multidrop System for Interfacing Smart Transducers*

This standard defines the communications interface between the TIM and the NCAP through the distributed communication architecture (multiple sensors). It also explains how transducers are attached to the smart TIM (STIM) and how they are connected to the NCAP through a multipoint connection.

The standard explains: the electrical interfaces, TEDS, channel identification protocols, protocols for rapid exchange of information, time synchronization protocols and functions for reading and writing used to access the TEDS and transducer data (Artaud et al., 2004).

In this case, each transducer is associated with a Transducer Bus Interface Module (TBIM). Multiple TBIMs are connected through multiple online connections to the bus driver specified in the standard (Ramos 2008).

### *IEEE 1451.4—Mixed-mode Transducer and Interfaces*

This standard defines a mixed-mode interface for analog transducers with analog and digital operation modes. A digital serial interface can be used to read information from the TEDS and the transducer set (Artaud et al., 2004). The analog electrical interface provides a signal (voltage, current) representing a physical phenomenon (temperature, humidity, speed, among others).

In addition, the model was refined to allow only TEDS with a minimal amount of data stored when using small sensors with a small amount of physical memory. It also defines additional TEDS classes for other types of sensors such as microphones and accelerometers. The standard primarily focuses on adding the TEDS function for analog sensors.

After power up, the TEDS of the transducer is sent to the instrumentation system of the communication interface, then the interface switches to analog operation and the same interface is used to move the analog signals from the transducer to the instrumentation system (Song and Lee 2008).

### *IEEE 1451.5—Wireless Transducer Networks*

This standard has been designed to define the interface from the transducer to the NCAP, along with TEDS for wireless transducers (NIST 2011). Developed to use wireless TIM modules known at this point as WTIM (Wireless Transducer Interface Module). The WTIM is composed of transducers, which are attached to the module with the required instrumentation for proper operation, the TEDS for each transducer and a radio. In this case, the NCAP has a radio to communicate and receive information from the WTIM. Even in this part of the standard, depending on the communication interface used, the WTIM can add more wireless devices to the NCAP.

Due to the fact that the standard provides easy "plug and play" connections, both devices simply use the same physical interface (the same wireless standard) to communicate and operate without any hardware or software reconfiguration. Therefore, there is no need to install new drivers, profiles or perform any other changes (Song and Lee 2010).

The wireless physical interface between the NCAP and WTIM may be based on any of the following protocols: IEEE 802.11 (WiFi), IEEE 802.15.1 (Bluetooth), IEEE 802.15.4 (ZigBee) and IPv6 over 802.15.4 (6LoWPAN). Obviously, each protocol has its own configuration and communication rules. In this case, only a radio link with the characteristics of each protocol will be used, so it is the user's responsibility to ensure that communication between the WTIM and NCAP is carried out and that it conforms to the guideline established by the selected communication protocol.

### *IEEE 1451.6—CANopen-based Network for Multi-channel Transducer*

The standard defines the format of TEDS and the interface between the transducer and the NCAP, using the interface of a high-speed CANopen network. This standard defines the mapping of the IEEE 1451 TEDS to the

CANopen dictionary, as well as communication messages, processing data, configuration settings and diagnostic information.

***IEEE 1451.7—Transducers to Radio Frequency Identification (RFID) Systems Communication Protocols and Transducer Electronic Data Sheet***

This standard is under development, and refers to the interface between smart transducers and radio frequency identification (RFID) systems. Currently, manufacturers of similar systems use their own interfaces and do not have the advantages of using a defined standard for interoperability with other devices.

The purpose is to specify standard methods of communication and data format for transducers, sensors or actuators and how they communicate with RFID tags, which use the standard ISO/IEC 24753. The IEEE 1451.7 also defines the format of family-based TEDS standards and IEEE 1451 protocols to access TEDS and transducer data (Ramos 2008).

## Using IEEE 1451 and WSN

The family of IEEE 1451 has been designed to support smart transducers networking. It has allowed important advances on defining and classifying interfaces, sensors and actuators. This family provides a set of interfaces for smart transducers working within a network, providing "plug & play" interoperability in commercial and industrial applications.

The IEEE 1451 standards add freedom and intelligence to the WSN, these features allow the network to grow or shrink, fitting user needs, while eliminating the technical restraints because it allows the connection or disconnection of any sensor that meets the standard, integrating or removing instrumentation without major complications.

The joining of IEEE 1451 and WSN forms a network that has the knowledge and ability to be distributed. This application can be used, for example, in a crop field, where the user has a disproportionate distribution of points of interest; this junction allows for the measurement of environmental variables in a field. Because of the complete portability of sensor devices and actuators, they can be installed throughout the field, allowing to work in the selected area without doing anything that involves moving a normal wireless network, like locate, implement and calibrate every movement of the transducers or the nodes in the network.

**Applications of IEEE 1451 with WSN**

Some IEEE 1451 and WSN applications can be found in the literature, mostly concerning research projects. At present, only a few commercial companies provide information about smart sensors, although the information is not very extensive and lacks specific details. These applications are related mainly to medicine, industry and agriculture. Some representative examples are described on next paragraphs.

A physical interface using Bluetooth communication is implemented between an NCAP and several WTIM through IEEE 1451.5 (Sweetser et al. 2006). The NCAP module was simulated from a computer through an USB-Bluetooth interface while the WTIM card was implemented on development boards, due that the project's purpose was to describe software architectures and modular hardware for rapid development of smart sensors which complied with the aforementioned standard. The firmware used for the microcontroller was developed in ANSI C to allow portability to a wide range of microcontrollers. If a commercial design is required, a new TIM card should be developed in order to meet the new requirements.

A Wireless Sensor Network utilizing the IEEE 1451.0 and IEEE 1451.5 standards with an interface for communication based on the IEEE 802.11 standard is described in (Lee and Song 2007). The authors used two wireless nodes, the NCAP and WTIM, communicated wirelessly via the WiFi protocol, while NCAP can access a server via the Ethernet network. The aim of the paper was to lay the foundations and some guidelines for deploying wireless intelligent monitoring with sensors based on standards already mentioned.

The design and implementation of IPv6 on wireless personal area networks (6LoWPAN) and how they use the IEEE 1451 for sensor networks based on IEEE 802.15.2 are explained by Higuera and Polo in (Higuera and Polo 2010a). The paper proposes an application using web services to transmit commands between the NCAP and WTIM over a client-server architecture. There use various WTIM devices attached to a NCAP controller, which, in turn, is attached to a computer via IPv6.

Service discovery protocols such as UPnP, which is integrated to a network based on IEEE 1451, will allow the sensor network to be discovered and accessed as a general UPnP device through the gateway (Sung et al. 2010). Integrating both technologies enable sensor self-identification and their ability to "plug & play." This paper explains that the proposed architecture uses the IEEE 1451 ZigBee-based standard, which provides bulk transfer modes for the application to interact with the devices. Basically, it maps the discovery applications, configuration and data exchange between sensor nodes (WTIM) and base station (NCAP), and behaves like the UPnP protocol between the base station and the control point.

The authors of this proposal (Zhaochun et al., 2009) describe two efficient solutions to develop a wireless sensor network, based on the IEEE 1451 standard. It is important to note that the hardware structure of the WTIM has been designed in detail. In addition, the authors configured NCAP and WTIM software architecture. The project makes use of other standards for transducers attached to WTIM based on IEEE 1451. For the communications with the NCAP through the IEEE 1451.5, a ZigBee interface is used.

Smart sensors to monitor health of people have been reported by the authors (Kim et al. 2010). The project considers the interoperability of IEEE 1451 and HL7. HL7 is a standard developed to exchange health information data generated by medical devices. The user would carry an entire sensor network (WSN) in his/her body that allows real-time information to be transmitted to the hospital personnel. The aim is to use IEEE 1451.4 and IEEE 1451.5 to place wireless smart transducers in patients' bodies and then send the information to an NCAP and finally to an internet connection.

A new electronic device, which is a transducer and controller at the same time, is described in (Xiangyin et al. 2009). The structures of hardware and software components of this device behave as a node in a Wireless Personal Area Network. The IEEE 1451.2 standard is used for communication between the STIM and NCAP, but has an external network using ZigBee, so the radio has a built-NCAP that is compatible with this technology. They use a gas station as a test site where there are several sensors installed to monitor the operation of the plant. Despite being based on the ZigBee project, it uses a star configuration to communicate devices with a gate bond central.

The combination of WSN and IEEE 1451 is used to build equipment to measure various physical parameters of soil. IEEE 1451.3 and IEEE 1451.5 were used to perform configuration and communication tasks (Skierucha et al., 2004). The wireless standard operates in the ISM (Industrial, Scientific and Medical) frequency of 433 MHz. The purpose of the project was to have multiple devices connected together as indicated by the IEEE 1451.3, where one of them had an Ethernet connection to the external network to control the internal network, based on point-to-point configuration. Communication between the STIM and NCAP uses the standard serial interface. This project aims to build hardware and complex software that can adapt to larger areas, including other communications to serve as a reference source for field measurements.

The authors of this article (Higuera and Polo 2010b) propose the design and implementation of an Intelligent Ultrasonic Anemometer based on the IEEE 1451 standard that operates with a Wireless Sensor Network. The wind speed and direction are important parameters in weather forecasting and air pollution. Even this is closely linked to agriculture and the damages to crop field. They worked with the IEEE 1451.5 standard on the interface using ZigBee. The anemometer was attached to a WTIM that established communication with an NCAP using the same wireless technology. In turn, this is connected to a database through a web server where data is stored.

To measure variables in a crop field, the authors (Morais et al., 2008) used ZigBee-based wireless sensors and a physical interface for communication, using part of the IEEE 1451 standard for such implementation. They proposed to create a node for interacting with different sensors in the field as well as using a ZigBee coordinator that, in this case, would be a standard module for the IEEE 1451.2 NCAP. This, in turn, sends the information gathered to the barn via a GSM interface, and then sends it to another antenna by either GSM or Internet, where both can be consulted.

## Conclusions

The information presented shows the relationship between IEEE 1451 and Wireless Sensor Networks (WSN), although this relationship has only reached the interface stage for communication between the NCAP and WTIM. We hope the reader can see the great versatility that the WSN provides to the IEEE 1451 standard.

The use of WSN also allows the connection between transducers and controllers with different stacks and protocols, which in addition is done without modifying IEEE 1451.

Any system that integrates IEEE 1451 benefits from the use of supported formats. When the user wants to connect a new sensor to a measurement system that does not accomplish the IEEE 1451 standard, it is usually necessary to perform a device reconfiguration on both hardware and firmware to provide an adequate communication with the sensor as well as a proper interpretation of data. By other side, using IEEE 1451, any TIM can be connected to any NCAP because the NCAP will recognize it without any problems as it needs no prior knowledge of the device, a physical reconfiguration or programming in order to communicate. Sensors and actuators can be simply connected or disconnected, hence the name "plug & play" transducers.

For example, a wireless TIM module does not require any other action than have an NCAP to work with, the same wireless protocol and, obviously, the same workspace. This ability illustrates how it is possible to use sensors or actuators in different procedures or situations. This versatility allows for the devices installed in the workspace to be removed or reconnected in an easy and fast manner.

The IEEE 1451 standard specifications allow separate communications in a network, which can be used to connect transducers and implement systems and networks. Using this standard, transducer manufacturers and users eliminate the requirement of using different protocols to support multiple networks. This standard has the ability to connect the transducers to either wired or wireless external networks.

Usage of electronic datasheets (TEDS) in IEEE 1451, provide benefits, such as: enabling the identification and description by themselves, eliminating human errors generated by configuration mistakes, making unnecessary the process of recalibration by the user, containing smart transducers with sufficient and necessary information to properly function, reducing the costs and problems that occur due to calibration and configuration (when compared with normal transducers), reducing training time to handle and allowing them for field installation, upgrades and maintenance, among others.

The incorporation of WSN with the IEEE 1451 standard offers very interesting advantages. For example, the network becomes more reliable, although a transducer may fail for reasons such as interference, physical damage or a lack of energy, among others, which does not affect network performance. Reliability in a WSN is the ability of the network to remain functional despite the failure of some nodes.

These kinds of networks are scalable and support a large number of connected nodes within the same network. The hardware used in WSN is the most basic and accurate for any deployment. A WSN can operate in hospitals, but also in harsh environments like in farming or precision agriculture. A sensor node is usually equipped with a limited amount of energy, so nodes contain many ways to save energy, which allows them to have a much longer and more productive life cycle than other RF devices.

When the transducers are wireless, they provide good interoperability, meaning they provide smooth operation and communication between each component of the system.

Utilizing the same communication module based on IEEE 1451.5 brings specific advantages to smart transducers: the WTIM from one manufacturer can be connected (plug & play) with an NCAP, which has been provided by another vendor. Furthermore, different WTIM from different manufacturers can be connected (plug & play) with an NCAP. Different WTIM from diverse manufacturers can be connected (plug & play) with other NCAP from different manufacturers. Without a doubt, this represents a great advantage for users.

While presented applications perform a connection between the NCAP and an external network via different protocols (most of the authors present variants for HTTP), this connection is not defined by the standard, which considers the external connection only to an arbitrary network. Consequently, the development and research for new external networking solutions should be considered. One feature that has not been solved yet is the use of an extended WSN to manage an external network, creating a network using many NCAP modules, based on WSN. Solving this will add all the features of a WSN to the external layer of IEEE 1451; in other words, a smart network with wireless smart sensors will be obtained. The intelligence would be provided by the interconnection of the NCAP to the TIM on the internal network and even for the connection from the NCAP to the external network.

As mentioned, there are presently only a few commercial applications. In fact, although some manufacturers offer wireless communication systems for use with sensors, they cannot be considered WSN. Something similar happens with the IEEE 1451 standard, where only a few applications can be found, from which available information is very limited.

Undoubtedly, applications with the IEEE 1451 standard and WSN are certainly useful in many sectors. For example, this technology could be applied in agriculture (farming or precision farming), where there is no need for high-speed links to collect the data measured. This is due to the fact that evolution of field variables is slow compared to other systems; therefore the use of WSN in agriculture is appropriate for managing information collected from required variables. The benefits offered by the IEEE 1451 standard include low staff training, a use-friendly system and measurement flexibility, and provide a highly reliable, configurable and manageable device to carry out the measurements requested by the user; no technical expertise is required to set up and use such devices in a crop field.

The material presented in this chapter gives the reader a clearer idea of how WSN can be used with IEEE 1451. The examples show the many possibilities for these new advances. The combination of WSN and IEEE 1451 grants the developer of embedded systems many opportunities for research and application development.

## REFERENCES


Artaud, G., P. Patrick, G. Furano, R. Magness and C. Plummer. 2004. IEEE 1451: Transducer networking. *DASIA, European Space Agency.* (pp. 1–5).

Baronti, P., P. Pillai, V. Chook, S. Chessa, A. Gotta and Y. Hu. 2007. Wireless sensor networks: A survey on the state of the art and the 802.15.4 and ZigBee standards. *Computer Communications.* **30(7):**1655–1695. doi: 10.1016/j.comcom.2006.12.020.

Bencini, L., F. Chiti, G. Collodi, D.D. Palma, R. Fantacci and A. Manes et al. 2009. Agricultural Monitoring Based on Wireless Sensor Network Technology: Real Long Life Deployments for Physiology and Pathogens Control. 2009 Third International Conference on Sensor Technologies and Applications (pp. 372–377). IEEE. doi: 10.1109/SENSORCOMM.2009.63.

Higuera, J. and J. Polo. 2010a. Understanding the IEEE 1451 standard in 6loWPAN sensor networks. 2010 IEEE Sensors Applications Symposium (SAS) (pp. 189–193). IEEE. doi: 10.1109/SAS.2010.5439427.

Higuera, J. and J. Polo. 2010b. Tiny and autonomous IEEE1451 Sonic Anemometer to deploy in environmental Wireless Sensor Network. 2010 Seventh International Conference on Networked Sensing Systems (INSS) (pp. 183–186). IEEE. doi: 10.1109/INSS.2010.5572161.

Kher, S., J. Chen and A. Somani. 2006. IEEE 1451 standard and wireless sensor networks: An over view of fault tolerant algorithms. 2006 IEEE International Conference on Electro/Information Technology (pp. 227–232). IEEE. doi: 10.1109/EIT.2006.252141.

Kim, W., S. Lim, J. Ahn, J. Nah and N. Kim. 2010. Integration of IEEE 1451 and HL7 exchanging information for patients' sensor data. *Journal of medical systems.* **34(6):**1033–41. doi: 10.1007/s10916-009-9322-5.

Lee, K. and E. Song. 2007. Wireless Sensor Network Based on IEEE 1451.0 and IEEE 1451.5-802.11. 2007 8th International Conference on Electronic Measurement and Instruments (pp. 4-7-4-11). IEEE. doi: 10.1109/ICEMI.2007.4351239.

Lee, P.G. 1995. A review of Automated Control Systems for Aquaculture and Design Criteria for Their Implementation. *Aquacultural Engineering.* **14(3):**205–227. doi: 10.1016/0144-8609(94)00002-I.

Lee, W.S., V. Alchanatis, C. Yang, M. Hirafuji, D. Moshou and C. Li. 2010. Sensing technologies for precision specialty crop production. *Computers and Electronics in Agriculture.* **74(1):**2–33. doi: 10.1016/j.compag.2010.08.005.

López-Riquelme, J.A., F. Soto, J. Suardíaz, P. Sánchez, A. Iborra and J.A. Vera. 2009. Wireless Sensor Networks for precision horticulture in Southern Spain. *Computers and Electronics in Agriculture.* **68:**25–35. doi: 10.1016/j.compag.2009.04.006.

Meijer, G.C.M. and C.M. Meijer. 2008. Smart sensor systems. Wiley Online Library.

Morais, R., M.A. Fernandes, S.G. Matos, C. Serôdio, P.J.S.G. Ferreira and M.J.C.S. Reis. 2008. A ZigBee multi-powered wireless acquisition device for remote sensing applications in precision viticulture. *Computers and Electronics in Agriculture.* **62(2):**94–106. doi: 10.1016/j.compag.2007.12.004.

NIST. 2011. 1451 Family. National Institute of Standards and Technology. Retrieved May 18, 2011, from http://www.nist.gov/el/isd/ieee/1451family.cfm.

Ramos, H.G. 2008. IEEE Standard 1451 and a proposed time synchronization approach. IEEE Instrumentation & Measurement Magazine. **11(2)**:29–37. doi: 10.1109/MIM.2008.4483731.

Rivera, J. 2009. Unidad V. CONTROL DE PROCESOS. *Reporte de Materia*. 1–30. Chihuahua, MX.

Sammarco-JJ, P. R. and K.V.K. Fries-EF. 2007. A Technology Review of Smart Sensors With Wireless Networks for Applications in Hazardous Work Environments. cdc.gov, 50. Pittsburgh, Pennsylvania, US.

Skierucha, W., A. Wilczek and R. Walczak. 2004. Application of smart sensors in the measurement of soil physical parameters. *Research in Agricultural Engineering*. **50(3)**:96–102. Lublin, Poland: Citeseer.

Sohraby, K., D. Minoli and T.F. Znati. 2007. Wireless sensor networks: Technology, protocols, and applications. Wiley-Blackwell.

Song, E.Y. and K.B. Lee. 2010. Interoperability Test of IEEE 1451.5 Standard-Based Wireless. 2010 International Conference on Measuring Technology and Mechatronics Automation (pp. 510–515). IEEE. doi: 10.1109/ICMTMA.2010.846.

Song, E. and K. Lee. 2008. Understanding IEEE 1451-Networked smart transducer interface standard—What is a smart transducer? IEEE Instrumentation & Measurement Magazine. **11(2)**:11–17. doi: 10.1109/MIM.2008.4483728.

Sung, J., T. Kim and D. Kim. 2010. Integration of IEEE1451 Sensor Networks and UPnP. 2010 7th IEEE Consumer Communications and Networking Conference (pp. 1–2). IEEE. doi: 10.1109/CCNC.2010.5421678.

Sweetser, D., V. Sweetser and J. Nemeth-Johannes. 2006. A modular approach to IEEE-1451.5 wireless sensor development. Proceedings of the 2006 IEEE Sensors Applications Symposium, 2006. (pp. 82–87). IEEE. doi: 10.1109/SAS.2006.1634242.

Wang, N., N. Zhang and M. Wang. 2006. Wireless sensors in agriculture and food industry—Recent development and future perspective. *Computers and Electronics in Agriculture*. **50(1)**:1–14. doi: 10.1016/j.compag.2005.09.003.

Wiczer, J. and K. Lee. 2005. A Unifying Standard for Interfacing Transducers to Networks–IEEE 1451.0. Proceedings of ISA Conference, Chicago, IL (pp. 1–10).

Wobschall, D. 2007. IEEE 1451—a universal transducer protocol standard. 2007 IEEE Autotestcon (pp. 359–363). IEEE. doi: 10.1109/AUTEST.2007.4374241.

Xiangyin, M., X. Shide, X. Ying and H. Huiping. 2009. ZigBee based wireless networked smart transducer and its application in supervision and control system for natural gas gate station. 2009 4th International Conference on Computer Science & Education (pp. 301–306). IEEE. doi: 10.1109/ICCSE.2009.5228440.

Zhaochun, L., H. Jin and C. Yuzhu. 2009. Approach to Wireless Sensor Network Development Based on IEEE1451. 2009 International Forum on Information Technology and Applications (pp. 443–446). IEEE. doi: 10.1109/IFITA.2009.563.

# 12

# Application Management in Low Power Distributed Embedded Systems

*F.M. Castanheira, A. Espírito-Santo and B.J.F. Ribeiro*

## Introduction

Embedded systems are nowadays widely found throughout the entire society. We can find them across all kinds of activities. The application field is so wide that it goes from industry, passing by biomedical applications, or by home automation, without forgetting its use in personal assistance applications. This dissemination has required, at different levels, a high increase of the structural complexity of these systems. In this context, we can forecast that several processing units could cooperate jointly to perform one or more tasks simultaneously. These tasks have a distributed characteristic, demanding, at the same time, the availability of real-time processing capacities executed by a single processing unit, or shared by a pool of cooperating processing units. This scenario is possible only if an infrastructure that supports this mode of operation is available.

This chapter presents the requirements of a distributed system, where all processing sub-units could share data and services among them. After an introductory section, the chapter presents a new infrastructure proposal designed to support the cooperation of several processing sub-units through the sharing of data, functions and services. The proposed infrastructure concept is illustrated through the application in the development of a modular energy management system built around the MSP430 low power family devices from Texas Instruments.

The advantages achieved by this concept are discussed through this example.

Information shared among all the processing units that are part of the distributed system must observe the localization of each one of them. Some processing sub-units can be placed very close, even a few centimetres from each other; in this situation, a wired communication line is a reasonable solution. Typically, neighbourhood situations can be managed by one of the existing low level protocols, I2C or SPI. These two protocols only specify how bytes are transmitted and received. The communication protocol must be implemented by the user. Almost all low power microcontrollers commercially available support I2C or SPI interface. The transceiver is implemented in hardware, or can be easily implemented in software.

More advanced protocols can be found, as for example de Controller Area Network (CAN). These communication protocols are almost totally specified. Unfortunately, they also require a lot of processing power, and their implementation, in a low power device, is rather cumbersome. Besides this, these protocols are not widely available in most of the low power microprocessors commercially available.

We must also consider the possibility of some of the processing units being geographically apart. This being said, some, or even all the processing sub-units, can only be accessed by a radio frequency link. The study of wireless protocols is actually one of the most active research branches in the area of embedded systems development, particularly in the field of low power consumption. Although there are several protocols that support wireless communications, in general, they focus on communication management, but the implementation of the application layer is left to the developer's responsibility. The computational resources required by these protocols are high. This results, at times, in an inability to share the resources available at the processing unit between the user application and the communications management task.

Special attention must be paid to the applications with restrictions at the power level. As system distribution increases, energy availability becomes a concern. Typically, for each processing unit, an independent power source must be provided. This requirement is fulfilled by a battery. Furthermore, low power devices also have small memory spaces to store data and a program code. Therefore, the development of applications must take into consideration the following aspects: application distribution, low power requirements, low computation power and reduced memory spaces.

## A Brief Review of the State-of-the-Art

Several examples can be found in the literature where the concept presented in the introductory session is, undoubtedly, a differentiating factor of the performance and quality of a distributed system. In the development of an infrastructure to support the operation of the distributed system as described, we must also consider a set of restrictions as the ones that will be discussed next.

A dynamic process to reconfigure a distributed system in real-time is described in (Brinkschulte et al., 2005). This issue is very important to systems that cannot be stopped. The reconfiguration process is required each time a new module is added to the system or the operation of an existing module must be changed. Two different methods can be used to reconfigure the distributed system: (1) static reconfiguration demands that the system go off-line, while (2) dynamic reconfiguration is performed in run-time.

The International Standard IEC61499 is proposed by the International Electrotechnical Commission to help the design of distributed systems (IEC61499 2004). This standard defines a function block to support application functionalities. This entity interacts with the external world through events that collect information related with the processes under supervision. The automatic reconfiguration topic is addressed in (Khalgui et al., 2011), where the authors propose a multiagent distributed architecture. The acquisition of information from the environment where the processing unit is inserted can be performed by different kinds of sensors. A non-electrical signal, such as temperature or pressure, for example, is converted by the sensor into an electrical parameter as resistance, voltage or current through analogue and digital electronics. The information collected by the processing sub-unit can be used by itself or sent to another one, belonging to the same distributed system. This task can only be performed if the infrastructure provides services to support data exchange among processing sub-units. The information acquired by the processing sub-unit can be processed locally or require additional information acquired by other processing sub-units. This task can be executed by the same processing sub-unit or can be distributed by several processing units. The communication infrastructure must be robust if information acquisition is performed by a distributed architecture. An illustrative example can be found in (Capatina et al., 2006), where industrial standards, such as: Profius, Modbus and OPC, are employed to acquire and manage data in a hydro-power plant.

Distributed systems operation should be optimized at different levels. The algorithm inside the COSYN tool developed by (Dave et al., 1999) meets performance, power, and cost goals. The distributed system's flexibility is due to the hardware's characteristics as well as the firmware which makes it work. This flexibility is important for the system's success and acceptance, seeing as it allows an effective adaptation to various situations. In functional terms, flexibility should translate into an easy extension of the system's capacities and should also contemplate quick functional modifications. The characteristics already identified permit a low production cost as well as an efficient performance. In the bibliography we can find papers which complement the problem addressed in this chapter.

The manufacturing process behind the Universal Robot Bus (URB) which helps to quickly integrate any sensor and heterogeneous actuator into a sole processing unit is described in (Saranli et al., 2011). The advantage of this technology resides in its ability to support heterogeneous collections of communication protocols and to include components with different sizes and capacities in the same system.

A communication methodology between devices making use of the CONIC language is presented in (Sloman et al., 2011). This programming environment, described in (Magee et al., 1989), provides a language based on an approach to distributed system construction which combines a simple and secure language with the flexibility and accessibility of the operating system. This system is designed for architectures of embedded systems with a distributive nature.

Research by (Brinkschulte et al., 2001) focused on the development of middleware which supports heterogeneous distributed real-time systems and simultaneously allows for the use of small microcontrollers in computational nodes. Heterogeneous systems can use OSA+[1] architecture to support communications between them. The proposed platform can be used not only at workstations, but in embedded systems, which are low cost, have little memory and possess reduced processing capacity.

Sharing the same perspective (Merrett et al., 2008) describes a simplified architecture for structures with multiple interfaces such as, for example, sensorial processes, and energy and communication management. Each sensor node has multiple stacks linked by a shared application layer.

In addition to the papers presented, research was also carried out regarding already existing task scheduling algorithms. Beyond the architecture itself, another important consideration is the priority given to tasks. In (Pop et al., 2004), a new methodology for task optimization

1 OSA+: Open System Architecture—Platform for Universal Services.

is presented. The approach must satisfy two conditions: first, it must minimize the disturbance of the applications already underway and, possibly in the near future, easily add new features to the system. In the same scope, another paper, presented by (Luo and Jha 2007), describes a new algorithm to reduce energy consumption in real-time heterogeneous distributed systems. This algorithm is based on optimizing the execution sequence of scheduled tasks. An alternative analysis of the consumed energy in real-time processes is discussed in (Mahapatra and Zhao 2005). It also introduced a new operation model for networks that allows energy consumption optimization. The proposed approach uses a detailed description of network load in the nodes and provides adequate information about traffic behavior, in the worst case scenario, anywhere in the distributed network. The system is thus enhanced with features that allow it to operate with greater energy efficiency.

## System Architecture Components

In order to develop a system that fulfils the compromise between flexibility and low power consumption, a modular architecture was defined, based on a heavily distributed structure supported on two distinct communications mediums: (i) I2C communications, which forms the backbone of the cluster unit; (ii) wireless communications in the 2.4GHz band, which allows the cluster unit the means that permits geographic dispersion beyond the electric cabinet where it resides. Figure 1 summarizes the view of the proposed system.

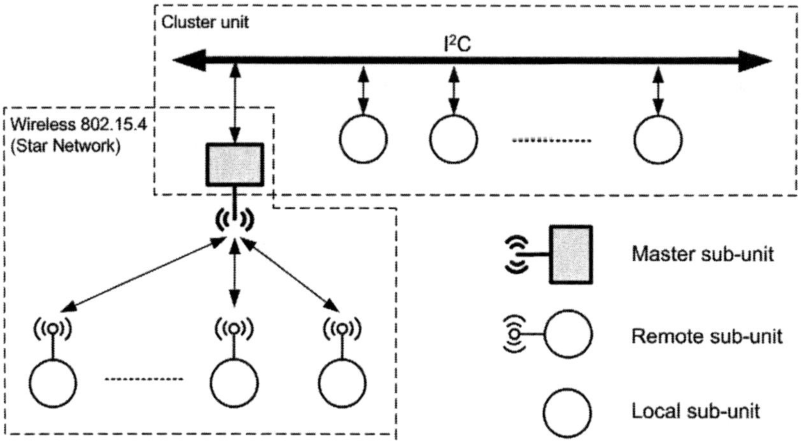


**Figure 1:** General cluster architecture.

## Communication Medium

A distributed system can be seen from various points of view. The sub-units that compose it may reside within the same box, or, alternatively, may be dispersed over a geographical area with a few dozen square meters. The adopted configuration depends on the type of distributed application selected. In any one of the above scenarios, there may be need to change the system configuration at anytime, which is possible either by changing the options of individual processing sub-units present, or through its substitution. Therefore, the communication medium should allow users to change its settings at any time which, in turn, changes its behaviour.

When the sub-units that make up the distributed system are located in the same physical space, the I2C bus becomes a very attractive solution. If there is a physical dispersion of sub-units, a connection through a radio link is often the most practical option to adopt.

Therefore, in a flexible distributed system, two physical communication mediums should coexist. Regardless of the medium chosen, it is important that the features supported by the distributed system are not affected. This means that the functionalities provided by the infrastructure should not be limited due to the physical connection. The following describes the operating modes of each form of communication used by the proposed distributed system, with particular emphasis on its features and how they contribute to increase system's performance.

### *Cluster incorporation—Bus I2C*

The adopted solution is based on a serial bus, and the main criteria behind this choice are: the physical dimensions, length of the bus, the rate of data transmission and availability of series communication interfaces in the MSP430 family of microcontrollers. It is important to note that the medium access mechanism is part of the "based Packet/Collision free" protocols, although the I2C bus supports multi-master and enables the implementation of a "Packet based/Collision avoidance" protocol, CSMA type. This case was chosen because it provides the master control over bus access, and, consequentially data exchange always takes place in this format (Gonçalves et al., 2011).

### *Cluster expansion—Wireless network*

The adopted solution establishes a low power star network topology in the 2.4 GHz radio band. The main selection criteria: the geographical area to cover, the rate of data transmission and the power required during the connection. In order to minimize energy consumption, message exchange in the network is done indirectly through the master sub-unit, which is also, the wireless network coordinator.

*Cluster sub-units*

A cluster unit is an aggregation of several independent sub-units that are located in the same space. By sharing resources, they can create a functional system that implements a complex task. The accomplishment of the task is achieved through cooperation between the various sub-units. In relation to the action performed by the system, this can be seen as a single entity that assumes different configurations, depending on what is required.

*Master sub-unit*

There are permanent and indispensable functions within the system to ensure its normal operation and the system would be inoperable without them. These functions include: (i) power management of sub-units which are incorporated in the cluster and (ii) communication management between the sub-units (medium access and message routing). The sub-unit master ensures these basic functions and in this way becomes the system's fundamental sub-unit. Since this sub-unit manages all communications, it needs substantial data memory and processing capacity to perform these functions as required.

*Local sub-unit*

The system's flexibility is achieved by allowing the possibility, theoretically unlimited, of adding new functions to the system. Functions that are strictly part of the system's functions range, in the distributed application context, are performed by single sub-units. From the network's viewpoint, these sub-units act as slave elements without autonomy to access the communications medium.

*Remote sub-unit*

To address the limitations inherent to the geographical location of the cluster unit, the use of wireless links is a convenient and effective solution to achieve greater geographic dispersion. The sub-units assist achieve a greater autonomy; they do not share the power supply because they have their own power. Their functions fall within the range of functions performed by simple sub-units.

# Network Operation Model

## Network Layer

The need to coordinate two different communication mediums led to the need to integrate a network management layer in the protocol. As shown in figure 2, this entity appears in two distinct ways: (i) Master Network Manager, whose primary mission is to start the network, coordinate the various communication mediums, perform message routing and check the network's vitality, and (ii) Slave Network Manager, which assumes the slave role in the network startup and participates in the network vitality verification.

### *Network startup/assembly*

The use of wireless sub-units introduces the need to establish rules that enable the coexistence of multiple networks operating in the same area. Thus, it is imperative that the master sub-unit and the simple sub-units know their potential partners in the system.

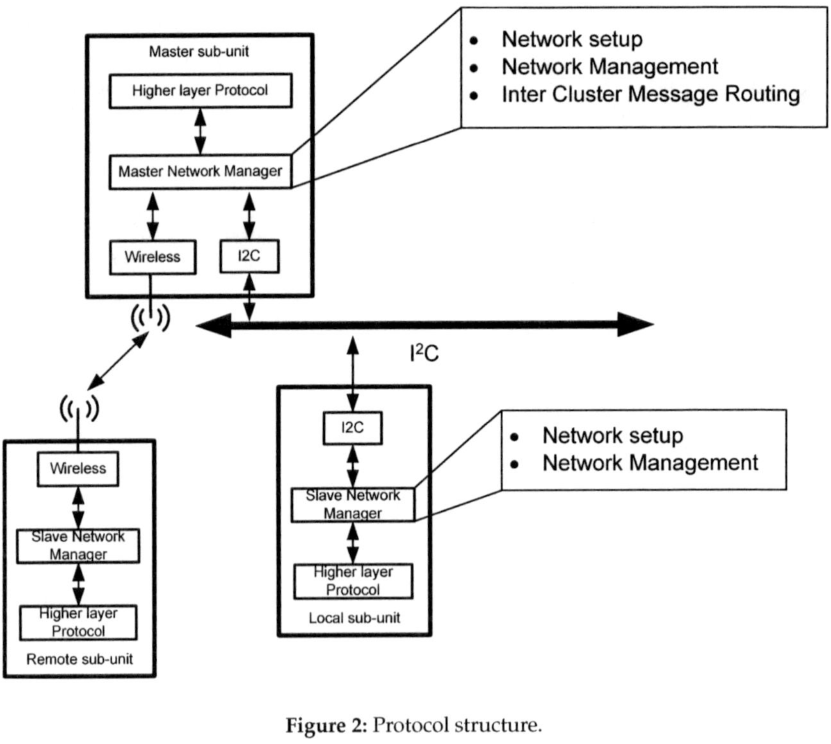


**Figure 2:** Protocol structure.

The network startup involves two distinct phases: (i) initialization, which is the responsibility of the master network, is essential to start the wireless network, and (ii) association, whose goal is to extend network capacity, binding its various sub-units by using an association procedure. It is important to note that the routing table is created in this stage.

## Message Routing Operation

As previously mentioned, two contingencies were created that resulted from using two different communication technologies, which forces users to employ message routing: (i) the impossibility of delivering messages directly between the cluster sub-units and wireless sub-units, (ii) the communication model proposed for the wireless sub-units to ensure effective energy management, and (iii) medium access control applied to the network cluster.

The routing function is unique to the master; the other sub-units do not play any role in forwarding messages. The routing algorithm used is presented in the flowchart shown in figure 3 and is based on finding the location of the message's destination address by means of the routing table.

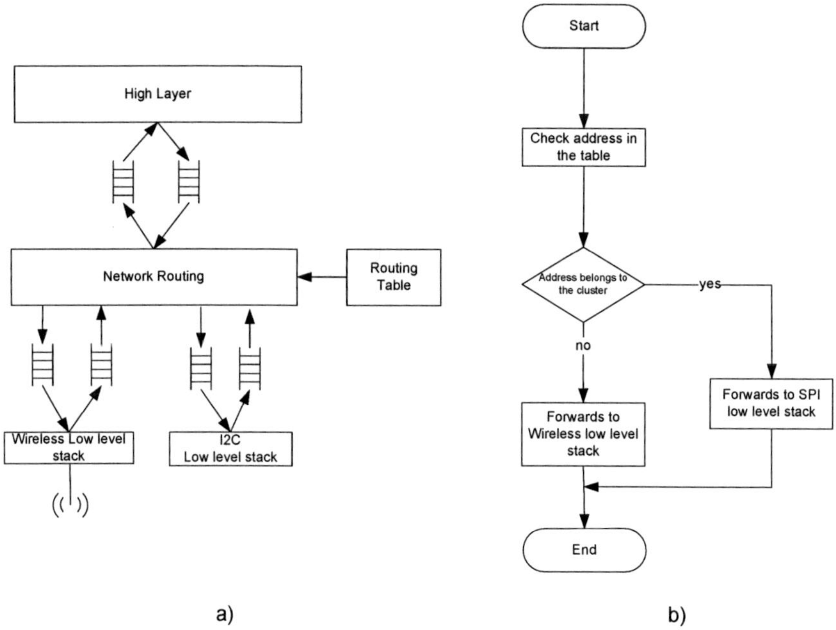


**Figure 3:** Routing operation.

## Data Exchange Model and Application Protocol

The communication protocol described below will help build the distributed applications. It is based on the invocation of remote actions RPC (remote procedure call), allowing virtual connections between various elements of the system. The message transmission mechanism is based on the classic Request-Response form (see figure 4). If, by chance, a message is not delivered or there is no response from the interlocutor, the protocol ensures that the application produces an error indication instead of the expected response. This response indicates the cause of error, which will be treated later by the application.

Each sub-unit has one or more skills in the distributed application context. These skills are known as services, as depicted in figure 5. Any service encapsulates one, or more, objects. The objects can live all in the same sub-unit, or can be distributed by several sub-units inside the cluster. A special service "Device" is encapsulated by the info object class that will be described next. It's mandatory that this object class must be present in

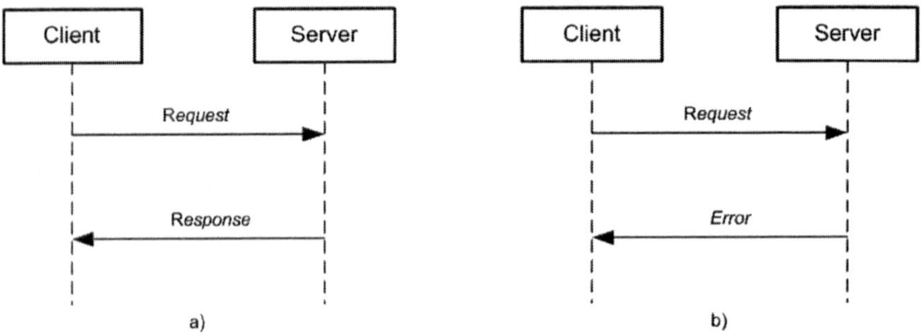


**Figure 4:** Message transmission mechanism (a) success, (b) failure.

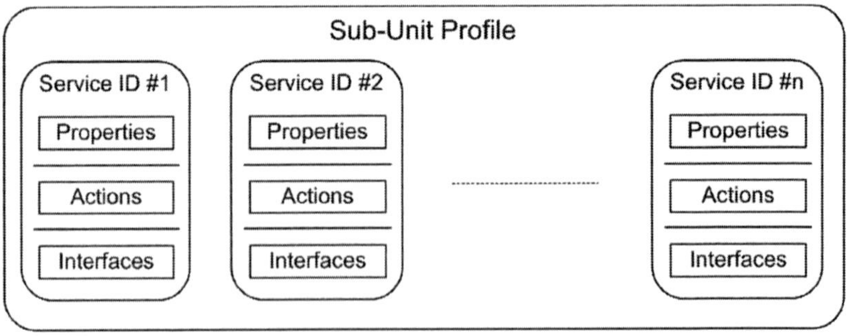


**Figure 5:** Example of a device (sub-unit) that supports several services.

all computation nodes. The information collected by the service "Device" is used by the distributed application's setup.

In the distributed system under consideration, there is no device which works as a service administrator and, therefore, the devices do not know the neighbouring network services. Each device only knows the services available therein. The identification of a service is done through an 8-bit identifier extension, allowing a total of 255 different services to exist in the system. During the distributed application setup, the application configuration, executed in a personal computer connected to the distributed system, performs a search of the provided services by all of the system's sub-units. The search produces a database which contains the services provided by these sub-units. The link configuration between the services present in each sub-unit produces the distributed application.

The protocol's package format is shown schematically in figure 6. It has five fields: (i) the "Control: field identifies the type of message and the need to acknowledge (ii) the "Address" field identifies the source and destination address of the data (iii) the "Seq. Number" field allows to relate a request and its answer, both messages should present the same sequence number—so, this field is added to each message sent (iv) the "size" field refers to the number of bytes contained in the message, and (v) the "data/command" field represents the portion of the message reserved for application data.

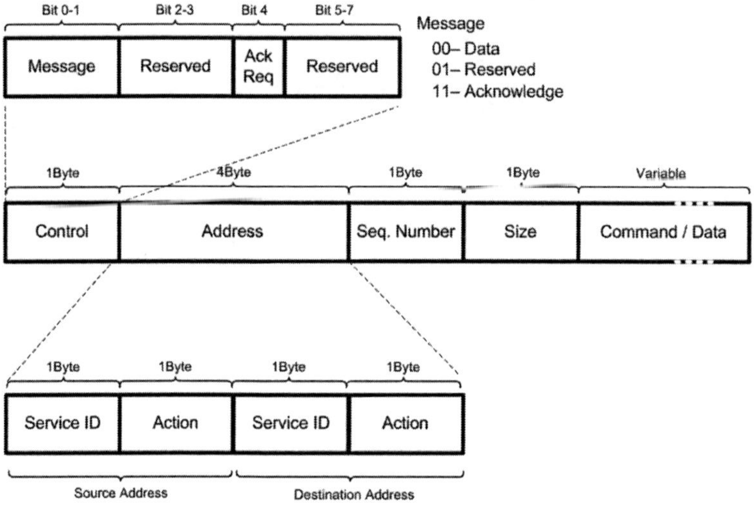


**Figure 6:** Application Layer frame format.

## Electric Energy Management Profile

The proposed infrastructure can support different applications. Following the previously described steps, users can quickly development different applications. This characteristic is supported by the ease with which new modules can be developed. Therefore, an existing infrastructure can be adapted by introducing new specific modules to develop new distributed products, including systems in the areas of: surveillance and security, agricultural plantation monitoring; industrial process supervision and monitoring and energy management, among others. The distributed infrastructure previously described is used to perform energy management actions in an electrical power system.

The current concern about the rational use of energy has contributed to the development of solutions that more efficiently manage energy resources. However, efficient management of these resources nearly always requires a system to collect information to support decision making. Simultaneously, the system should allow different strategies that permit the desired energy performance. Load shedding can be performed on heating, ventilation, air conditioning, and lighting appliances, when a threshold of power consumption is reached. This strategy allows controlling energy consumption pecks. The maximum power consumption criteria, at a given time period, promotes energy consumption deviation to more economically favourable time frames.

The electrical network has a distributed nature. This configuration is imposed by the characteristics of the productive process served by the electrical installation. The introduction of an energy management system should follow the distributed nature of the electrical system that will monitor/manage the system. This approach represents an important contribution to raising the overall performance of all processes involved.

The energy management system described below allows users to achieve the established objectives of managing energy resources by installing a distributed system cluster, consisting of distributed sub-units. The sub-units can be classified as local or remote, depending on the communication medium (I2C or wireless). The distributed system's features are established through each sub-unit's settings. Normally, local sub-units, including the cluster management sub-unit, are installed in an electric cabinet. There is the possibility of adding more local sub-units via the I2C bus if necessary. The cluster can also have sub-units that allow remote control operation of the state of loads through relays and push buttons. The remote sub-units are connected to the cluster by radio frequency.

In order to explain the distributed application programming methodology, we introduce concepts associated to object-oriented programming. The objects defined in each of the sub-units set the behaviour of this sub-unit within the distributed system. An object always has an identifier field, unique within the sub-unit. Each sub-unit also has a unique identifier within the cluster. The objects have a set of methods, public or private, that may be linked to other methods, belonging to objects that exist within the same or other sub-units. Before the cluster can perform the assigned task, the distributed system must be configured. This operation is performed by using a specific configuration method, called *Config()*, that establishes the links between the object in question and the various objects to which it relates. The *Proc_MSG()* method that belongs to the objects processes the message exchange between objects within the cluster. The following sections present the classes that provide the features of the application to each sub-unit.

**Distributed Energy Management Cluster Profile Overview**

The cluster consists of several sub-units capable of operating by sharing data and functions in a distributed fashion. The cluster system, which is described below, has both local and remote sub-units. The cluster's management is executed by a special sub-unit (master sub-unit) which, in addition to managing the cluster, also provides the energy required for local sub-unit operation. This sub-unit is built around a MSP430F5438A microcontroller that has greater computing capacity. With exception to the energy measuring sub-unit, which is built around the MSP430F47197 microcontroller, the remaining sub-units are built on the MSP430F2231 microcontroller. The MSP430 microcontroller family allows developing applications whose energy consumption represents a significant constraint, since it is a device with low-power operation characteristics. In the illustrated example detailed in the following sections, two energy management sub-units, an interface sub-unit with the user and a USB/PC interface sub-unit are used. These local sub-units are installed in the electric cabinet that powers the loads to be managed. Two remote control sub-units are located near the user and allow direct control over the two charges operating status in an independent form. A remote relay is located next to the loads to be managed and determines, independently, the operating state of each load.

**Sub-unit Description Class**

The object Info class (see figure 7) is present in all cluster sub-units, which serves to describe the sub-unit's content and nature. This information is

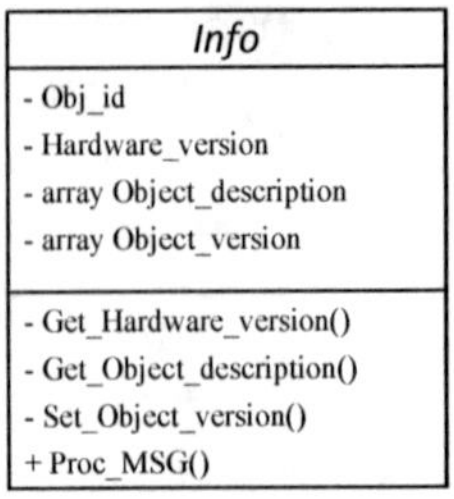


**Figure 7:** Sub-unit description class.

required so that the sub-unit may be integrated into the cluster: hardware version; the type of objects present, description and version. This information is defined when the firmware is compiled. In addition to this information, the object class provides methods that allow access to each of the parameters in an individual or global manner.

## Energy Measurement And Energy Management Classes

The energy measurement is carried out by an object belonging to the Measurement class. The objects found in this class should work with the MSP430F47197 microcontroller and have the ability to perform measurements on both single phase and three-phase installations, depending on how their acquisition channels are connected. The measurement process consists of two different tasks: One, with a fixed duration, which runs along the fundamental harmonic cycle is called *sampling_period*, which collects samples of voltage and current so as to elaborate a pre-processing measurement; the other, with a variable length, but multiple of the previous, determines the average values computed during the *Measurement_period*.

The sampling process is controlled by the *Timer_ADC_Start()* event that, at regular intervals, initiates the ADC conversion, the *ADC_EOC()* event, triggered by the end of the conversion performed by the ADC, collects samples and performs a pre-processing measurement. Calculation algorithms are then applied to the information gathered that update the *Sum_period* data structure of the acquisition period. Finally, this data structure is copied for processing to the *Sum_period_shadow* structure, thus enabling the start of the new acquisition. The *Acquisition_period_process()* method determines the measured value corresponding to the sampling period. These values then update the *Sum_measurement_period* structure.

At the end of the measuring period, the values of the *Sum_measurement_period* structure are placed in the *Sum_measurement_period_* *shadow* structure to allow the measuring process to continue by using the *measurement_period_process()* method, which determines the average values that corresponds to this measurement period .

The Measurement class of objects can be simultaneously connected to two different classes of objects. The object from the *DataStructure* class shown in figure 8 has the ability to store the measured data. This object is identified by the *Data_Unit_Link_id* and the *Data_Obj_Link_id* properties. Data gathered during the sampling period is stored by the *Data_adquisition_period_Method_Link_id()* method. Data related to the measurement period is stored by the *Data_measurement_period_Method_Link_id()* method. The other object that it can be connected to is the Relay class that allows control over the operating state of the load, shown in figure 11. The maximum power allowed is observed during each acquisition interval. If there is a violation of the *Max_active_power* value, a message is sent to the class object identified by *Load_Unit Link_id, Load_Obj_Link_id* using the method identified by *Load_Method_Link_id*. At the end of the time established in *Disabled_time,* the load is reactivated using the same procedure.

Associated to the features of the objects of the class described earlier, there is a class of DataStructure objects where measured values are stored. This class of objects has two distinct fields: one field has arrays that store data related to the last completed period of acquisition; the other field saves structures where the successive values related to the measurement intervals are stored since the beginning of the measuring process. This object class provides a set of methods that allow reading, writing and managing the status of the data structures of the objects of this class. The classes described are shown in figure 8.

**Interface Classes**

The interface classes provide the ability to interface with the distributed system. The interface can be divided into two distinct types: the user interface and the interface with a personal computer.

The UserInterface class of objects allows the user to view information gathered by the distributed system through the objects of the measurement class. The interaction is done by hardware with a LCD and four buttons. Since the latter are capable of causing events. The information is displayed on the 16 × 2 alpha numeric LCD, taking into account the context in which the interface is at the refreshing time. The context is defined by the LCD page being exhibited, identified by the *pag_id* parameter, and the cursor position stored in the object fields *cursor_x* and *cursor_y*. By using the context, it is possible to implement a current state machine which allows users to browse the available information. When a page is entered, the

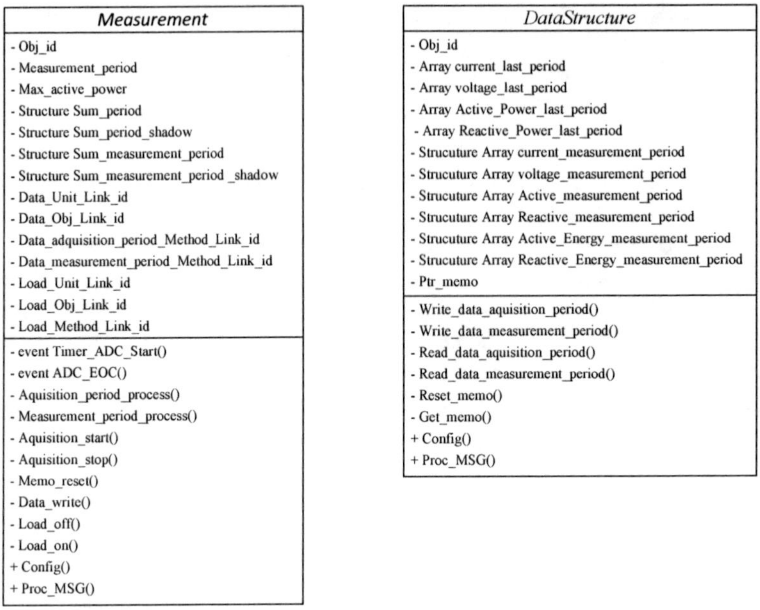


**Figure 8:** Interface classes: Measurement and DataStructure classes.

context is used by the *Get_Data()* method to request the information necessary for visualization. The page is actualized on a time basis. The location information of data to be displayed is stored in an object class *Data_Link*, identified in the array of parameters: *unit_Link_id* and *Obj_Link_id*. Using arrays allows object classes to be linked to various data structures of the class DataStructure simultaneously.

Whenever a push button is triggered, it generates an event that responds according to the keyboard state machine, taking into account the context of the object. Each button is associated to an event: *up_press()*; *down_press()*; *next_press()*; *enter_press()*.

The personal computer interface is made of a class of objects called *USB_PC*. This class always resides in a sub-unit that supports USB interface hardware. The class allows personal computer access to the public method *config()* objects, used to build the network. Access to the data structure as well as the objects of the UserInterface class is also possible. The three classes described above are shown in figure 9.

The user interface state machine shown in figure 10 allows multiple data structures simultaneously. In the main menu, and using the next button, the user chooses the data structure to view. Upon pressing Enter, the user interface state machine will be taken in such a way that some of the quantities acquired can be viewed, including: voltage, current, active power and reactive power, based on the sampling period. Apart from

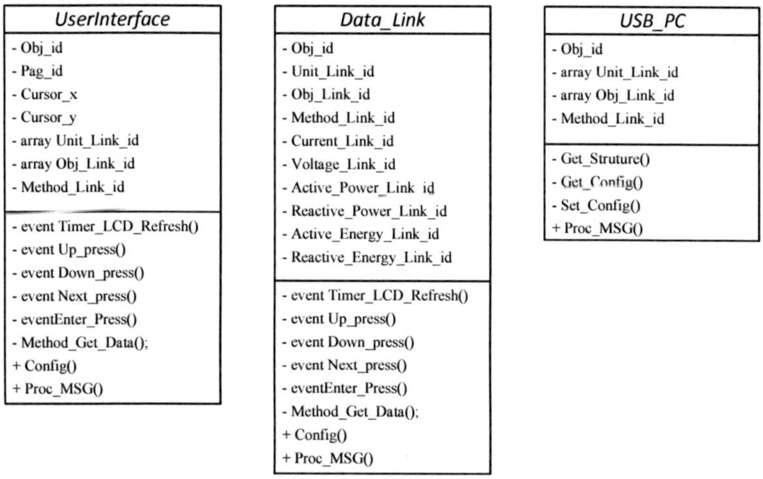


**Figure 9:** Interface classes: *UserInterface, Data_link* and *USB_PC.*

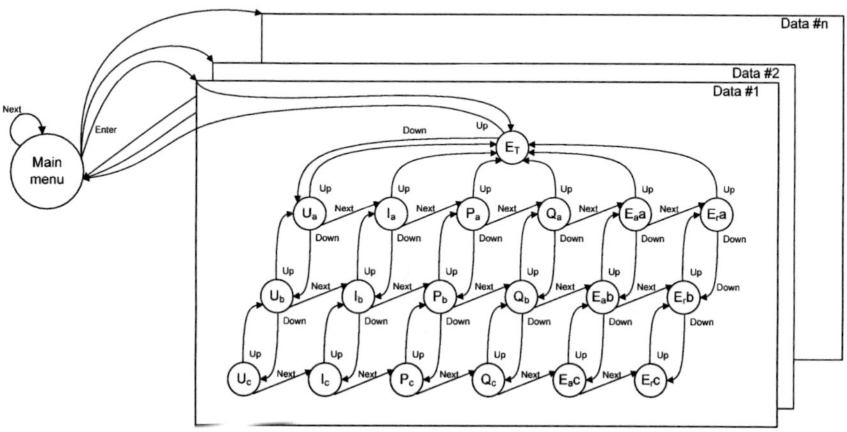


**Figure 10:** User interface state machine: LCD and Keyboard.

these, the user can still see the value of active and reactive energy totals from the beginning of the measuring process.

## Load Management Classes

The main activity of the energy management sub-unit is to control the load status. Whenever a power limit is reached, the power management sub-unit commands a shutdown of the corresponding load for a pre-determined period of time. The load's operating status can also be controlled directly by the user via a pushbutton. This function is performed by objects that belong to the following classes: LED, Button and Relay.

The *Toggle_Load()* method from the Relay class, identified by the *method_Toggle_Load_id* property, enables switching the relay state of associated hardware. Whenever this action is performed, a message is sent by the *ACK_MSG()* method, identified by the *method_ACK_MSG_id* property, informing the object from the LED class that belongs to the *Unit_Link_id* unit and identified by the *Obj_Link_id*, soliciting the execution of the method identified by the *Method_Link_id*. Whenever the operating status of the load is to be changed, the user must press the buttons that command this desire. When the network configuration is takes place, the object of the Button class that has the *Obj_id* identification is connected to the object from the Relay class, linked to the *Unit_link_id* sub-unit, with the *Obj_Link_id* identification. When the *Press()* event of the Button class object takes place, a request to execute the method identified by *Method_Link_id* is sent, belonging to the object identified as *Obj_Link_id* of the sub-unit and identified as *Unit_Link_id*. The method pertaining to the previous reference should match the *Toggle_Load()* method of a Relay class object. The LED class allows for the indication of the operating status of objects to which they are connected. This class has the LED_Status property which indicates the LED status. The *LED_Toggle()* method, identified by the *Method_LED_Toggle_id* property, can toggle the LED status. The three classes described above are shown in figure 11.

## Application Example

An illustrative example of a distributed cluster to perform power management is given by way of example. The previously described sub-units are programed with the objects classes. After the physical installation of the sub-units, as shown in figure 12, the configuration's functionality is used, through a computer application running on the

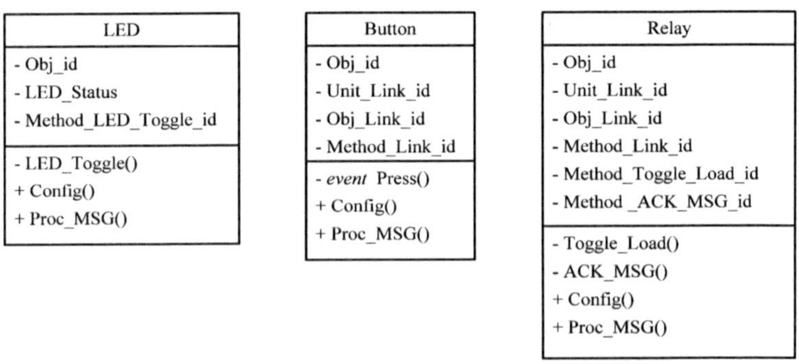


**Figure 11:** Load management classes: LED, Button and Relay.

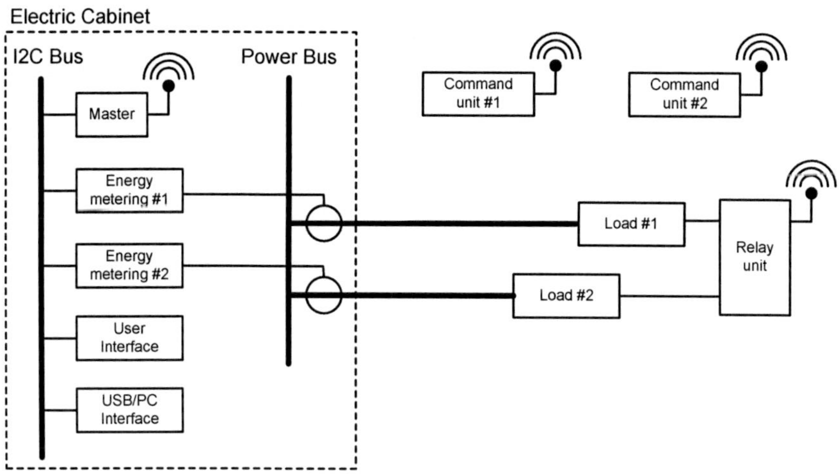


**Figure 12:** Energy Management Cluster: hardware layout.

personal computer, to establish the functional connections between objects residing in multiple sub-units. The application setup is connected to the cluster through the USB/PC interface sub-unit. This module allows access to *Config()* configuration methods of the classes being used to establish the effectiveness of the distributed network.

The operation strategy is defined as follows:

- Through the features described above, the USB/PC interface sub-unit allows the clusters to communicate with a computer. The configuration of each of the sub-units in the cluster is carried out through this channel of communication that may require any of the available services by the configuration application on the computer.
- The direct modification of the operating status of the two electric charges is possible through the remote control sub-unit. The equipment's operating status is indicated by the LED status included in these sub-units.
- The Relay sub-unit controls the load #1 and load #2 operating status; therefore, it has two relays.
- The measurement sub-units 1 and 2 monitor the energy consumption of loads 1 and 2, respectively. The information is stored in memory for later analysis. If the threshold value of the load's power is exceeded, the corresponding measurement sub-unit orders the temporary shedding of the load in question.
- The interface sub-unit and the user allow observation of the values measured by the two sub-units of measure.

A detail of the distributed system is shown is fig. 13. Each command sub-unit has one object of the Button class and one of the LED class. The relay sub-unit has two objects of the Relay class. The links between the objects establish the operation; these connections can be seen in fig. 14.

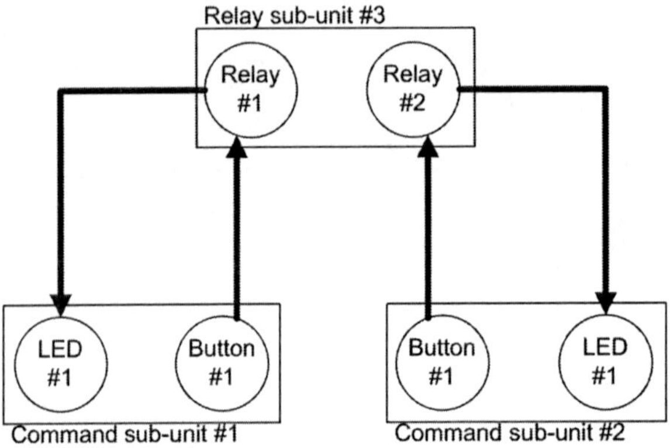


**Figure 13:** Distributed application detail: buttons, LED and relays operation.

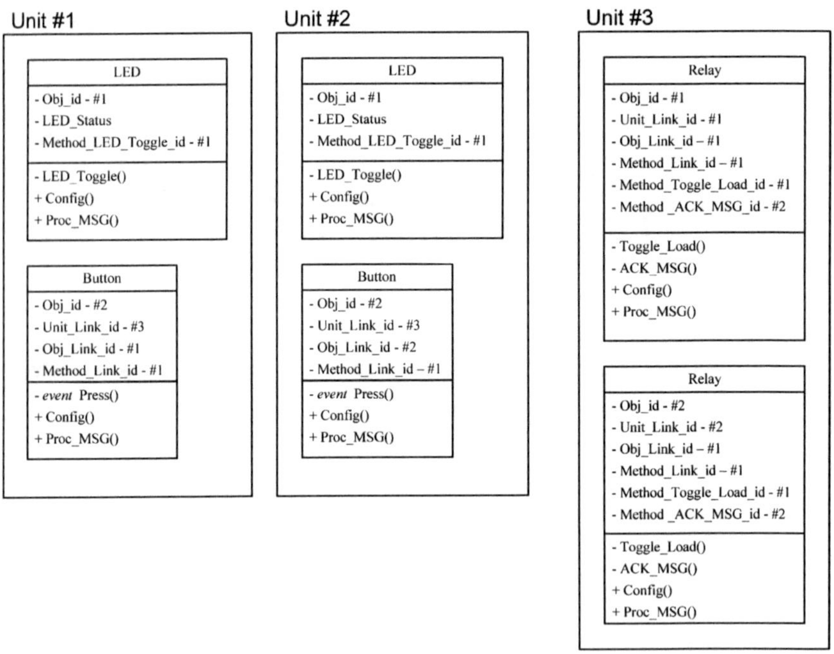


**Figure 14:** Distributed objects configuration.

## Conclusions

A design system that would carry out functions in the broad sense of instrumentation and automation systems was briefly described. Its key capabilities were presented, emphasizing its highly distributed nature and its main operating mechanism, based on virtual connections that result from the dynamic definition of parameters in the various sub-units that belong to the cluster.

The implementation of a distributed system following the methodology proposed in this chapter will achieve the following: (i) distributed architecture—high reliability, modularity and flexibility, which are inherent to distributed systems; (ii) in the wake of virtual instrumentation—ease of installation with the distributed application features scattered throughout the various sub-units, the system may be modified at any time with the addition of new modules, easy maintenance, virtual connections established between sub-units that regulate the functions of the distributed system; (iii) practical aspects—complete absence of cables and performance located in the electrical cabinet.

## REFERENCES

Brinkschulte, U., A. Bechina, F. Picioroaga, E. Schneider, T. Ungerer, J. Kreuzinger and M. Pfeffer. 2001. "A Microkernel Middleware Architecture for Distributed Embedded Real-Time Systems", Proceedings of the 20th IEEE Symposium on Reliable Distributed Systems, October.

Brinkschulte, U., E. Schneider and F. Picioroaga. 2005. "Dynamic Real-time Reconfiguration in Distributed Systems: Timing Issues and Solutions", Proceedings of the Eighth IEEE International Symposium on Object-Oriented Real-Time Distributed Computing, May.

Capatina, D., I. Stoian, T. Sanislav, O. Ghiran, E. Stancel and I. Filip. 2005. "Integration Techniques of the Embedded Distributed Systems Using Programming Environments and Industrial Standard Communication Protocols", Proceedings of the 2006 IEEE International Conference on Automation, Quality and Testing, Robotics. May.

Dave, B.P., G. Lakshminarayana and N.K. Jha. 1999. "COSYN: Hardware-Software Co-Synthesis of Heterogeneous Distributed Embedded Systems", IEEE Transactions on Very Large Scale Integration (VLSI) Systems, Vol. 7, No. 1, March.

Gonçalves T., A. Espírito-Santo, B.J.F. Ribeiro and P.D. Gaspar. A Scalable Hardware Enviroment for Embedded Systems Education. pp. 233–245. In: Sio-Iong Ao and Len Gelman [eds.] 2011. Electrical Engineering and Applied Computing, Lecture Notes in Electrical Engineering, Volume 90. Springer Science.

IEC61499—Industrial Process Measurements and Control Systems, I.S. IEC61499, 2004.

Khalgui, M. and Olfa Mosbahi, Zhiwu Li, and Hans-Michael Hanisch. 2011. "Reconfiguration of Distributed Embedded-Control Systems", IEEE/ASME Transactions on Mechatronics, Vol. 16, No. 4, August.

Luo, J. and N. K. Jha. 2007. "Power-Efficient Scheduling for Heterogeneous Distributed Real-Time Embedded Systems", IEEE Transactions on Computer-Aided Design of Integrated Circuits and Systems, Vol. 26, No. 6, June.

Magee, J., Jeff Kramer and Morris Sloman. 1989. "Constructing Distributed Systems in Conic", IEEE Transactions on Software Engineering. Vol. 15, No. 6, June.

Mahapatra, R.N. and W. Zhao. 2005. "An Energy-Efficient Slack Distribution Technique for Multimode Distributed Real-Time Embedded Systems", IEEE Transactions on Parallel and Distributed Systems, Vol. 16, No. 7, July.

Merrett, G.V., A.S. Weddell, N.R. Harris, B.M. Al-Hashimi and N.M. White. 2008. "A Structured Hardware/Software Architecture for Embedded Sensor Nodes", March.

Pop, P., P. Eles, Z. Peng and T. Pop. 2004. "Scheduling and Mapping in an Incremental Design Methodology for Distributed Real-Time Embedded Systems", IEEE Transactions on Very Large Scale Integration (VLSI) Systems, Vol. 12, No. 8, August.

Saranli, U., A. Avci and M. C. Öztürk. 2011. "A Modular Real-Time Fieldbus Architecture for Mobile Robotic Platforms", IEEE Transactions on Instrumentation and Measurement, Vol. 60, No. 3, March.

Sloman, M., J. Kramer, J. Magee and K. Twidle. 1986. "Flexible communication structure for distributed embedded systems", IEE Proceedings, Vol. 133, Pt. E, No. 4, July.

# 13

# Embedded Systems and Applications in Telecommunications

*Sorin Zoican*

## ABSTRACT

The purpose of this chapter is to illustrate how an embedded system for telecommunications can be developed using digital signal processors (DSP). The generic DSP architecture is presented and real-time applications of digital signal processing in telecommunications domains are exemplified.

This chapter provides an insight into the implementation of real-time applications using digital signal processors. The presentation is done gradually, starting from general principles to the practical realm to exemplify the programs that implement different applications.

The principles and examples presented are especially useful in developing telecommunications applications and other digital signal processor applications.

The structure of the chapter is as follows:

1. Introduction—presents general characteristics of a signal processor, parallel processing mechanisms and how DSP can be used to developing of telecommunications applications.
2. Background—illustrates the embedded system architecture for signal processing, the unified micro-architecture, two DSP processor architectures (ADSP21xx and Blackfin BF5xx) with comparison and benchmarks and a real-time operating system kernel to support complex telecommunications applications.
3. General framework—for applications development employing DSP processors.
4. Examples of real-time signal processing applications—adaptive echo cancellation, discrete multi-tone (DMT) modulator and demodulator and the development of network applications.

5. Future research directions.
6. Conclusions.

## 1. Introduction

The purpose of this chapter is to illustrate how embedded systems with applications in telecommunications can be developed, both in hardware and software, using digital signal processors (DSP). The motivation of using DSPs in telecommunications is quite obvious: most telecommunications applications, especially 3G wireless communications, require significant computational power and operating system support. DSPs have the required computational resources in contrast to general-purpose microcomputers. Furthermore, DSPs have sufficient memory to support a real-time operating system kernel. This kernel has a crucial role in the rapid development of very complex telecommunications applications.

This section introduces the main features of a digital signal processing processor (DSP) and emphasizes the way in which parallel processing is achieved (Zoican 2002), (Berkeley Design Technology 2000).

The design elements, including the type of processor architecture (Harvard or von Neumann), computational and command units, memory and input-output port organization provide the differences between DSPs and other processors (MCUs).

DSP processors are designed to achieve real-time execution of specific applications (wireless communications, audio and video processing, and industrial control) which have many arithmetic computations. The real-time signal comes to the DSP as individual samples from an analog-to-digital converter (ADC).

To carry out the application in real-time, the DSP must complete all the calculations and operations required for processing samples, without missing the next samples.

This section explains the important differences between DSP and MCU, and focuses on features that a system designer should examine to find the processor that best fits the specific application.

First of all, a telecommunications application implements a specific algorithm that implies signal processing operations: delay lines, sum of products, multiplications and addition/subtractions. Second, most of the applications are infinite loops that process input sample flows received from an input device (commonly a codec). It is essential that the communication and synchronization between the DSP processor and I/O device be fast enough to achieve real-time functioning. There are two essential parts to a DSP processor: multiple specialized computational units and flexible IO ports. If multiple computational units exist, it is desirable to have multiple busses and addressing units in order to be able to fetch two or more units simultaneously. As the DSP processors are designated to implement infinite loops, they will have specific hardware mechanisms to control these loops without software involvement. Furthermore, circular addressing necessarily considers that most algorithms have convolution functions in their structure.

The main features of a generic DSP architecture are the following (Zoican 2002):

1. Fast and flexible arithmetic
2. Large dynamic range for multiplication and accumulation
3. Loading of two operands in one cycle
4. Hardware circular buffers
5. Hardware loops and conditional jumps without additional cycles

The arithmetic used by computational units in the core is very important because most DSP algorithms require a large dynamic scale for number representation. Most DSP use *fixed point* arithmetic, where numbers are represented as fractions in a fixed range of -1.0 to +1.0. Other processors use floating-point arithmetic, where values are represented by a mantissa and an exponent. The mantissa is generally a fraction in the range -1.0 to +1.0, while the exponent is an integer that represents the number of places that the binary point must be shifted left or right in order to obtain the value represented.

The floating-point DSP processors are generally easier to program than fixed point processors, but they are also more expensive and consume more power.

Embedded applications use fixed point processors because the priority is on low cost and, often, low power consumption. It is possible to perform general-purpose floating point arithmetic on a fixed-point processor by using software routines that emulate the behavior of a floating point device. However, such software routines are usually very expensive in terms of processor cycles.

The DSP algorithm should be implemented without any modifications, thus flexible computational units are required.

The differences in instruction sets make MIPS comparisons inaccurate measures of processor performance. A better measure to compare the processors is the instruction set and the mechanism that concurs with a high level of parallelism (hardware loops, functional units work simultaneously, DMA channels, dynamic range).

The organization of a processor's memory subsystem can have a great impact on its performance. The DSP processors have separate memories for the program and data. The memory should be large enough to accommodate all the data and instruction codes, and it should be accessed through DMA channels. Depending on the specific application, the amount of memory can be from hundreds of Kbytes (for audio) to thousands of Kbytes (for multimedia) for each of the programs and data memory. For program memory, these amounts are sufficient because the processor architecture is characterized by a high level of parallelism and the instruction set is powerful and, consequently, the code density is high. For data memory, the above amounts are sufficient because a DSP application does not require very large data buffers (that is, the application works in real time).

Parallelism is achieved using two or three computational units in parallel. Moreover, the command unit (or program sequencer) has specialized functional blocks that allow hardware control of circular addressing and loops. Accordingly, along with the operations' types in a digital system (sums, delays and sums of products) a digital signal processor also has specialized arithmetic units (arithmetic and logic unit—ALU, multiply and accumulate unit-MAC and shifter unit-SHIFTER).

The main differences between DSP and MCU processors are summarized in Table 1.

**Table 1**

| *Feature* | *DSP* | *MCU* |
|---|---|---|
| Architecture | Harvard (separated data and instruction busses) | Von Neumann (common data and instruction busses) |
| Computational units | More than one unit and specialized computational units (ALU—arithmetic and logic operations, MAC—multiply with accumulate and SHIFTER – shift and normalize operations) | A single computational unit for all arithmetic and logic operations (ALU) |
| Addressing units | Generally two data address generators DAG1 and DAG 2 | One single addressing unit with support of ALU |
| Instruction cycles | Most DSP has a one processor cycle instruction | The MCU instructions have more than one processor cycle |
| Parallelism | Physical (spatial)—many units can work simultaneously | Temporal (pipe-line) |

Tools that make system development as simple as possible are necessary. These tools include both software tools (assemblers, linkers, simulators, debuggers, compilers, code libraries, and real-time operating system kernels) and hardware tools (development boards and emulators).

Because DSP applications have substantial computational necessities, the compilers often produce assembly code that executes slowly. In most applications, it is better to write the DSP programs in assembly language or mixed (C and assembler).

## 2. Background

### 2.1. Embedded System Architecture for Signal Processing

As a rule, a real-time system must maintain a well-timed response to both internal and external data. Real-time systems are characterized by a deadline for the completion of the overall task. If the hard time limit is not met, the task has failed. In a real-time application, running multiple parallel functions is desirable for the following reasons (Zoican 2002), (Analog Devices 2011):

- The application contains a lot of processing to be logically linked in a specific order
- Troubleshooting and maintenance is made easier if the program is modularized
- CPU hardware resources can be used more efficiently if the program is divided into distinct functions so that sections of code may be reused among different applications

The general architecture for a DSP system is illustrated in Figure 1a.

Besides a DSP microcomputer, digital signal processing systems consist of a block of data acquisition (sensors), a display block and a communication control unit. The system development is illustrated in Figure 1b.

The system is characterized by a very high level of integration (only a few chips form the whole system—the DSP chip, codec (sample and hold, analog digital, digital analog convertors in a single chip), boot memory and small analog interface circuits (few operational amplifiers, resistors and capacitors).

Figure 2 illustrates the common architecture of an embedded DSP system.

### 2.2. Unified Micro-architecture

Embedded systems are usually a part of multipart systems and are typically implemented on dedicated hardware with related software that accomplishes a specific function. Unlike general-purpose microcomputers (MCU), an embedded system is a specialized microcomputer system. Embedded processors are used in cellular phones, personal digital

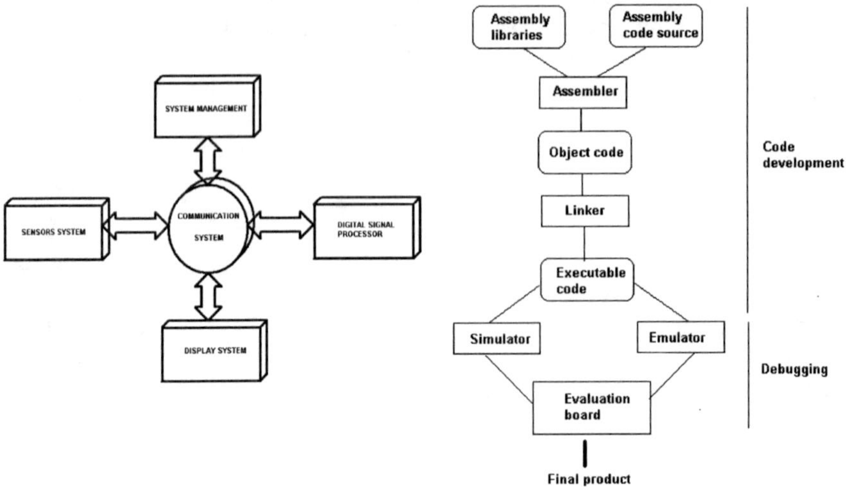


a) b)

**Figure 1.** (a) Generic digital signal processing system, (b) Development of a digital signal processing system.

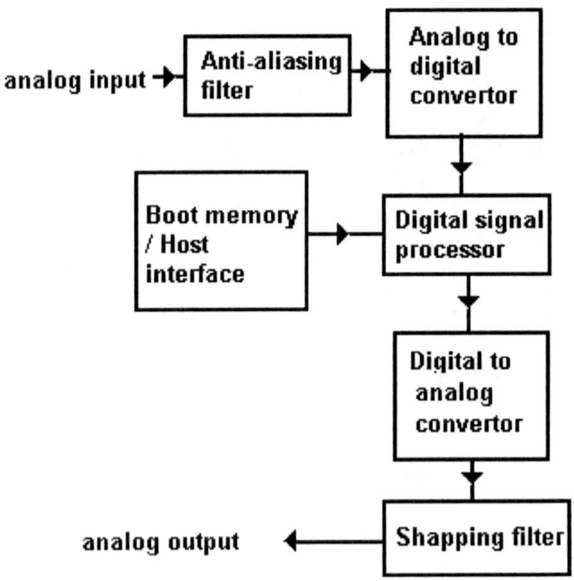


**Figure 2.** Common architecture of a DSP system.

assistants, portable gaming devices, digital versatile disc players, digital cameras, fax machines, scanners etc. Along with the embedded systems, digital signal processors are becoming indispensable for processing signals such as speech, audio, images and video in real time (Analog Devices 2003).

The control functions (typically implemented by microcontrollers) involve many conditional operations, with a lot of changes in program flow. These programs are frequently written in a high level language (such as C or C++) and code density is important, making variable-length instructions a key characteristic.

A typical embedded system (digital signal processing system) in a classical approach is shown in Figure 3.

The MCU core is used to do networking, memory management and control function and the DSP core is used to carry out control signal processing functions. Hardware interfaces to the MCU and DSP cores include:

- Internal memories such as read-only memory (ROM) for boot loading code
- Random-access memory (RAM) and cache for storing code and data
- Direct memory access (DMA) controller
- System peripherals that contain timers, clocks, and power management circuits for controlling the processor's operating conditions

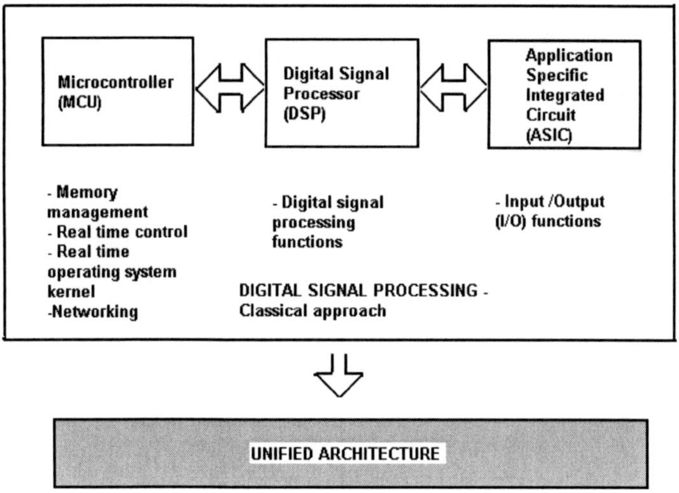


**Figure 3.** Embedded system (classical approach and unified approach).

- I/O ports that allow the embedded core to monitor or control some external events and process incoming media streams from external devices

These underlying hardware units and the processing core are the representative building blocks that comprise an embedded system.

The unified architecture is designed to achieve high-speed DSP performance and top power efficiency. This is different from other cores that require a separate DSP processor and microcontroller. A unified architecture benefits from both DSP and MCU architectures and is more efficient and faster than a dedicated multiprocessor system.

An example of such a unified architecture is Micro Signal Architecture (MSA) which achieves high-speed digital signal performance and the best power efficiency (Gan and Kuo 2007). This architecture combines the capabilities of a microcontroller and DSP processor into a single programming model. The main advantage of the MSA is the integrated feature that combines multimedia processing, communications and the user interface on a single, easy-to-program platform.

A dynamic power management circuit continuously monitors the software running on the processor and dynamically adjusts both the voltage delivered to the core and the frequency at which the core runs. This results in optimized power consumption and performance for real-time applications.

In the next section, an example of the MSA architecture, Blackfin BF5xx, will be presented.

## 2.3. DSP architecture examples

### *2.3.1. The ADSP21xx digital signal processor architecture*

This section illustrates a flexible and powerful DSP architecture that follows the general characteristics shown in the above section.

All the general features for a powerful DSP processor, presented in Section 1, are found in the ADSP21xx architecture (Analog Devices 2001).

The ADSP-21xx has the following set of hardware elements:

- Arithmetic-Logic Unit (ALU)
- Multiplier-Accumulator (MAC)
- Barrel Shifter
- Two Data Address Generators (DAG)
- Program Sequencer.

Efficient data transfer is achieved with the use of five internal buses:

- Program Memory Address (PMA) Bus
- Program Memory Data (PMD) Bus
- Data Memory Address (DMA) Bus
- Data Memory Data (DMD) Bus
- Result (R) Bus.

The architecture of the ADSP21xx is a modified Harvard architecture. There are also there are a number of processor extensions such as program and data memories, serial ports, parallel DMA ports and timers. All of these resources are designated to ensure the connectivity and synchronization of the ADSP21xx with other processors or devices (including the codec) in order to implement an efficient DSP system (as illustrated in Figure 2). The internal memory is sufficient to store the necessary instruction codes and data for medium complexity applications (such as telephony and audio applications).

Figure 4. Illustrates the general architecture of the ADSP21xx family from Analog Devices.

The ALU and MAC architectures, in processor ADSP218x, are illustrated in Figures 5 and 6.

It should be noted that the ADSP21xx architecture has a high degree of parallelism obtained by the three functional units that can operate in parallel. Also, the ALU and MAC are flexible and are specialized to perform fast fixed-point calculations.

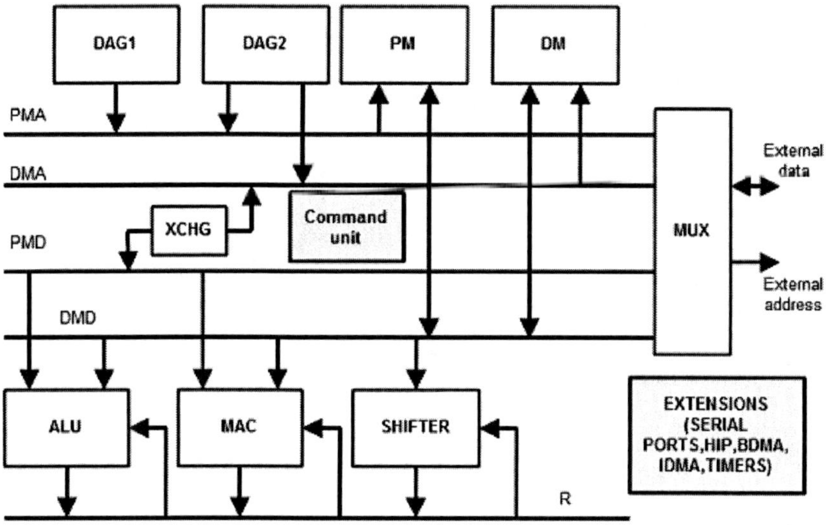


**Figure 4.** ADSP 21xx general architecture.

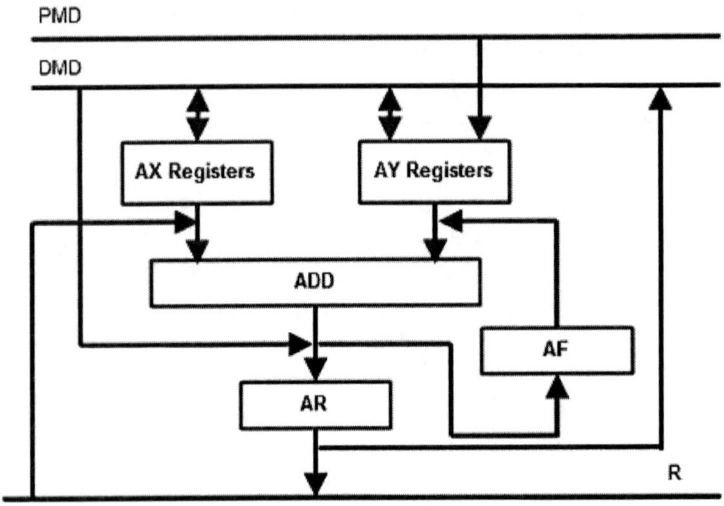


**Figure 5.** The ADSP 21xx ALU architecture.

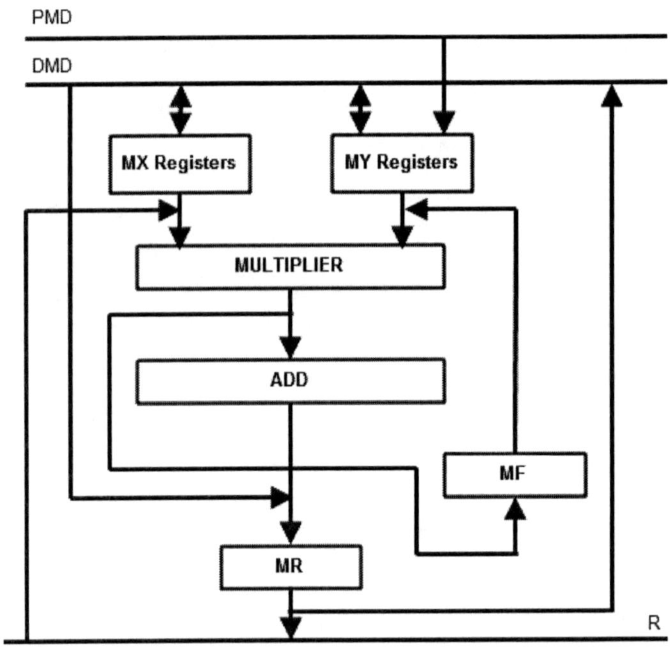


**Figure 6.** ADSP 21xx MAC architecture.

Processing parallelism is achieved in particular by:

- Harvard architecture
- Ability to simultaneously load two operands (from data memory and program memory)
- Possibility of generating two addresses simultaneously
- Multifunction instructions
- Possibility of using a result of one computational unit as input for another computational unit (on the results bus, R)
- Fast response to interrupts (fixed vector)
- Program loop hardware control
- Conditional jumps without overhead

The family of ADSP 21xx consists of various chips which differ by core extensions. The remarkable members of this family are the ADSP2181 and the ADSP219x (dual core).

### *2.3.2. The architecture of the Blackfin digital signal microcomputer*

The Blackfin microcomputer is a 16-bit fixed-point processor that is based on the MSA core, developed in cooperation by Analog Devices and Intel (Analog Devices 2007). Because of its low cost and high performance, this processor is suitable in power-sensitive applications (portable audio players, cellular phones, digital cameras) and computationally intensive applications (video equipment, third-generation cellular phones).

The Blackfin processors provide an embedded fast Ethernet controller that offers the capability to connect directly to a network.

The Blackfin family can operate between 350 MHz at 1.6 V. and 750 MHz at 1.45 V. The clock rate and operating voltages can be switched dynamically for specified tasks via software for reduction power.

The Blackfin core combines dual multiply-accumulate (MAC) units, an orthogonal reduced instruction-set computer (RISC) instruction set, single instruction, multiple data (SIMD) programming capabilities and multimedia processing features, into a unified architecture. The Blackfin BF5xx processor includes system peripherals such as a parallel peripheral interface (PPI), serial peripheral interface (SPI), serial ports (SPORTs), general-purpose timers, universal asynchronous receiver transmitter (UART), real-time clock (RTC), watchdog timer, and general-purpose input/output (I/O) ports.

The Blackfin Processor is optimized to powerfully execute the DSP applications code.

Blackfin processors have high peripheral supports, a memory management unit (MMU) and RISC-like instructions, which are usually found in many high-end microcontrollers. These processors have high-speed buses and highly developed computational units that support variable-length arithmetic hardware operations. These features make Blackfin processors appropriate to replace other high-end DSPs and MCUs.

The Blackfin processor uses a modified Harvard architecture which allows multiple memory accesses per clock cycle.

The Blackfin processor instruction set is optimized so that 16-bit operation codes represent the frequently used instructions. Complex DSP instructions are encoded into 32-bit operation codes like multifunction instructions. Blackfin microcomputers bear a limited multi-issue facility, where a 32-bit instruction can be issued in parallel with two 16-bit instructions. This allows the programmer to use several of the core resources in a single instruction cycle.

The Blackfin architecture supports instructions that control vector operations. We take advantage of these instructions to carry out concurrent operations on multiple 16-bit values, as well as add, subtract, multiply, shift, negate, pack and search (Zoican 2008).

Figure 7. Illustrates the Blackfin general architecture (Analog Devices 2007).

Figure 8 shows that the core architecture consists of three main units: the address arithmetic unit, the data arithmetic unit and the control unit.

The data arithmetic unit contains the following hardware blocks:

1. Two 16-bit multiplier and accumulate (MAC0 and MAC1 each with 40-bit accumulators).
2. Two 40-bit arithmetic logic units (ALU0 and ALU1).
3. Four 8-bit video ALUs (AV0-AV3)
4. A 40-bit barrel shifter (SHIFTER).
5. Eight 32-bit data registers (R0 to R7) or 16 independent 16-bit registers (R0. L to R7.L and R0.H to R7.H).

Computational units acquire data from data registers and execute fixed-point operations. The data registers collect data from the data buses and move the data to the computational units for processing. Computational results are transferred to the data registers before transferring to the memory using data buses. The data registers get data from the data buses and move the data to the computational units for processing.

The arithmetic unit supports SIMD operation and has Load/Store architecture. The assembler instruction syntax is algebraic; it is insightful and makes it simple to understand what the instruction does. Figure 9

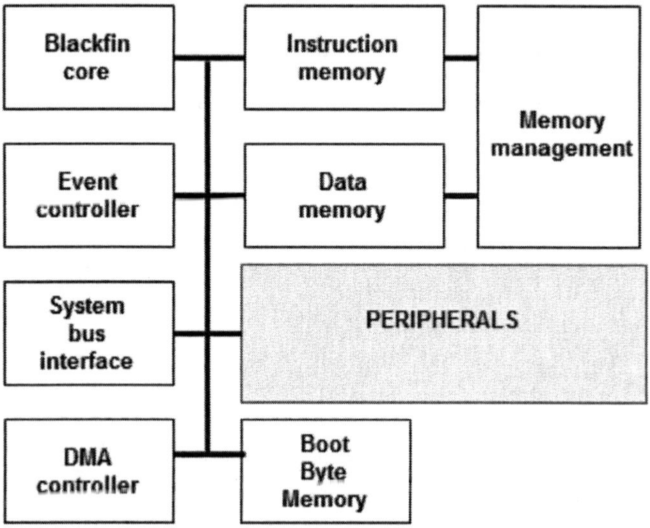


a)

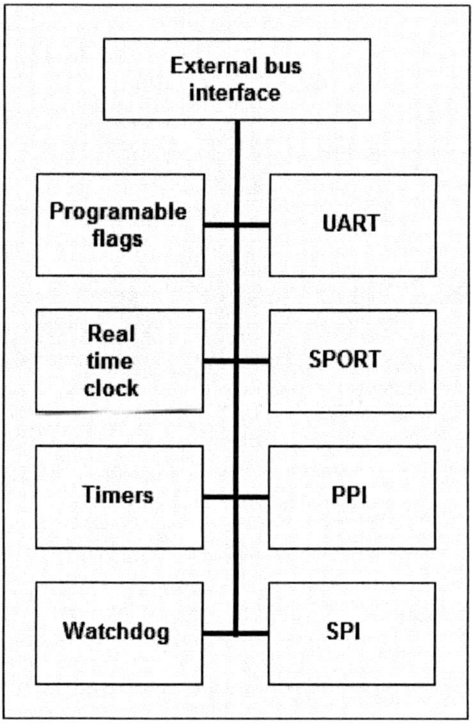


b)

**Figure 7.** (a) Overall Blackfin architecture; (b) The Blackfin peripherals.

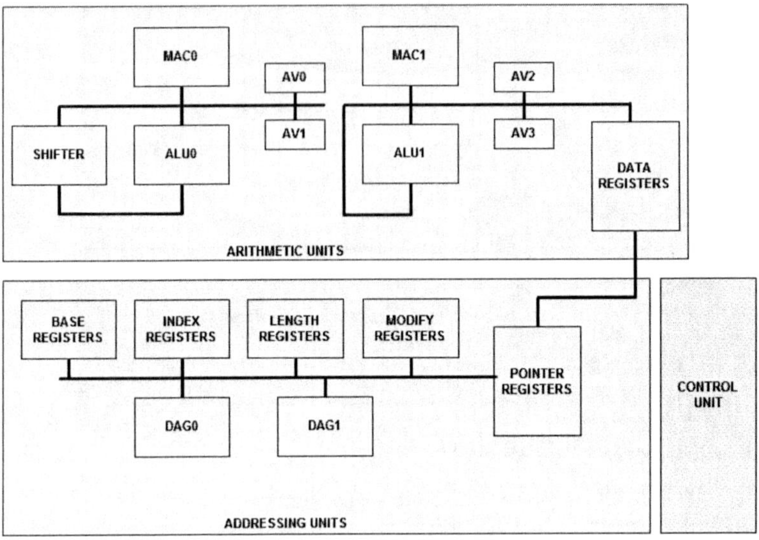


**Figure 8.** The Blackfin core.

illustrates several arithmetic instructions that are efficiently executed in a single cycle. The video ALUs offer parallel computational power for video operations (quad 8-bit additions and subtracting, quad 8-bit average, Subtract-Absolute-Accumulate). All these instructions take one cycle to complete.

The address arithmetic unit consists of the following hardware blocks (Figure 8):

1. Two data address generators (DAG0 and DAG1) that generate addresses for data transfer to and from the memory. The advantage of using two data address generators is to permit dual-data fetches in a single instruction.
2. Six 32-bit general-purpose address pointer registers (P0 to P5).
3. One 32-bit frame pointer (FP) pointing to the current procedure's address.
4. One 32-bit stack pointer (SP) pointing to the last location on the current user stack.
5. A set of 32-bit data address generator registers:
   a. Indexing registers, I0 to I3, have the effective addresses
   b. Modifying registers, M0 to M3, hold values used for updating index registers
   c. Base address registers, B0 to B3, contain the starting addresses of the circular buffers

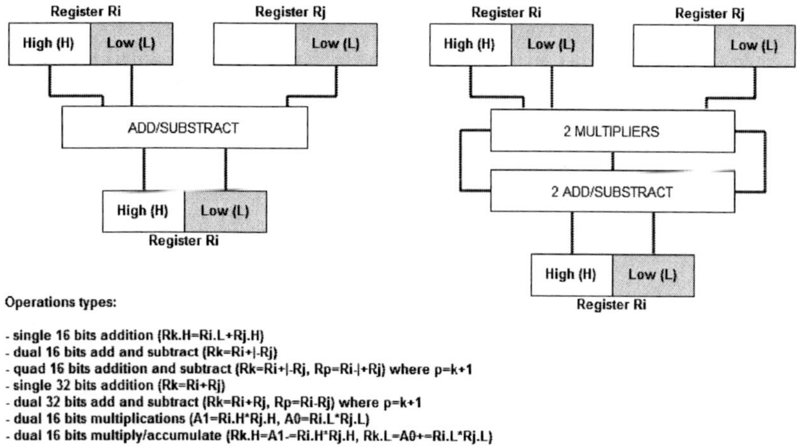


**Figure 9.** Blackfin arithmetic instructions.

d. Length value registers, L0 to L3, hold the lengths of the circular buffers

The main purpose of the address arithmetic unit is to produce addresses to access the memory (byte addressable) of the Blackfin processor. However, data can be accessed in 8, 16, or 32 bits using the pointer registers (P0–P5); 16-bit and 32-bit accesses use index registers (I0–I3) and 32-bit access employs the stack and frame pointer registers.

The control unit shown in Figure 8 consists of the following blocks:

1. A program sequencer controls the instruction execution flow, which includes instruction alignment and instruction decoding. The address generated by the program sequencer is a 32-bit memory instruction address. A 32-bit program counter (PC) is used to point to the current instruction being fetched.
2. The loop buffer is used to support zero-overhead looping. The Blackfin processor supports two loops with two sets of loop counters (LC0, LC1), loop top (LT0, LT1) and loop bottom (LB0, LB1) registers to handle looping. Hardware counters are used to calculate the loop condition. Loop unit 1 has a higher priority than loop unit 0. Consequently, loop unit 1 is used for the inner loop and loop unit 0 is used for the outer loop.

The program sequencer controls all program flow, which includes maintaining loops, subroutines, jumps, idles, interrupts and exceptions. Figure 10 and Figure 11 illustrate the different program flows and the pipeline stages, respectively.

The Blackfin processors have pipeline architecture. This pipeline is used to allocate best the workload among the processor's units, which leads in efficient parallel processing among the processor's hardware.

Overlapping different stages of instructions allows different functional units to work concurrently in the same clock cycle. Each instruction can be executed in a single clock cycle when the pipeline is full, giving a throughput of one instruction per clock cycle. On the other hand, every non-sequential program flow will decrease the processor's throughput.

The Blackfin sequencer component manages events which include interrupts (hardware and software) and exceptions (error situation or service related).

Event handling is divided between the CEC (Core Event Controller) and SIC (System Interrupt Controller). CEC has 16 levels and deals with requests based on priority. There are nine general purposes (lowest levels) that are used for Peripheral Interrupt Requests. Each level has a 32-bit Interrupt Vector Register that points to the beginning of the ISR associated with that level. SIC enables Peripheral Interrupt Requests, maps them to specific CEC levels, and sets peripheral request priorities.

The Blackfin processor uses a modified Harvard architecture which allows multiple memory accesses per clock cycle. However, the Blackfin processor has a single memory map that is shared between data and instruction memory. Instead of using a single large memory, the Blackfin processor supports a hierarchical memory model as shown in Figure 12.

The L1 data and instruction memory are placed on the chip and are generally smaller in size but faster than the L2 external memory, which has a superior capacity. As a result, the Blackfin memory is organized in a hierarchy from the slowest memory (L2) to the fastest memory (L1).

The justification of memory hierarchy is based on three principles: (1) code and data that have to be accessed frequently are stored in the fastest memory; (2) program locality (instructions and data in use are close to each other in memory.); and (3) smaller memory has a faster access time.

Table 2 compares the main characteristics of the ADSP218x and Blackfin BF53x microcomputers.

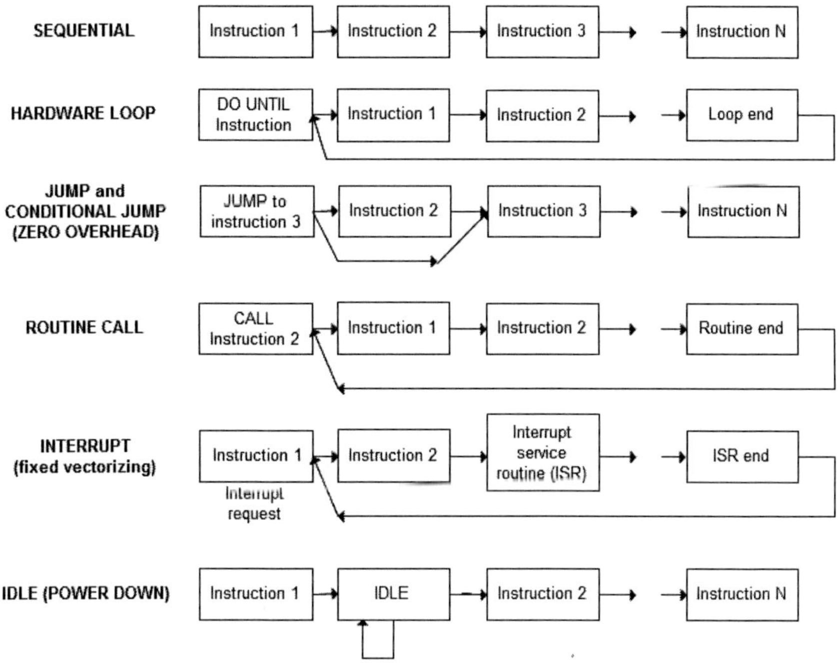


**Figure 10.** The Blackfin program flows.

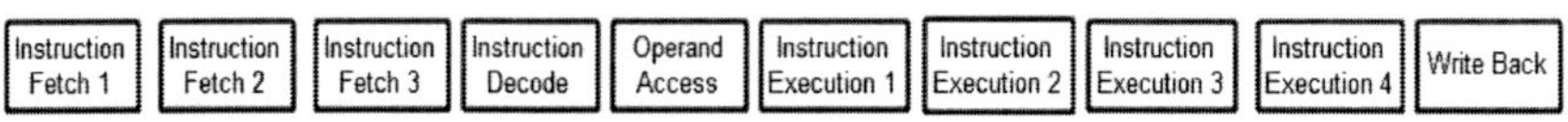


**Figure 11.** The Blackfin pipeline.

**Table 2**

| *Feature* | *ADSP 21xx* | *Blackfin* |
|---|---|---|
| Number of arithmetic units | 3 (ALU, MAC SHIFTER) | 9 (2 ALU, 2MAC, 1 SHIFTER, 4 ALU for video) |
| Addressing modes | Indirect register, post-modified, no memory management, circular buffers | Indirect register, post-modified, general pointers, memory and cache management, circular buffers |
| Pipeline | Low (3 stages) | Moderate (10 stages) |
| Memory | 80 Kbytes, internal, overlaid | 132 Kbytes, internal, non overlaid |
| External devices communication | Moderate (serial ports, IDMA, BDMA and HIP ports) | Intensive(serial ports, DMA, USB, COM,GPIO) |
| Typical applications | Audio, process control and industrial automation | Multimedia, image processing |

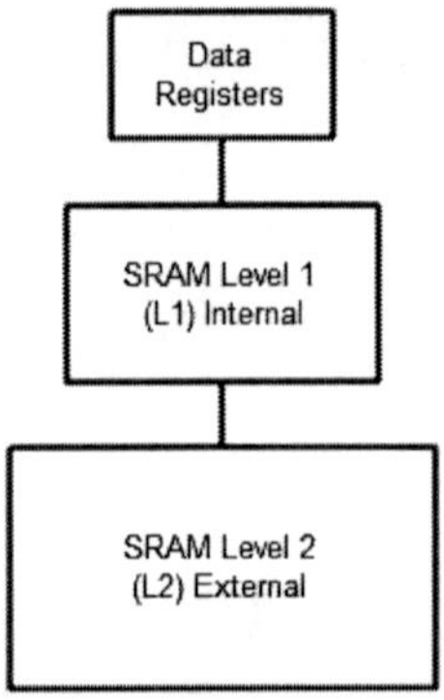


**Figure 12.** Blackfin memory model.

The benchmarks for Blackfin processors are illustrated in table 3 (Analog Devices 1990).

**Table 3**

| *Benchmark Description* | *Blackfin Processor (cycles)* |
|---|---|
| Delayed LMS Filter | 1.5 number of taps + 4.5 |
| Radix 2 Butterfly Kernel | 3 |
| 256 Complex FFT | 2324 |
| Block FIR Filter | (number of samples/2) (2+number of samples) |
| Complex FIR Filter | 2 number of taps +2 |
| Max Search | 0.5 number of samples |
| Max Index Search | 0.5 number of samples |
| Biquad IIR | 2.5 number of biquads +3.5 |

The benchmarks for ADSP21xx processors are illustrated in table 4 (Analog Devices 1997), (Analog Devices-web1 2011).

**Table 4**

| *Benchmark Description* | *ADSP 21xx Processors (cycles)* |
|---|---|
| FIR Filter | number of taps + 4 |
| IIR Filter (direct form) | 2 filter order + 9 |
| 1024 decimation in time real radix 2 FFT | 52911 |
| Standard ADPCM | 892 |
| 64 x 64 two dimensional FFT | 633701 |
| RPE-LTP LPC Encoder | 49300 |
| RPE-LTP LPC Decoder | 14400 |
| Voice Activity Detector | 02141 |

The ADSP21xx and Blackfin processors are used in industrial telecommunications products such as the SoftFone AD6758 (GPRS/GSM multimedia baseband processor) and the Digital Media Device (DMD).

The AD6758 provides a commercial solution for essential multimedia handsets for GSM/GPRS networks. Based on the Analog Devices ADSP-218x processor, it provides sufficient processing power for multimedia algorithms such as MP3 music playback and MPEG-4 video encode/decode for QCIF-resolution displays. Other industrial products include the Digital Media Device (DMD) 2000, DMD3000 and DMD2500 from TransVideo, based on the Blackfin microcomputer. These products support a proprietary CODEC based on the new H.264/MPEG4-AVC standard. The IPTV terminals feature support for the H.264 decode format at 25–30 frames per second, a 500–1,500 Kbps audio bit rate, multiple output and network connections, MP3 stereo, S-Video, and RCA connections. The products also support cable modems, ADSL, Ethernet, and WLAN. All of these devices support Video-On-Demand (VOD) and multi-channel broadcast.

From the previous tables, it can be observed that both DSPs are suitable for telecommunication applications. Of course, there is a great difference in speed, but this is a consequence of technology not of architecture. On the other hand, the Blackfin architecture is a unified architecture, as discussed in the previous section. As a result, the digital signal processing system has a higher integration level and eliminates the need to implement input-output functions using a microcontroller.

### 2.4. A Real Time Operating System Kernel Example

Supervisor software is needed for more complex applications. This supervisor provides support for applications development both at lower and higher levels of program organization.

This section presents the main characteristics of a real time kernel for DSP processors. The real-time kernel, Visual DSP Kernel (VDK), produced by Analog Devices, is illustrated (Analog Devices 2011). The VDK supports both the ADSP219x and Blackfin architectures.

Running multiple parallel tasks is achieved by using a real-time operating system kernel. Real-time operating systems (RTOS) perform a wide variety of tasks, from managing and processing high-level I/O, to tracking and managing files, to controlling several concurrent processes, ensuring they do not affect each other. The RTOS have to resolve the following problems associated with embedded systems: scheduling, exclusive access to shared resources, inter-process communications, and process synchronization.

Signal processing applications can be divided into processes, with each process carrying out a specific function of the global system.

Execution of the processes is based on the priority associated with each process. For DSP applications, the process scheduling is preemptive (higher-priority processes will interrupt lower-priority process).

Real-time operating systems for DSP applications manage signal processor resources and carry out the following functions:

- Process scheduling, according to their priority (the processes having all resources available, except CPU, are placed in a ready queue)
- Inter-process communication
- The synchronization between processes and external devices
- Mutual exclusion for shared resources

The real-time kernel state diagram is shown in Figure 13.

Priority associated with each process is allocated dynamically and the operating system maintains a minimum priority maintenance process.

The priority assignment is imposed by the process deadline. Furthermore, interrupts that can trigger the execution of a process are considered (such as the analog to digital convertor interrupt). The operating system chooses the next process, in order of priority, from the ready queue and this process will be run.

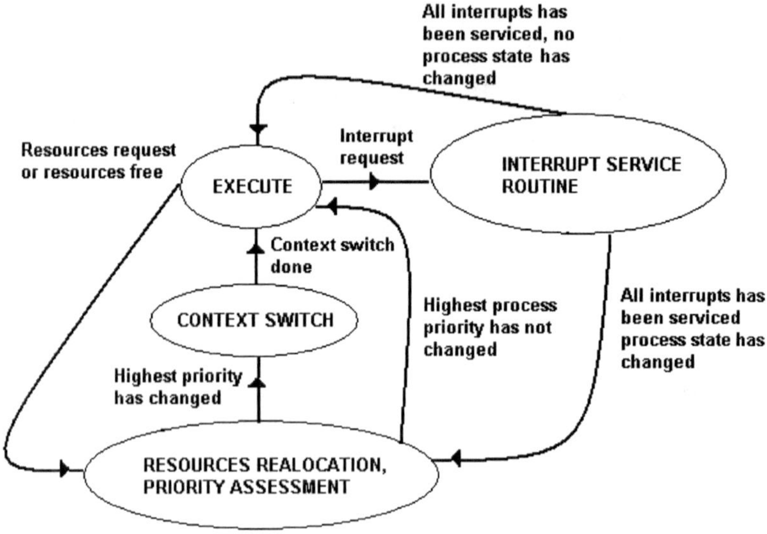


**Figure 13.** Diagram of the real-time kernel state.

The operating system cannot perform a process context switch as long as the process that is running has a higher priority—even if it frees the resources or determines, by appealing to the primitive system, the loading ready queue for other processes with equal or lower priority.

Typically, signal processing system resources are accessed atomically, meaning that resources must be protected. There are two categories of protection: unscheduled regions and critical regions.

Unscheduled regions are portions of code that do not call scheduling primitive of the real-time kernel, which is necessary when one wishes to execute a sequence of instructions without switching context, even if there is a priority change. However, interrupts are allowed in unscheduled regions.

Critical regions are portions of code that can inactivate both as context switching and interrupts.

Unscheduled regions are useful, for example, when multiple system resources are freeing and critical areas are required when global variables can potentially be modified by an interrupt service routine.

In DSP applications, the processes have basic information about the system hardware structure. Physical control is achieved by using additional software called drivers or device drivers. This structure ensures independence from both the DSP processor (if processes are written in high level languages) as well as at the input—output interfaces.

The processes must minimize time spent in unscheduled regions or in critical regions to avoid blocking the scheduler or interrupt services.

The process scheduler role is to run the process which has the higher priority in the shortest time possible. The process scheduler is not called directly by a process; it will be executed whenever a real-time kernel primitive is called (in a process or a routine interrupt service routine), provided that this primitive can change priorities. The scheduler will not be executed in unscheduled or critical regions, but will be called as soon as these protected regions are left.

Signal processing applications have three methods of scheduling: cooperative scheduling, uniform time division (round robin) and preemptive scheduling.

Cooperative scheduling is involved in processes that have the same priority. Each process takes the DSP processor and passes it, on its own initiative, to the next process. The running process will yield the processor by calling a primitive system; the suspended process is then placed in the ready queue in the last position.

Round robin scheduling assigns processes with an equal priority, an equal length time quantum for execution. When the execution time allocated to a process is past, it is placed in the last position in the ready queue of its associated priority.

If the current running process is interrupted by another process, the duration of its execution will not be updated until the interrupted process does not run again.

Preemptive scheduling gives the processor to the highest priority process as soon as possible, after it has entered the ready queue.

A DSP application can use all three scheduling methods: processes without time-critical requirements (lower priority) are scheduled according to a round robin scheme and processes with time-critical requirements (higher priority) are scheduled using a preemptive method. In this way, low priority processes do not interfere with higher priority processes. Moreover, low priority processes may not wait for their quantum expiration, but to run the next low-priority process on their own initiative—using cooperative scheduling.

In signal processing applications, there are three ways to achieve communication and synchronization between processes: semaphores, events and I/O flags.

A process waiting for a signal may continue executing if the signal is available or can enter the blocking state if the signal is not available. The process will unlock when the signal becomes available or when a timer expires.

Semaphores are global variables in the system and are therefore available to any process. A semaphore is a data structure that can be used to:

- Control access to shared system resources
- Allow synchronization of processes
- Periodically execute processes

The semaphore may be posted or it may be pending. During pending, the semaphore value is testing with zero. If the semaphore is zero, the process increases the semaphore value and accesses the shared resource, otherwise the process waits.

Processes can wait at semaphores if they are not taken by another process. If a semaphore is taken by a process, the execution process will resume; if not, the process is suspended until a predetermined time interval expires. In this situation, the process will run the associated error function. It is possible that a process to wait continuously for a semaphore.

When the process has finished using a shared resource, it posts the semaphore. Posting a semaphore means that its value is decreased and the resource is made available.

In signal processing applications, semaphores are not objects that belong to a particular process. All semaphores can be accessed by any process created in the system. If a process that takes the semaphore is destroyed before releasing it, the semaphore will remain locked.

In addition to their use for exclusive access to shared resources and synchronization between processes, semaphores can be set periodically by the operating system. In this way, they can schedule processes.

Processes interact with semaphores using real time kernel routines.

Events are signals that control process executions. They define some bits indicating a logical condition in the process. An event is a logical operation performed on the state of bits in a system variable. The logical operation is performed between the event and a mask. The event is true if all mask bits correspond to the event or the event is true if at least one bit in the mask matches the corresponding bit of the event bits. If the event is true, the process that is waiting for this event will be placed in the ready queue; otherwise, it will wait until the event occurs or it will call the associated error function. Specific operating system primitives control the events. Any change in the event leads to a reevaluation that event.

As with semaphores and events, processes can wait for I/O flags. Processes waiting I/O flags are immediately blocked if the I/O flags are false. I/O flags switch to be true only after a special function named the I/O driver is called. In this way, processes are synchronized with the I/O device. The core of DSP processors has special primitives for I/O devices: initialization, device control, channel opening, reading, writing and channel closing.

## 3. The Framework for Applications Development with DSP Processors

This section illustrates a general framework for telecommunication applications' development. This framework is based on the DSP hardware presented in section 2 with the support of the VDK operating system kernel.

A real-time system processes data at a customary and well-timed rate. A real-time system must suit firm response time constraints. The response time is defined as the time between the arrival of input data samples and the output of processed data samples. The sample-by-sample processing mode requires all operations be completed within the given sampling period. A new input sample is read, then it is processed and the result of processed sample is written in every sampling interval. Figure 14 illustrates how sample-by-sample processing works.

The advantages of using sample-by-sample processing are:

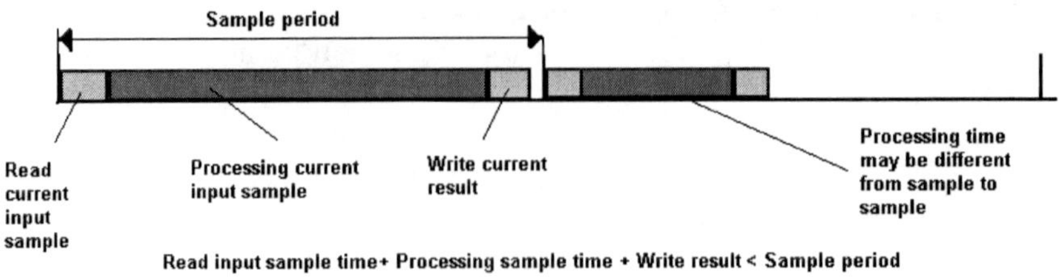


**Figure 14.** Sample by sample processing.

1. The delay between the input and the output is small (less or equal with one sampling interval).
2. In some applications, multiple channels could be acquired, processed, and output within the sampling interval. Memory increases to store the multichannel data samples, but it remains low.
3. Results are obtained within the sampling period and, therefore, they are simple to handle.
4. The method is simple to put into practice. No supplementary memory buffers or memory administration code are required.

An achievable method to reduce the processing speed requirement is to group a chunk of input samples and perform processing on this block of data as a batch.

In the block mode, input samples are stored into an input buffer of $N$ samples and a whole block of samples are processed after the buffer is full. At the same time, a new chunk of $N$ samples are acquired, using interrupts from an analog to digital converter. Figure 15 shows block processing for a block of samples.

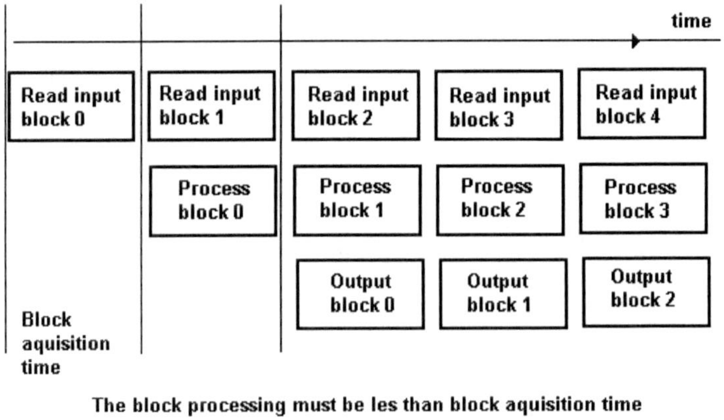


**Figure 15.** Block processing.

To perform the block processing shown in Figure 15, the input samples must save constantly in different memory buffers, thus increasing memory requirements. A memory buffering scheme known as the switching buffers technique is illustrated in Figure 16.

The switching buffers approach solves the input samples missing if the acquisition time of the whole input buffer is greater than the processing time of the corresponding buffer. The disadvantages of using block processing include:

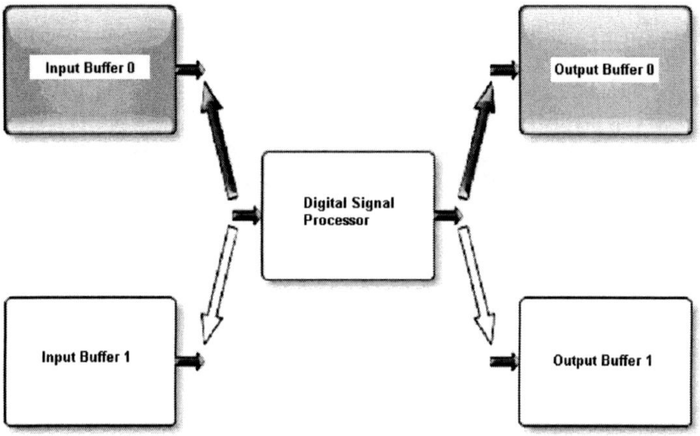


**Figure 16.** Switching buffer technique.

1. Four memory buffers of length *N* are required for storing input and output data samples.
2. The delay incurred in block processing is greater than in sample-by-sample processing (it is twice the acquisition time).
3. Programming needed to supervise switching between buffers becomes more difficult.

Complex applications may be benefit by operating system (like VDK, presented above) support. Using the VDK, leads to a faster developing of the applications.

All the necessary inter-process communications, exclusive access to shared resources and memory and input—output device management are done by VDK primitives. Users can focus on writing the applications for the specific processes to be integrated into the system using the operating system. Both of the above presented techniques may be implemented more easily by using VDK.

The following section presents some application examples in the area of telecommunications.

## 4. Examples of Real-time Signal Processing Applications

This section presents real-time implementations of some telecommunications applications including: adaptive acoustic echo canceller, discrete multi-tone modulation, and networking applications framework. These applications have been implemented on Blackfin processors, but some of them may be implemented similarly with ADSP21xx.

### 4.1 Adaptive Acoustic Echo Canceller Implementation with the Blackfin Processor

Figure 17 illustrates a typical acoustic echo canceller system (Tretter 2003).

The main problem regarding acoustic echo implementation is the long delay line of the adaptive filter. This may lead to overflows due the limited register length of computational units in the DSP processor. This problem may be solved by a proper scaling of the input signal. Most DSP processors (including ADSP21xx and Blackfin, presented in the above section) have special mechanisms (additional bits in internal MAC registers and hardware logic for overflow detection and saturation) to counteract this issue.

Figure 18 illustrates the adaptive LMS algorithm (Haykin 1993).

An efficient FIR filter implementation, based on the Blackfin architecture, is illustrated in Figure 19. At each clock period, two FIR outputs are calculated in the two MACs of the Blackfin architecture.

The input data delay line and coefficient buffers are declared as circular buffers. The loop in Figure 19 is hardware controlled.

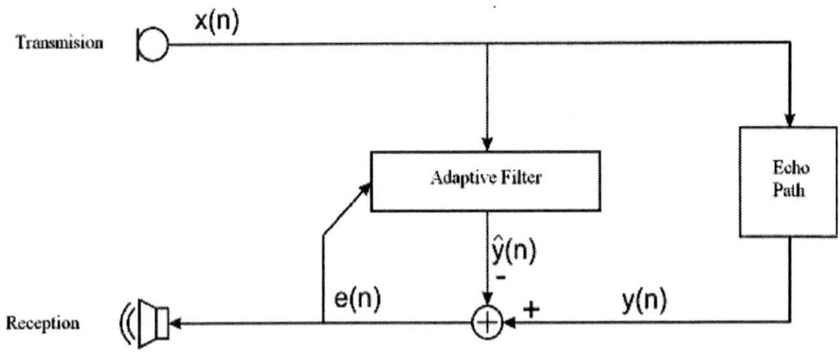


**Figure 17.** Echo canceller.

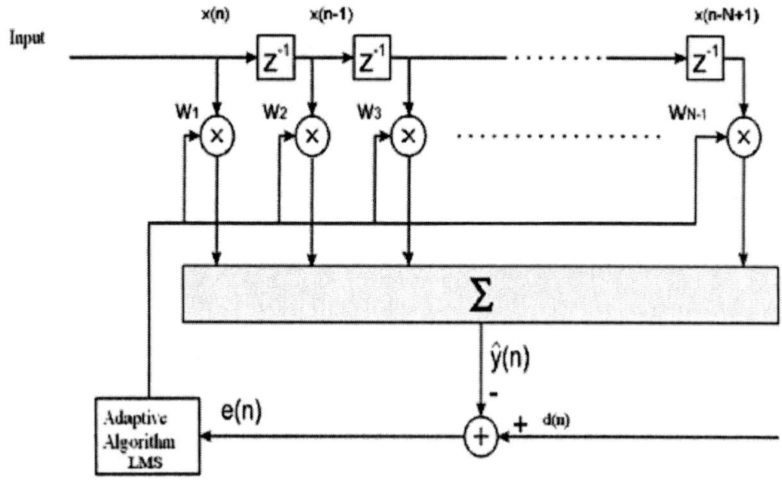


$$\hat{y}(n)=\sum_{i=0}^{N-1} w_i(n)x(n-i)=X^T(n)\cdot W(n)$$

$$e(n)=y(n)-X^T(n)\cdot W(n)=y(n)-\hat{y}(n)$$

$$W(n+1)=W(n)+2\mu\cdot e(n)X(n)$$

**Figure 18.** Adaptive LMS algorithm.

The acoustic echo canceller is implemented on an EZ-KIT-LITE BF 533 (Analog Devices-web1 2011) evaluation board.

The following implementation problems have been identified (Zoican 2008):

**Problem #1**: The vector length (for FIR delay line) increases the memory requirements and execution time.

**Solution**: Use the external SDRAM evaluation board.

**Problem #2**: The evaluation board codec has a 48 KHz sampling frequency

**Solution**: Decimation filter by 8

**Problem #3**: The FIR routine in C (floating point) takes too many processor cycles

**Solution**: Use the fract16 format (assembly language FIR routine)

**Problem #4**: Real time implementation

**Solution**: Employ the switch buffers technique

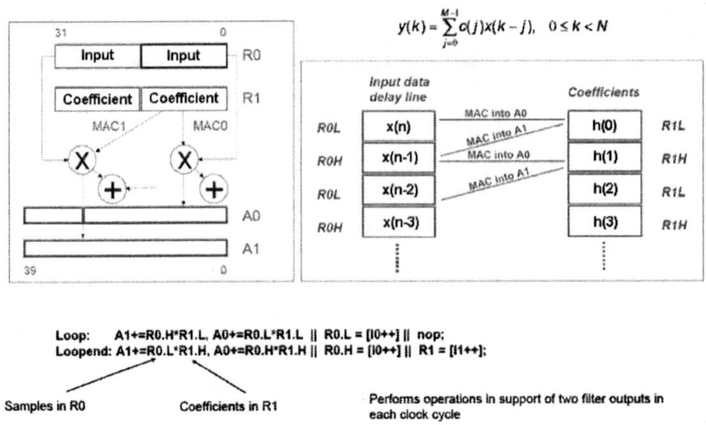


**Figure 19.** Efficient FIR implementation.

*(Color image of this figure appears in the color plate section at the end of the book.)*

Figure 20 shows the acoustic echo canceller functional description. The switching buffer technique was used. Several data buffers were defined: *Buffer_in_0, Buffer_in_1, Buffer_out_0, Buffer_out_1* – the input and output switching buffers, *Buffer_aux_d* – the auxiliary buffer, *Buffer_in_0_sec* and *Buffer_in_sec_1*- the secondary buffers for the real echo signal, and *d* – the desired signal buffer.

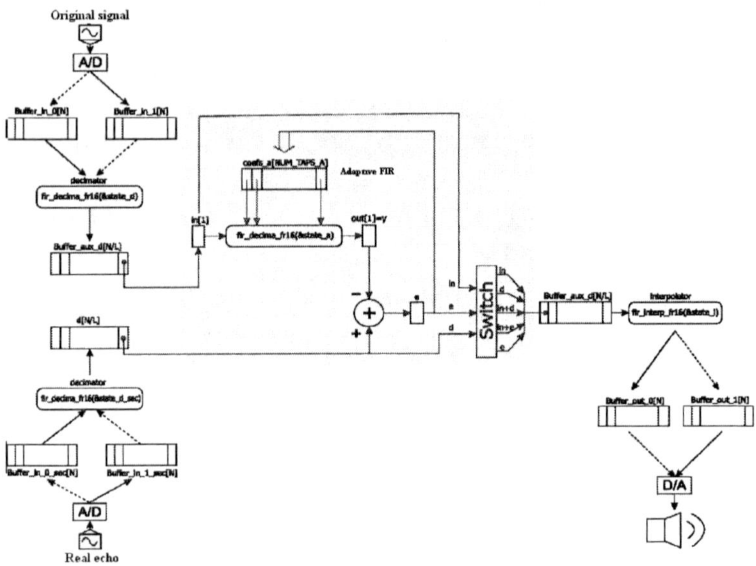


**Figure 20.** Acoustic echo canceller.

*(Color image of this figure appears in the color plate section at the end of the book.)*

The FIR filters were implemented using C-able functions *fir_decima_fr16* and *fir_interp_fr16* from the VisualDSP C Library.

Figure 21 provides the program flowcharts. Two software modules have been written, in C and C-able functions from Visual DSP++ for Blackfin processors: the interrupt request service routine (used to fill in and fill out the input and output buffers) and the main program (which implements the adaptive filter algorithm). Also, the interrupt routine is used to switch the input and output buffers when the current input buffer is full. An interrupt request (from serial port SPORT0) is generated when a new sample is acquired from the codec.

The main results are illustrated in Figure 22.

The adaptive echo canceller has been implemented in real time for long path echo (about 500 ms). The convergence speed can be improved using the normalized LMS (NLMS) algorithm or the Variable Step Size (VSS) algorithm (Zheng and Lynch 2009).

### 4.2 DMT Modulation

Discrete multi tone (DMT) is considered the preferred modulation technique for DSL systems (Harte 2005).

The DMT transmitter and receiver are illustrated in Figures 23 and 24.

The modulator and demodulator of a DTM system are implemented for use by the Fast Fourier Transform (FFT).

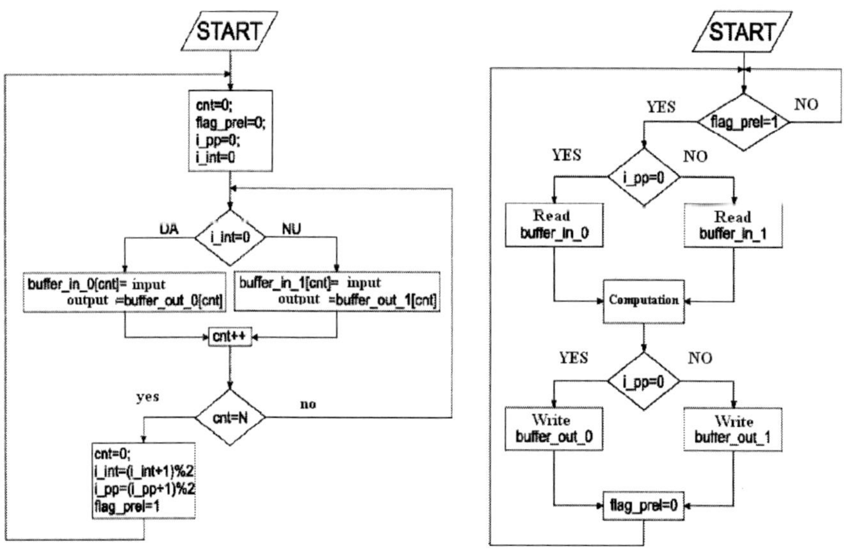


**Figure 21.** Echo canceller program flowchart.

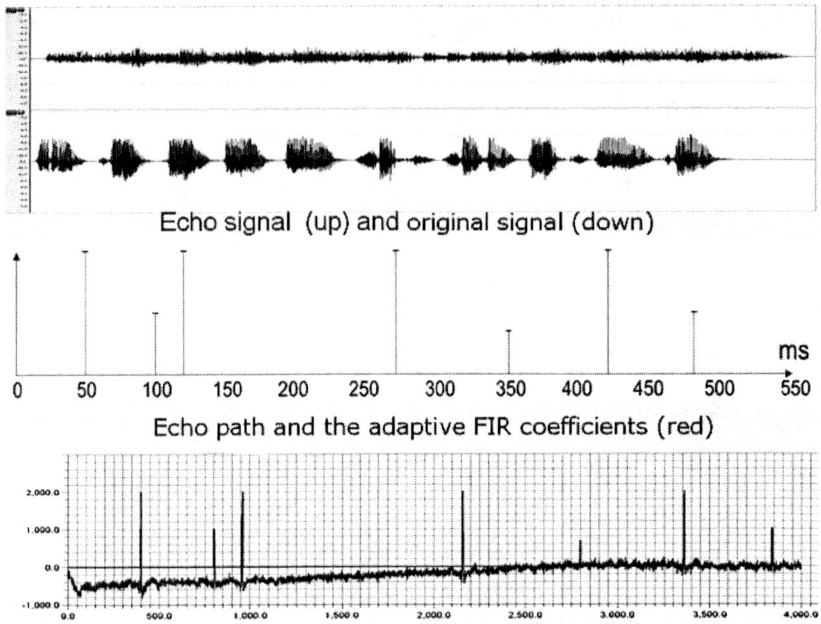


**Figure 22.** Results for the acoustic echo canceller.

*(Color image of this figure appears in the color plate section at the end of the book.)*

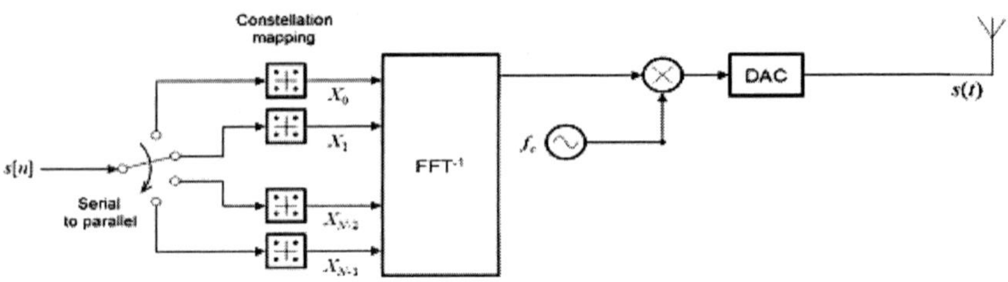


**Figure 23.** DTM transmitter.

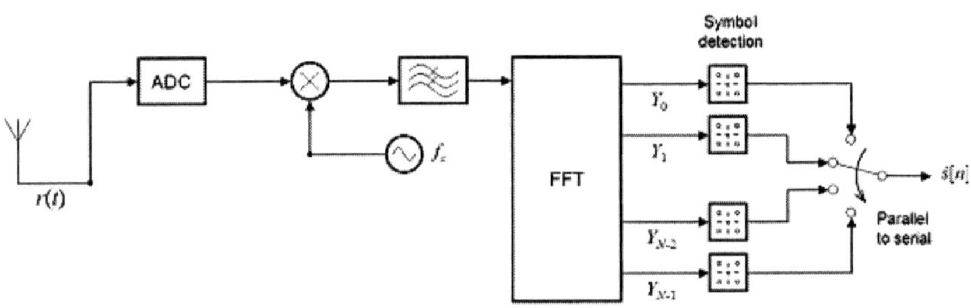


**Figure 24.** DTM receiver.

The inverse FFT (IFFT) is used as the modulator and the FFT is used in the demodulator.

The stream input is divided in symbols of k bits each.

Each symbol is mapped in a constellation and a sequence of complex numbers is obtained. Using the symmetry properties of the FFT transform, a real sequence can be obtained at the IFFT block. For example, if the FFT sequence has N=8 points, a number of 3 symbols must be mapped in a complex sequence S1, S2 and S3. The input sequence in IFFT block is: 0, S1, S2, S3, 0, S3*, S2*, S1* (where * means complex conjugate).

The obtained real sequence is modulated and transmitted in the channel.

At the receiver, the transmitted signal is demodulated and filtered. The real sequence obtained after low-pass filter is the input of FFT block. An estimate of the original sequence at the IFFT input at the output of FFT block will be obtained. Then the estimates of symbols S1, S2 and S3 are obtained as output of detection block.

### 4.3 Networking Applications

Embedded networking applications are now increasingly being used to transmit multimedia content over wired or wireless networks. Control applications and sensor networks are additional areas where adding network means is desirable.

More or less, all systems connected to the Internet communicate using the IP protocol stack (Estrin 2001).

Owing to the dominance of IP networks, many networked embedded systems are connected to such networks and must be able to communicate using such protocols. However, the IP protocol suite is quite heavy because the protocols often require significant resources in terms of memory and processing power.

In this section, a lightweight IP (lwIP) stack implementation for Blackfin processors is illustrated. A VoIP system based on adaptive multirate (AMR) codec is exemplified.

The lwIP IP stack package on the Analog Devices Blackfin family of processors uses a standardized driver interface to permit it to be used with different Ethernet controllers. All of the drivers are implemented as part of the Analog Devices Inc. (ADI) driver model and use the System Services Libraries (SSL). The stack provides the standard BSD socket API to the application and has been intended to separate it from both the working environment and the particular network interface being used.

The stack is connected to the nearby environment by means of two standardized interfaces (TCP wrapper and LwIP library) which are implemented as a different library for each separate environment shown in Figure 25.

The kernel API hides the complexity of operating system services.

The kernel abstraction simplifies the movement of the lwIP stack to different operating system environments. The fundamental services contain synchronization services, interrupt services, and timer-based callback services (Analog Devices 2007).

Blackfin processors can benefit from the system service library (SSL) that provides reliable, easy C language access to Blackfin features such as the interrupt manager, direct memory access (DMA), and power management units. Clock frequency and voltage can be changed without difficulty at run time through a set of simple APIs. Interrupt handling is done at the time of the event, or delayed according to the specific application's requirement.

A device manager integrates device drivers for on-chip and off-chip peripherals. The SSL is operating system neutral and can be run as separate, or in conjunction with a real-time operating system (RTOS).

The Visual DSP kernel (VDK) is a strong real-time operating system kernel. It provides vital kernel features as illustrated in the previous section (pre-emptive scheduler, thread creation, semaphores, interrupt management, events, memory administration and inter-thread messaging).

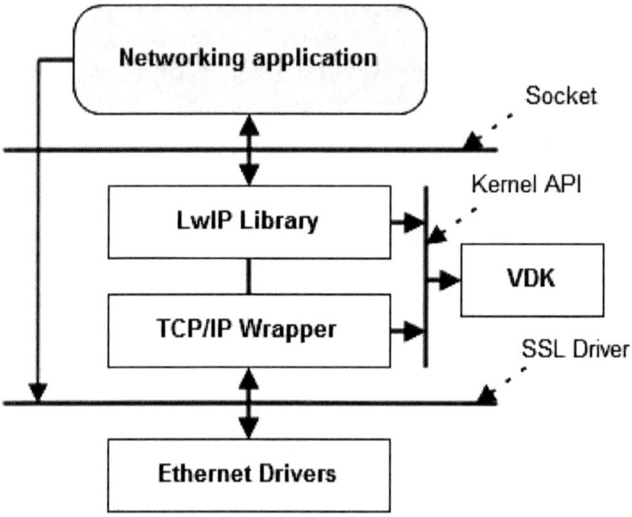


**Figure 25.** The LwIP system architecture.

The kernel API hides the primitive operating system services and simplifies the migration of lwIP stack to different operating system environments.

The basic services include synchronization services, interrupt services, and timer based callback services.

The stack supports the following protocols: TCP (Transport Control Protocol), UDP (User Datagram Protocol), IP (Internet Protocol), ICMP (Internet Control Message Protocol), ARP (Address Resolution Protocol) and DHCP (Dynamic Host Configuration Protocol)

The operating system used as support for the LwIP stack is Visual DSP kernel (VDK), as presented above (Analog Devices 2010).

### 4.4. A real-time Implementation Approach and Case Study: VoIP Based on AMR Codec

This section defines a framework for networking applications based on the following assumptions (Third Generation Partnership Project 1998), (Zoican 2011):

- There are two groups of tasks: the first one deals with the digital processing of analog signals (such as audio or video samples) and the second one deals with network packet processing (receiving and transmitting packets).
- The first group has very complex tasks and the second group has less complex tasks (block processing may be necessary).
- The system should work in real time.
- There are two types of information: signals, represented by samples of analog signals, and data, represented by binary results which are transferred over the network.

Two independent applications may be implemented using core A and core B of the dual core Blackfin BF561. The application in core A implements the tasks from first group and the application in core B may implement some simple tasks in the first group and the networking tasks (these tasks are certainly less complex).

In the first group (tasks A), consider an adaptive multi-rate (AMR) encoder; in the second group (tasks B), consider an AMR decoder, codec rate setting and the packets' processing, receiving and transmitting over network using UDP protocol).

A switching buffers technique is involved to ensure real-time functioning of the entire communication system. In this technique, there are pairs of input and output buffers that are switched periodically. One pair of buffers is used for processing and the other pair is used for receiving inputs and transmitting previous results.

For the specific case of the AMR-VoIP system, the following buffers are defined:

*Signal_RX[0]* and *Signal_RX[1]* – for input frames

*Signal_TX[0]* and *Signal_TX[1]* – for output frames

*Data_RX[0]* and *Data_RX[1]* – for coder results

*Data_TX[0]* and *Data_TX[1]* – for decoder inputs

The input frames have N samples each, and the coder/decoder output buffer contains M results each.

Two flags, *flag_A* and *flag_B*, are defined in order to control the program flow in cores A and B.

Also, a variable named counter is defined to manage the sample frames acquisition. An interrupt is generated by the audio analog to the digital converter (ADC) for every input speech sample. The interrupt routine service acquires the new sample, stores it in the current input signal buffer and transmits the results from the current output buffer to the digital to the analog converter (DAC).

After N signal samples have been acquired, the signal and data buffers are switched. A necessary condition for real-time operation is that the acquisition time to input the current frame must be greater than the processing time of the frame.

The above defined buffers, flags and counter are shared resources of the two cores in Blackfin microcomputers.

Figures 26 to 30 illustrate the switching buffer technique as well as flowcharts of core A, core B and interrupt service routine.

Networking processing is illustrated in Figure 31.

The application code, written in C, may be optimized for speed. Several techniques for C optimizations were used (Analog Devices 2010):

- Built-in functions for fractional data
- Circular buffers and hardware controlled loops
- Inter-procedural analysis (IPA)—the compiler optimizes across translation units instead of within individual translation units. The compiler sees the entire source file used in a final link at the time of compilation and uses that information while optimizing.

The most computationally expensive block in voice over IP (VoIP) based on the AMR codec is the AMR encoder block (implemented in core A). The rest of the computations (AMR decoder, setting AMR rate at each frame and network processing) require fewer computational resources.

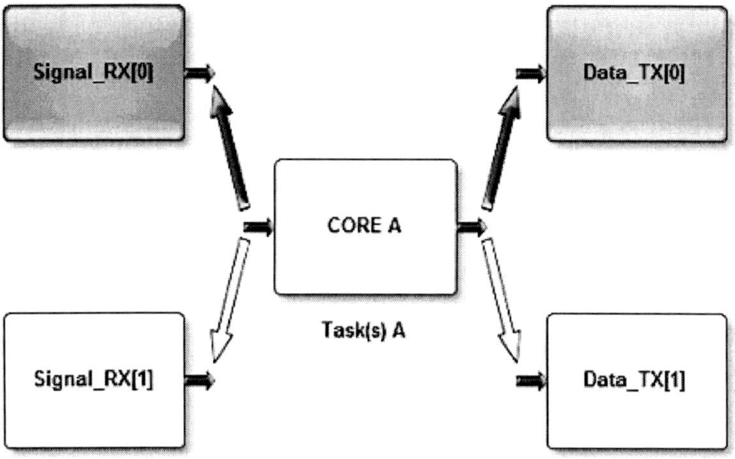


**Figure 26.** Switching buffer technique—core A.

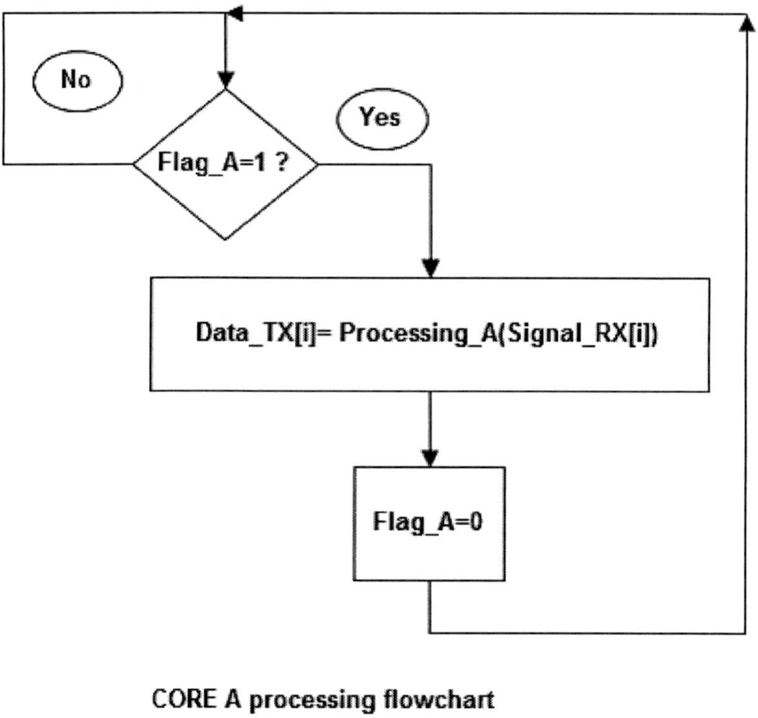


**Figure 27.** Flowchart—core A.

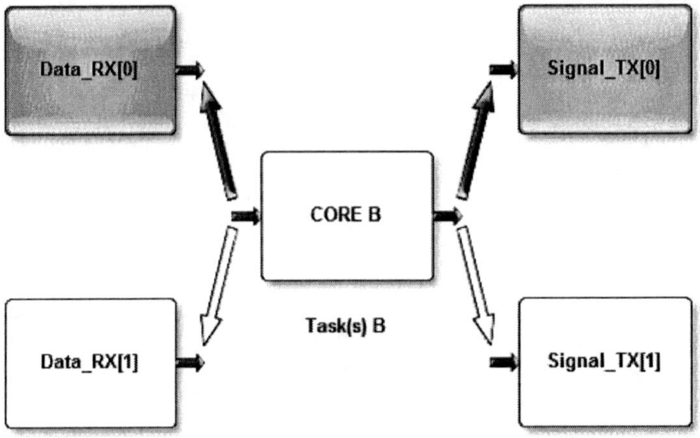


**Figure 28.** Switching buffer technique—core B.

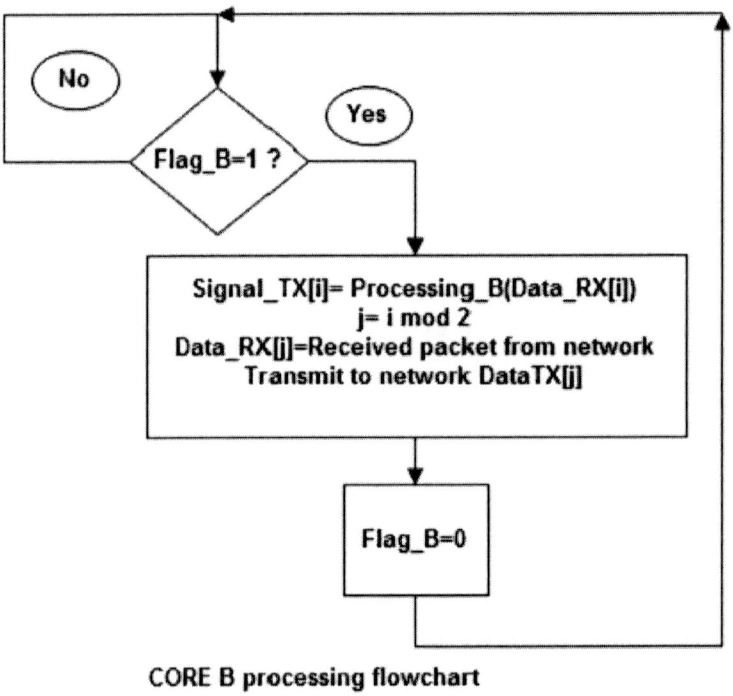


**Figure 29.** Flowchart—core B.

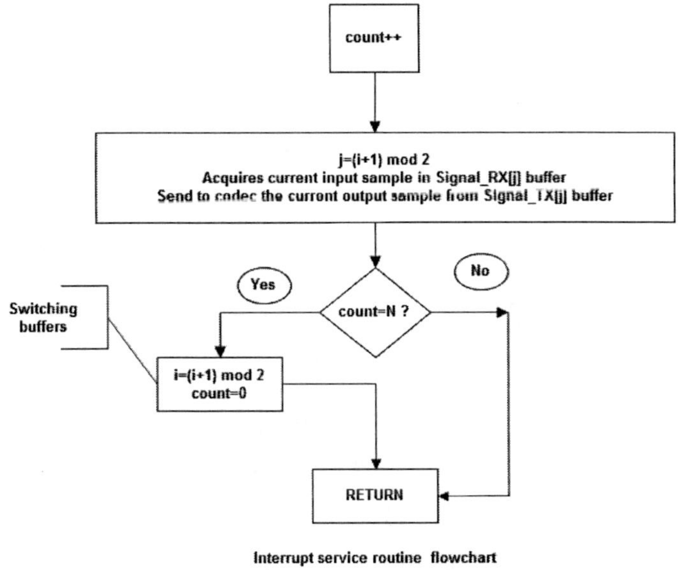


**Figure 30.** Interrupt service routine.

The optimized execution time is shown in table 5. This table illustrates how the processing time is shorter than the frame acquisition time of 20 milliseconds. Therefore, the VoIP system works in real time.

The proposed strategy may be used as a framework for various applications, especially for sensor network applications that require fast computation, lower power consumption and network (fixed or mobile) connectivity.

## Future Research Direction

Future work will more closely research the development of real image processing (noise elimination from video clips) as well as multimedia applications (real-time implementation of audio and video compression algorithms, improved voice over IP systems), using the Blackfin processor family.

## Conclusions

This chapter suggests how DSP may be used in telecommunications applications and to provide some implementation results for various applications with low and high level of complexity. The motivation for using DSP microcomputers in telecommunication applications resides in the characteristics of such applications: the need for real-time input signal

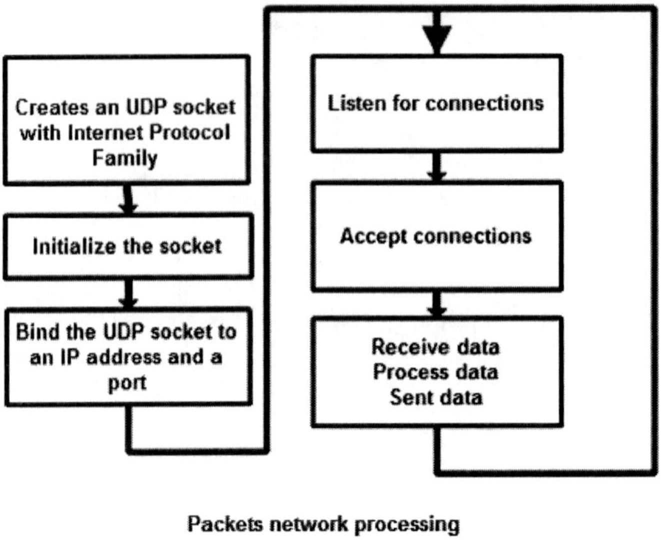


**Figure 31.** Networking processing.

**Table 5**

| | *AMR Codec—Execution time (BF High Speed Processors)* | | | | | |
|---|---|---|---|---|---|---|
| | *CODER* | | *DECODER* | | *TOTAL CODEC* | |
| *MODE* | *Cycles* | *Milliseconds* | *Cycles* | *Milliseconds* | *Cycles* | *Milliseconds* |
| 4.75 Kbit/s | 11327744 | 15.10 | 1386462 | 1.85 | 12714206 | 16.95 |
| 5.15 Kbit/s | 8894915 | 11.86 | 1363597 | 1.82 | 10258512 | 13.68 |
| 5.90 Kbit/s | 10118267 | 13.49 | 1353256 | 1.80 | 11471523 | 15.30 |
| 6.70 Kbit/s | 11980754 | 15.97 | 1354639 | 1.81 | 13335393 | 17.78 |
| 7.40 Kbit/s | 11320627 | 15.09 | 1360085 | 1.81 | 12680712 | 16.91 |
| 7.95 Kbit/s | 11687914 | 15.58 | 1425358 | 1.90 | 13113272 | 17.48 |
| 10.20 Kbit/s | 11833369 | 15.78 | 1410845 | 1.88 | 13244214 | 17.66 |
| 12.20 Kbit/s | 12238264 | 16.32 | 1473111 | 1.96 | 13711375 | 18.28 |

acquisition and output signal generation, intensive numeric calculation and the need for fast communication with external devices. Based on this motivation, the necessary tools (DSP hardware architectures, the real-time operating system kernel) and a general framework for developing DSP telecommunication applications are illustrated. Particular results are shown in order to validate this framework.

The conclusion is that DSP microcomputers and its associate software (for programming development and the operating system kernel) represent a useful tool for rapidly and efficiently developing complex real-time telecommunications applications.

## REFERENCES

Analog Devices. 1990. Digital Signal Processing Applications using the ADSP-2100 Family Edited By Amy Mar, Prentice Hall, Englewood Cliffs, NJ 07632, 1990, ISBN 0-13-212978-7.

Analog Devices. 1997. Digital Signal Processing Applications Using the ADSP-2100 Family, Volume 2, Prentice Hall, ISBN 0-13-178567-2.

Analog Devices. 2001.ADSP-218x DSP Hardware Reference.

Analog Devices. 2003. Unifying Microarchitecture for Embedded Media Processing.

Analog Devices. 2006. ADSP-BF533: Blackfin Embedded Processor DataSheet, Rev. C. Gan, Woon-Seng and Kuo Sen. 2007. Embedded Signal Processing with the Micro Signal Architecture, Wiley-Interscience.

Analog Devices. 2007. Blackfin Processors Manual.

Analog Devices. 2011. VDK (Kernel) User's Guide, Revision 3.5.

Analog Devices-1. 2010. Lwip user guide.

Analog Devices-2. 2010. VisualDSP++ 5.0 C/C++ Compiler and Library Manual for Blackfin Processors.

Analog Devices-web1. 2011. http://www.analog.com/en/processors-dsp/content/blackfin_benchmark_comparison/fca.html

Analog Devices-web1. 2011. http://www.analog.com/static/imported-files/eval_kit_manuals/288418312ADSP_BF533_EZ_KIT_Lite_Manual_Rev._3.1.pdf

Berkeley Design Technology. 2000. Choosing a DSP Processor.

Estrin, Deborah. 2001. Embedded Everywhere: A Research Agenda for Networked Systems of Embedded Computers, National Academy Press.

Harte, Lawrence. 2005. Introduction to Digital Subscriber Line (DSL); Technologies, Operation and Systems, ALTHOS Publishing.

Skolnick, David and Noam Levine. 1997. Why Use DSP? Analog Dialogue 31-2.

Softfone-web. 2011. http://www.analog.com/static/imported-files/product highlights/51136650630ftfonelcr_plus.pdf

Third Generation Partenership Project. 1998. Adaptive Multi Rate (AMR) speech; ANSI-C code for the AMR speech codec (3GPP TS 06.73 version 7.6.)

TransVideo-web. 2011. http://www.analog.com/en/processors-dsp/blackfin/products/customer-case-studies/transvideos_blackfin-based_set_top_boxes/resources/fca.html

Zheng, Yahong and Robert Lynch. 2009. Performances of variable step-size adaptive algorithms in non-Gaussian interference environments, Aerospace conference, 7–14 March 2009, pp. 1–7, DOI 10.1109/AERO.2009.4839470

Zoican, Sorin. 2002. Digital signal processors. Applications, MatixRom, Bucharest, Romania.

Zoican, Sorin. 2008. The Role of Programmable Digital Signal Processors (DSP) for 3G Mobile Communication Systems, Acta Tehnica Napocensis, Electronic and Telecommunications, Vol. 49, No. 3, 2008, pp. 49–56.

Zoican, Sorin. 2011. The Adaptive Multirate Speech Codec: Deployment Strategy Using the Blackfin Microcomputer", Sped 2011—The proceedings of 6th Conference on Speech Technology and Human—Computer Dialogue, Brasov, Romania.

## ADDITIONAL READING SECTION

ADSP-218x DSP Hardware Reference (Rev 1.0, February 2001), www.analog.com

ADSP-BF537 EZ-KIT Lite Evaluation System Manual (Revision 2.4, April 2008) www.analog.com

Bentham, Jeremy. 2000. TCP/IP Lean: Web Servers for Embedded Systems, CMP Books, Lawrence, Kansas.

Bhattacharyya, Shuvra S., Leupers Ranier and Marwedel Peter. 2000. Software synthesis and code generation for signal processing systems. IEEE Transactions on Circuits and Systems—II: Analog and Digital Signal Processing. **47(9):**849–875.

Burns, Alan and Andy Wellings. 1997. Real-Time Systems and Programming Languages, Addison-Wesley.

Chassaing, Rulph. 2002. *DSP* Applications Using C and the TMS320C6x DSK, Wiley.

Haykin, Simon. 1993. Adaptive Filter Theory, Prentice Hall, Englewood Cliffs. AllanV. Oppenheim and Robert W. Shafer. 2001. Discrete—Time Signal Processing, Englewood Cliffs, NJ. Prentice Hall, edn. 2.

Mitra, Sanjit K., and Kaiser James F. 1993. Handbook for Digital Signal Processing, John Willey and Sons.

Parhi, Keshab K. 2009. VLSI Digital Signal Processing Systems: Design and Implementation, Wiley India Pvt. Ltd.

Smith, Steven W. 2002. Digital Signal Processing: A Practical Guide for Engineers and Scientists, California Technical Publishing, 2002.

Tretter, Steven. 2003. Communication system design using DSP algorithms: With laboratory experiments for the TMS320C6700, Kluwer Academic Publishers.

VisualDSP++ 3.5 C Compiler and Library Manual for ADSP-218x DSPs, (Rev. 4.1, October 2003), www.analog.com

VisualDSP++ 5.0 C/C++ Compiler and Library Manual for Blackfin Processors (Revision 5.4, January 2011), www.analog.com

VisualDSP++ 5.0 Kernel (VDK) Users Guide (Revision 3.5, January 2011), ww.analog.com

## KEY TERMS AND DEFINITIONS

AMR – Adaptive Multi Rate – standard for compression of voice signals

DMA – Direct Memory Access

DMT – Discrete Multi Tone (a modulation method for discrete signals)

DSL – Digital Subscriber Line (a wire data transmission technology)

DSP – digital signal processor—specialized processor with high level of parallelism and multiple computational unit

FIR – Finite Impulse Filter

GPRS/GSM – General Packet Radio Service/Global System Mobile – data transfer system in mobile communication

Hardware loops – a mechanism that control a loop in hardware, without necessity of software

Harvard architecture – processor architecture with separate instruction and data buses

IP – Internet Protocol

LMS – Least Mean Squared (algorithm used in system identifications and adaptive filters)

LwIP – Light weight IP – a reduced resources consuming IP protocol implementation

MCU – microcontroller unit (general purpose microcontroller)

Micro Signal Architecture (MSA) – unified architecture with digital signal and general purpose processor characteristics

MIPS – Million of Instructions per Second

Networking application – an application that require network connectivity and packet processing

Real time operating system kernel (RTOS kernel) – an operating system kernel that has specific primitives to support real time functioning

Real time systems – a system works with time restriction or deadlines

TCP – Transport Control Protocol

UDP – User Datagram Protocol

14

# Embedded System Using GNU/Linux for Automating Low Earth Orbit Satellite Tracking

*Omar Alvarez Cardenas, Miguel A. Garcia-Ruiz, Jesus Rafael Carrillo and Margarita Glenda Mayoral Baldivia*

**ABSTRACT**

There are different types of artificial satellites orbiting the Earth; most of them are used for communications, astronomy, atmospheric studies, navigation, space exploration, and weather forecast, among other specific functions. All these satellites need to be located and tracked according to their orbit and altitude, but many commercial tracking systems are expensive and some of them are difficult to use and maintain. This chapter describes a low-cost and efficient embedded computer system running on GNU/Linux, intended for automating the orbit calculation and tracking of satellites, as well as performing the orientation of an antenna system to receive and transmit radio signals from Low Earth Orbit (LEO) satellites used for amateur radio communications. Our embedded system performs the same basic functions of any commercial satellite tracking system but with a low cost and running under the free software development philosophy, making the tracking process of Low Earth Orbit satellites accurate and easy to maintain.

## Introduction

Amateur radio (also called ham radio) is a nonprofit, non-commercial and private recreation of exchanging messages via voice and data communication using designated segments of the radio frequency (RF) spectrum. Amateur radio enthusiasts generally use the RF spectrum to contact people from all over the world, carrying out wireless experimentation for self-training and for some types of emergency communications (ARRL 2010). Amateur radio operation is regulated and coordinated globally by the International Telecommunication Union (ITU), a branch of the United Nations (UN) (ITU 2008), and licensed by national governments that regulate operational and technical characteristics of radio frequency transmissions and issue individual and group (club) stations with an identifying call sign (an unique identifier composed of letters and numbers).

In order to obtain a radio amateur license (and its respective call sign), future amateur radio operators are tested for their knowledge of key concepts in analog electronics, electricity, RF, and the host government's radio regulations. For example, the government body that regulates the amateur radio operations and licensing in the United States is the Federal Communications Commission (FCC), who defines amateur radio service as "a voluntary noncommercial communication service, used by qualified persons of any age who are interested in radio technique with a personal aim and without pecuniary interest" (FCC 2011). In some countries, future amateurs must also pass a Morse code test. Currently, an estimated two million people worldwide are licensed amateur radio operators. They generally use commercially-available transceivers, antennas and other parts of their amateur radio station, but operators can build their own "homebrew" (homemade) equipment.

People involved in amateur radio communications use a variety of text, voice, and image communications modes and can communicate across a city, region, country, and even to outer space with astronauts, since a number of them hold a radio amateur license and communicate to the Earth using their radio amateur equipment installed in spacecraft. There are also specially-built artificial satellites for radio amateur operators to carry out radio communications.

Amateur radio operators can also use other resources to communicate regionally and even throughout an entire continent using outer space capacities. Some organizations and groups of radio amateurs have designed and built artificial satellites capable of this type of communications for decades. These satellites have been used for amateur radio communications using voice and data (mainly in the form of Morse code), using designated very-high and ultra-high frequencies (VHF and UHF), respectively (Davidoff 2009). Amateur satellites are generally used to communicate to two or more amateur radio operators by retransmitting an RF signal carrying voice or data. Amateur satellites may also transmit telemetry data that an amateur operator may pick up and analyze. In addition, some amateur satellites transmit a recorded greeting voice or data message using Morse code, which is transmitted continuously, working as electronic "beacons" for any amateur radio operator to pick up and decode. However, commercial satellite antennas and commercial satellite tracking systems for radio amateurs are expensive and difficult to maintain. One way of overcoming this is to build homebrew satellite equipment for amateur radio operation.

This chapter describes the development of an embedded computer system running GNU/Linux which is responsible for automating the orbit calculation and orientation of an antenna used to transmit radio signals to radio amateur satellites and receive radio signals from them (the satellite tracking process). The advantages of our embedded system are: Ease of construction, low cost, and use of a homebrew antenna. Our satellite tracking system performs the same basic functions of any of the commercial systems for tracking satellites and orienting antennas, but under the free software philosophy. With the automation of the tracking process, the operation of low Earth orbit (LEO) satellites becomes easy and accurate.

Other types of artificial satellites that amateur operators may use (and be tracked using our above mentioned embedded computer system) are the weather satellites used for atmospheric analysis and forecast (Taggart 1996). Some of these weather satellites were built by the National Oceanic and Atmospheric Administration (NOAA) of the United States (NOAA 2011). They transmit real-time images of the surface of the Earth that can be down linked and decoded using special software. It is interesting to note that no amateur radio license is required to receive data from the weather satellites or for picking up beacons from amateur satellites.

A typical amateur radio station for satellite communications consists mainly of specialized antennas, pre-amplifiers, transceivers for the VHF, UHF and HF frequencies, a low loss feed line and satellite tracking software (Newport n/d). For best results, amateur radio operators use directional (Yagi) antennas. The use of a directional antenna is an important aspect and there is a great range of choice in commercial systems. Table 1 shows the most popular of these. For most amateur satellites, radio operators have to transmit and receive in different bands; one for the uplink (a radio signal transmission from the operator to the satellite) and a different one for the downlink (the reception of a radio signal from the satellite to the operator). An amateur radio operator can use a dual-band antenna or two separate antennas to carry out the uplink and the downlink. An operator also can use several antennas to operate on various uplink/downlink bands as needed (ITU 2008). One good option is to buy commercial satellite antennas from different brands. Table 1 describes some of them. However, most of them are expensive, especially the ones with high gain.

**Table 1. Commercial Amateur Satellite Antennas (ARRL 2011).**

| *Model* | *Brand* | *Elements* | *Gain* |
|---|---|---|---|
| A-144S5 | Diamond (VHF) | 5 | 9.1dB |
| A-27010S | Cushcraft (VHF/UHF) | 5/5 | 10/10 dB |
| 2M/440L5 | Elk (VHF/UHF) | 5/5 | 8.9/9 dB |
| 2MCP | M2 (VHF) | 22 | 14.3 dB |
| 436CP30 | M2(UHF) | 30 | 15.50 dB |
| 146/437-10 | Arrow Antenna (VHF/UHF) | 3/7 | 0/0 dB |

If the cost of the satellite antennas is a problem, there are some "homebrew" (homemade) low-cost antennas that can be used with great results (Britian 2009). They have been built, tested and used by many ham radio operators in different countries. The most popular homebrew antennas are:

- Turnstile
- Moxon
- Eggbeater
- TPA
- The quadrifilar helicoidal
- The EZ Linderblad
- XE1MEX Yagi/UDA

Having selected a homebrew antenna, a way to greatly improve the downlink signal is to install a receive preamplifier in the satellite antenna system. It consists of a high gain and low-noise amplifier with a frequency response for a specific band. A well-designed preamplifier can give and extra gain in the order of 15 to 25 dB with a minimum noise figure of 0.5 to 2 dB (ARRL 2007).

With respect to amateur radio transceivers for satellite operation, there are many commercial amateur radios that cover the VHF, UHF and HF bands and some of them offer satellite-specific features like the ICOM IC-2820, which receives and transmits on 2 m and 70 cm band independently. One of the transceivers with full HF coverage, plus VHF and UHF with full duplex satellite capability, is the Kenwood TS-2000. Another solution is using a hand-held (portable) VHF/UHF transceiver like the Yaesu FT-60R.

To complete an amateur radio satellite station, it is necessary to include a very low loss feed-line coaxial cable. All amateur satellites transmit using low power to extend their operation life in space, and the downlink signals must arrive readable at the amateur radio operator's transceiver. Amateur radio transceivers are designed to work with 50-ohm coaxial cable. If the transceiver finds a different impedance, it reduces its output to protect itself from damage that than can result from standing wave ratio (SWR) variations (ARRL 2007).

One important aspect of amateur radio satellite communications is the ability to track and predict a specific orbit related to the radio station position. To do this, it is possible to find satellite tracking software for the Windows, Macintosh and Linux operating systems (see Table 2). Most of them can automatically correct for the Doppler Effect and move the azimuth and elevation rotors of a directional antenna to get the best download radio signal. In general, most of the professional tracking software have the following functionalities:

- Professional features intended for serious satellite users
- Satellite tracking and prediction
- Antenna steering through popular interfaces
- Radio tuning with automatic Doppler correction
- Simulations and graphical interfaces
- Extensive documentation and vendor support (AMSAT 2011).

**Table 2. Sampling of Satellite Tracking Software (AMSAT 2011; ARRL 2011).**

| *Name* | *Source* | *Operating System* | *Radio Control* | *Antenna Control* |
|---|---|---|---|---|
| Nova | www.amsat.org (store) | Windows | No | Yes |
| SCRAP | www.amsat.org (store) | Windows | No | Yes |
| SatPC32 | www.amsat.org (store) | Windows | Yes | Yes |
| SatScape | www.satscape.co.uk | Windows | Yes | Yes |
| MacDoppler | www.dogparksoftware.com | Mac OS | Yes | Yes |
| Ham Radio Deluxe | hrd.ham-radio.ch | Windows | Yes | No |
| Predict | www.qsl.net/kd2bd/predict.html | Linux | No | Yes |
| WinOrbit | www.sat-net.com/winorbit | Windows | No | No |

There are currently a number of Low Earth Orbit (LEO) satellites that can be used as digital/voice repeaters and receive weather images that amateur radio operators can use. In order to access them, an amateur radio operator needs to compute the orbit and the real-time position of an amateur radio satellite using one of many available computer programs for orbit calculation, adjust and point a special antenna towards the satellite, and use their radio communication equipment. Operators can manually set up their antennas, or use commercially-available systems to do it automatically.

However, these types of commercial systems are very expensive and difficult to maintain. To overcome the cost of commercial systems for satellite tracking, it is possible to build a homemade system running the Linux operating system (O.S.), which rivals expensive commercial equipment. The next pages describe the design and development of a homemade satellite tracking system that consists of an embedded computer system and state-of-the-art software, including a specially-adapted version of GNU/Linux O.S.

Our embedded system uses an adapted version of Slackware GNU/Linux, adapted to run from an USB Pen Drive. The core of the tracking system runs at the top of the O.S., which involves the pass scheduler, orbit calculation and the rotor control. Other system services such as SSH/Web also run on the Embedded OS, but with low priority.

The antenna orientation system uses a servo motor and a stepper motor to move the antenna. The servo motor is used to control the elevation of the antenna from zero to ninety degrees. The interface between the embedded system and the servo motor is the PicServo control board, which is an open-hardware design and uses a PIC Micro controller.

The stepper motor is used to move the antenna on the azimuth axis. The motor is connected to an adapted old dot matrix printer mechanism. This helps the stepper motor to gain the required strength to move the elevation mechanism and the antenna. The azimuth movement of the motor goes from zero to 360 degrees. This motor is controlled by a computer parallel port using a power interface that includes TIP120 transistors and opto-couplers (SHARP PC817) to prevent parallel port damage because the motor draws 1.5 Amperes.

The antenna system uses an antenna design proposed by two Amateur Radio Operators from Spain (with amateur radio call signs EB4DKA and EA4CYQ) which is popular among amateur satellite operators. This design is called The Canary Jail Umbrella (CJU) Antenna. This design has been discussed in a number of amateur radio magazines such as QST (www.arrl.org/qst), CQ Amateur Radio (www.cq-amateur-radio.com), AntenneX (www.antennex.com), CQ VHF (http://www.cq-vhf.com), and Onda Corta (Alvarez et al. 2010), among others. The CJU antenna has been praised worldwide because of its ease of construction and the good quality of signal reception. One of the key characteristics of its design is the weight of the antenna. Using light materials like Balsa wood or Bamboo in its construction can make it light enough to mount on a small servo motor.

Figure 1 shows the system and its components. To understand how it works, we need to first understand what a satellite is, what azimuth an elevation movements are, how the servo motor and stepper motor are controlled, how to modify a GNU/Linux distribution to run from an USB pen drive, and how satellite orbit calculations are done.

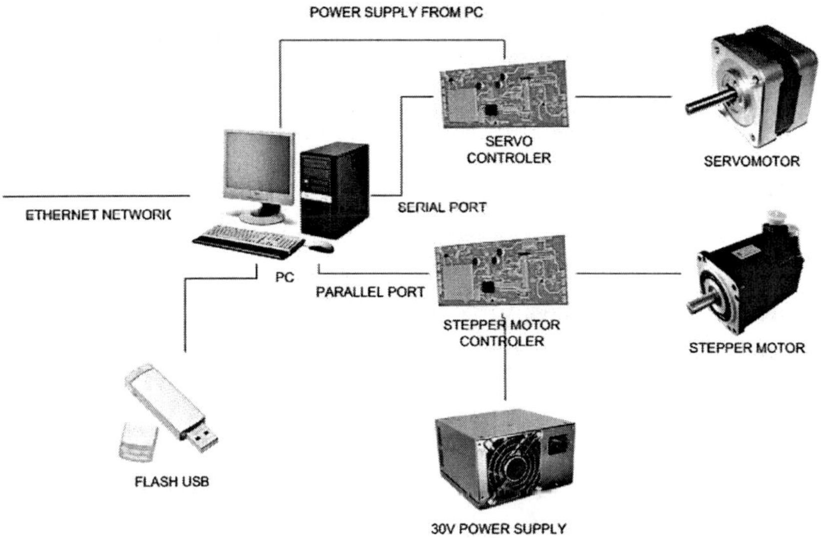


**Figure 1.** Elements of the embedded system.

To understand how amateur radio satellites work, it is necessary to describe some key concepts and definitions:

**Azimuth:** Azimuth refers to the rotation of the whole antenna around a vertical axis. It is a side to side angle measured clockwise with respect to the Earth's North Pole.

**Equatorial orbit:** The satellite spins around the Earth's equator in this kind of orbit.

**Elevation Angle:** This is the angle of the ground station from the horizon. If the angle is small, the distance the radio waves need to travel is longer.

**GEO (Geosynchronous Orbit):** In this type of orbit, satellites are at a distance of 35,848 Kilometers above the equator. At this altitude, the rotation period of satellites is 24 hours, which makes them appear in the same place from the planet surface.

**Inclined Orbit:** Any other kind of orbit is called inclined orbit because it goes from one side to another crossing the equator (Tomasi 1996).

**LEO (Low Earth Orbit):** These satellites orbit at 5,035 Kilometers above sea level or even lower, between 600 and 1,600 Km. (Roddy 2006). The amateur radio satellites are LEO Satellites. When a ground (amateur) station, A, wants to transmit a signal, it is first received by the satellite and then immediately amplified and retransmitted. After that, another amateur station, B, receives the signal and starts communication. To perform the communication, two types of angles from the ground station must be known (Tomasi 1996).

**MEO (Mid Earth Orbit):** The satellites with this kind of orbit are between 10,075 and 20,150 Kilometers from the Earth. This type of satellite is not geostationary.

**Polar orbit:** Type of orbit in which a satellite passes above or nearly above both poles of the Earth.

**Satellite:** A satellite is an object that spins around the earth; specifically, a communications satellite is a radio frequency repeater that orbits the Earth. A satellite system consists of a transponder, a terrestrial station to control the satellite, and a group of users. The correct relations between them provide the facilities to the correct reception of the radio signals in a satellite system (Cakaj 2009; Tomasi 1996). Orbital satellites spin around the Earth using different types of orbits.

**Trajectory:** A trajectory defines a geostationary orbit, which is a geosynchronous orbit above the Earth's equator that has an orbit period equal to the Earth's rotational period and an orbital eccentricity of approximately zero. An object in a geostationary orbit appears motionless to ground observers.

The satellites stay in orbit due to the centrifugal force caused by the Earth's rotation. LEO satellites have closer orbits to the Earth and their speed is around 7,800 m/s. At this velocity, satellites orbit the earth in approximately 1.5 hours. Because of this, the satellite can only be seen on the horizon for about 15 minutes.

In order to know when and where a satellite is going to pass over a specific point on the Earth, there are several tools to calculate the pass. With the help of a computer and these tools, we can predict the coverage, direction, and position of a particular satellite. To calculate these parameters, we need the information contained in the Keplerians. Keplerian elements are the inputs to a standard mathematical model of spacecraft orbits and are formed by seven numbers. This set of numbers defines an ellipse, orient it about the earth, and place the satellite on the ellipse at a particular time. With the Keplerians, the correct time, and your local station, it is possible to compute the exact satellite position in the sky and where to point a satellite antenna (Leick 2004).

## GNU/Linux for Embedded Systems

Having described the different types of satellites and their orbits, it is necessary to know about GNU/Linux embedded systems, which we used for our satellite tracking module. GNU/Linux is being used as a platform for embedded systems in diverse applications. The main advantages of using GNU/Linux for embedded systems are the following (Sally 2009):

- Source code portability.
- GNU/Linux has support in a wide range of architectures.
- Long Term Support.
- Total access to the Kernel Source code and the ability of changing it.

In order to develop an efficient and reliable Linux version for our embedded system, we followed five general steps and ten best practices.

The following are the general steps defined by (Chou et al., 2007):

- Choose the right hardware platform.
- Get an efficient development environment. (BuildRoot is recommended).
- Get the necessary packages for our embedded system.
- Once packages are obtained, it is necessary to integrate all of the above.
- Test the System.

The following are some of the best practices to develop an efficient and small embedded system, according to (Chou et al., 2007):

- Remove unused executables.
- Only install the needed libraries.
- Compile the software platform kernel only with the options needed by the system.
- Remove unused libraries.
- Use a lightweight library like uClib.
- Compile the libraries manually.
- Remove manual pages ("man") and documentation.
- Use a filesystem like CRAMFS.

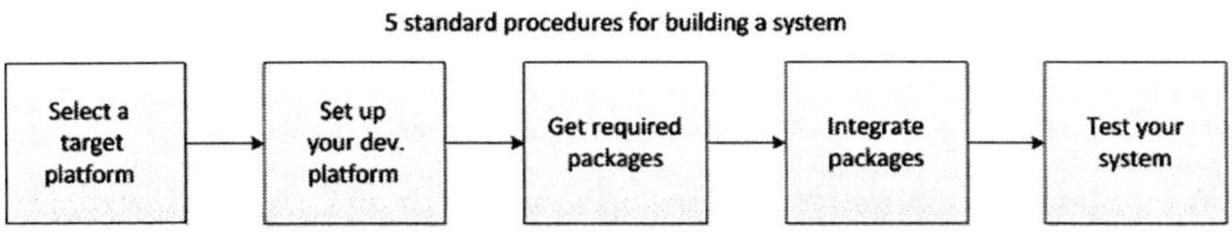


**Figure 2.** Steps for designing an embedded system (Chou et al., 2007).

After the kernel compilation is performed, one must decrease the "STATIC_MEMORY_ALLOCATION" parameter.

Do not compile unused kernel modules. Also, it is better to compile the kernel in a non-modular way.

## Preparation of the Embedded System

The option of creating a GNU/Linux distribution from scratch was discarded because for prototyping software development purposes it is better to have the support and applications right away (Yaghmour et al., 2008). We developed the software system for our embedded system using Slackware Linux O.S. In addition, we used Qemu to install the Linux distribution on a USB flash drive. This is an x86 compatible processor emulator. With this software we can run the installation from a computer running an O.S. other than Linux (Surhone et al., 2010).

## Installing GNU/Linux in a USB Flash Drive

For software development purposes, a 2 GB Flash drive it is recommended. We needed to delete the current flash drive partitions and creates a new one using all the space available. It also needs to have the bootable flag. To install Slackware on the flash drive we needed to follow the following steps:

## USB Support

To create the partition, we first need to identify the device file of the USB Flash drive. This can be done following the following command from the Slackware Terminal:

***root# dmesg***

*[118828.614766] sdb: assuming drive cache: write through*

*[118828.614769]* ***sdb: sdb1***

This command provides access to the kernel log, where it can be confirmed that Linux has detected the flash memory with the device name "sdb".

### Partitioning

After we have identified the name of the flash drive device, we started its partitioning with the following command:

***root# cfdisk /dev/sdb***

This command deletes all the partitions and creates a new one. Users must make sure to mark the partition as bootable.

Then the Qemu must initiate the installation (Fig. 3):

***root# qemu –hda /dev/sdb1 –cdrom slackware-10.1-disc1.iso –boot d***

The installer is run with the following command:

***root# setup***

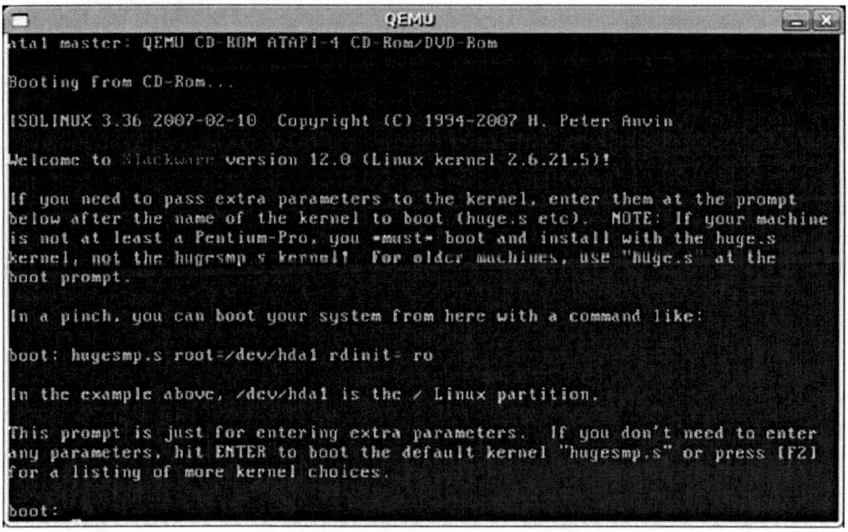


**Figure 3.** Slackware booting process.

Once placed in the installation menu, users should choose the "target" option and select /dev/hda1. Then the Ext2 should be selected as the file system for the partition (Fig. 4). Ext2 is an improved version of Fat32 File system. For a long time Ext2 was the default file system on most of the GNU/Linux distributions and it is suitable for this project since it is stable and does not have journaling (a file system that is used in the event of a system crash or power failure).

The installation process copies and install of all the packages into the flash drive. This process takes about 30 to 45 minutes. Importantly, Slackware uses LILO as the default bootloader. Finally, it was installed in the MBR of the flash drive. After that, the installation was completed.

## Kernel Building

The latest version of the Linux kernel is recommended because it provides support for newer hardware and avoids old bugs. It is also better to build

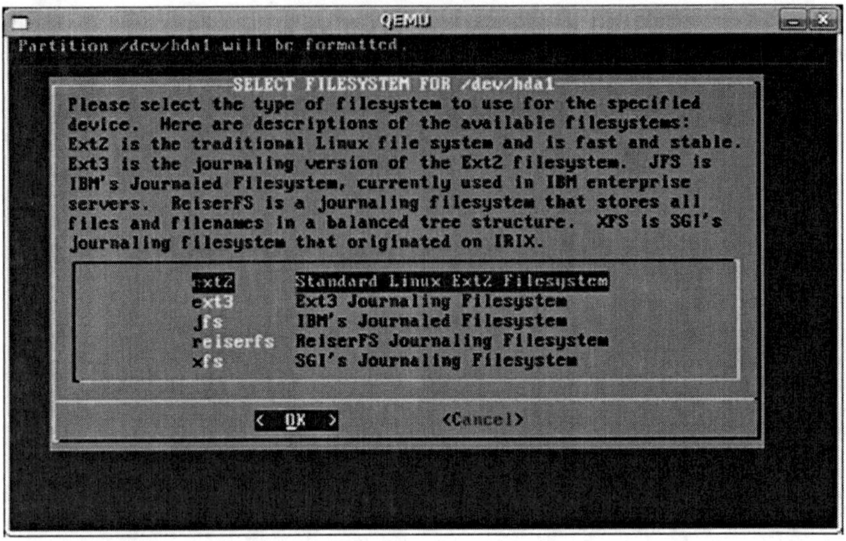


**Figure 4.** File System selection menu.

***root# make modules && make modules_install && make bzImage***

After this, users must select the modules they wish to install on the kernel. Do not to forget to mark them as static modules (*) and decrease the "STATIC_MEMORY_ALLOCATION" parameter. Users should also take into account the support for the ext2/ext3 file system.

The following must be executed to build the kernel:

***root# make modules && make modules_install && make bzImage***

The build process will then create a Linux kernel image. This image needs to me moved to the /boot directory using:

***cp arch/i386/bzImage /boot/linux-kernel***

After copying the kernel image, a series of changes needs to be done to the LILO configuration file in order to boot the new kernel. This time the changes need to be done in a live environment using Qemu again:

***root # qemu –hda /dev/sdb***

The first change needs to be done on the /etc/fstab file. The line that mounts the root / partition must be changed from:

*/dev/**hda1** / ext2 defaults 1 1*

to:

*/dev/**sda1** / ext2 defaults 1 1*

the kernel using the host computer rather than using Qemu since the build process is I/O, CPU, and memory intensive.

The steps to build the kernel are the following:

Have root access to the build computer.

Mount the flash drive.

Execute a chroot to the directory where the flash drive is mounted.

Download the kernel code from http://kernel.org

Decompress the source code.

Execute "make menuconfig"

Configure the kernel parameters.

Build and install the kernel.

The kernel building process will run natively using this procedure after executing chroot and running the following command:

***root# mount –t ext2 /dev/sdb1 /SlackwareMemoria***

After this, the working environment must be changed. The chroot command realizes this process:

***root# chroot /SlacwareMemoria***

The purpose of the previous steps was to create a Slackware terminal. The next step is to download the kernel source typing these commands:

***root# cd /usr/src/***

***root# wget http://www.kernel.org/pub/linux/kernel/v2.6/linux-2.6.25.4.tar.bz2***

The kernel source comes stored in a TAR file, so an un-TAR needs to be run:

***root# tar xjf linux-2.6.25.4.tar.bz2***

Next, the menuconfic command needs to be executed

***root# make menuconfig***

The File Systems section activates the support for both the USB file system and the support for the ext2 file system. After this, the following command is used to compile and install it.

The next modification is to the LILO configuration file located at / etc/lilo.conf. This step is important because it defines the kernel image to load in the boot process. An important detail to keep in mind is that the Linux kernel is multi threaded; therefore, it is mandatory to stop the kernel execution a few seconds to let the motherboard initialize the USB Ports. This can be achieved by setting up the rootdelay directive defined in the append section in the LILO config file:

*image = /boot//linux2.6*

*root = /dev/sda1*

*label = linux_SDA*

*append = "rootdelay=8 ramdisk_size=16000"*

*read-only*

Next, the LILO must be installed:

*root# /sbin/lilo*

## Design and Control the Elevation and Azimuth Mechanisms

### Servo motors for Elevation System

Servo motors, also called servos, are direct current motors that have the capacity to spin to a desired position within their operation range. To do this, the servo needs a control signal, which is a pulse modulated signal. The duration of the pulse tells the servo how many degrees to spin. Thus, the operation range for common servos is 0 to 180 degrees.

### Servo Motor Control Board

The control board has a PIC microcontroller, capacitors, and resistors. This board uses a PIC16F84 microcontroller from Microchip to optimally drive servo motors and digital outputs. The control board receives commands from the host embedded system via a standard RS232 serial interface. It was designed by Ashley Roll (an Australian engineer) who released its design to the public domain (Fig. 5).

To enable the controller and start moving the servo motor, it is necessary to send control commands to it. These commands have a 1-Byte length. The board has the following control commands:

#### *a) Reset*

The reset is needed when an emergency stop or any channel deactivation is required. This command is a zero ("0") succession and the board interprets them as a RESET and shut down all the outputs.

**Figure 5.** The PicServo controller (Roll 2000).

***b) Channel activation/deactivation:***

*Activation*

To move the servo, the channel that is connected to the motor must be enabled. The activation command is 48 Decimal + the channel number. For example, to enable Channel 4, we need to send the command 48 + 4 = 52.

*Deactivation*

To deactivate a channel, the command is 64 + channel number. In other words, to deactivate Channel 4, the number 68 needs to transmit to the control board.

***c) Servo movement 'position'/'offset'***

In order to achieve the full 180 degrees of movement, we need to handle two types of movements; the position which goes from 0 to 90 degrees and the offset, which is an extra movement that also goes from 0 to 90.

## Elevation Control Daemon

This daemon handles the elevation movement and the communication with the control board. It is possible to have 512 positions that cover 180 degrees. However, the system only requires 90 degrees. With a simple equation, we found a multiplication factor to move the motor. If users want to move the servo 60 degrees, they need to send the code to move the motor and the number 60 * 2.8444444 to the control board.

In order to use this daemon, the tracking software needs to write information to a UNIX Socket. The protocol is simple; the software only needs to send the degree to position the servo. For example, to move the servo 60 degrees, the tracking software only needs to open the socket and write 60.

**2) Azimuth movement of the Stepper Motor**

Stepper motors are used when precise movements are required. The main advantage of these kinds of motors is that they are pulse-driven and they move the same number of degrees each time. In addition, they remain locked on each position unless another pulse is applied.

There are two types of stepper motors: unipolar and bipolar. This project uses unipolar motors since they are easy to drive and they are widely available. This motor has five to six cables. Four cables are for the coils and the remaining ones are for ground. The basic setup in the control circuit for this kind of motors consists of an array of transistors which is found in the ULN2003 integrated circuit.

To drive these kinds of motors, it is necessary to activate their coils in a specific order. We use the war drive sequence which provides a smooth movement and good speed of rotation. The disadvantage of this sequence is that only one coil is activated each time and this produces a loss of power. This can be solved with the use of a lightweight antenna attached to it.

## Electromechanical Platform and Control Software

The mechanical platform was built using a stepper motor and a gear system taken from an old dot matrix printer (Figures 6 and 7). This gear demands the stepper motor to make 866 steps to complete a full spin. Using a simple mathematical procedure, a factor was calculated to obtain the number of steps required to move a certain number of degrees.

The parallel port of the PC used to communicate with our satellite tracking system did not deliver enough current to drive the stepper motor. Thus, a power interface, which uses transistors, was designed to isolate and protect the parallel port. The transistors we used are TIP120 NPN, which provide enough current to drive the stepper motor. Also, opto-couplers were used to isolate the parallel port from the high current of the transistors.

The parallel port control software responds to a series of codes that define the direction of rotation and the number of degrees to spin. It was

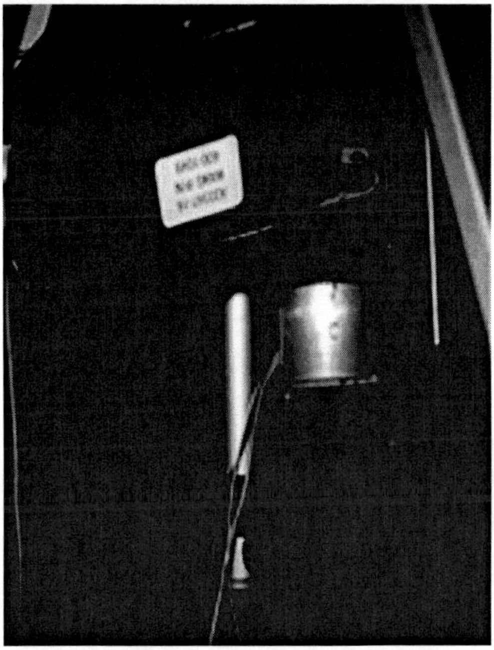

**Figure 6.** Stepper motor used for the Azimuth.

**Figure 7.** Gear system.

necessary to define memory allocation for the parallel port, where the allocation address was the hexadecimal value 0x378h.

In order to write to the software to communicate through the parallel port, the software needs to ask the kernel for special access rights and be executed as root user. The request for the access is issued by the *ioperm* function. This function establishes the access bits for the memory address on which the parallel port resides.

The process of writing data to the parallel port is done by the *outb* function.

## Prediction/Scheduling and Orbit Calculation

The most important software subsystems of our embedded system are: orbit prediction, orbit calculation and scheduling.

### Orbit Prediction

This subsystem consists of a series of web pages that are used to predict the orbits and schedule them. This is the main user interface of the embedded system and the user (the operator) interacts directly with it.

### Orbit Calculation

This is the most important subsystem of the project. It calculates the position of any satellite in real time. The information obtained from this subsystem is used to drive the azimuth and elevation subsystems.

### Scheduling

This subsystem is in charge of verifying if there is a scheduled pass in real time and start the orbit calculation subsystem.

To achieve orbit prediction, scheduling, and orbit calculation, this project uses a piece of software called "Predict" which is delivered under the GNU/GPL License and can be freely used.

### Antenna

To select or construct the antenna, we considered the following aspects:

- Easy to build.
- Light weight.
- Inexpensive.

The project uses the CJU (Canary Jail Umbrella) (Figure 8), which was designed by two amateur radio operators from Spain (the operators' amateur radio call signs are EB4DKA and EA4CYQ). This type of antenna has all the above mentioned construction aspects and a radiation pattern that is not as directive as a Yagi antenna.

## Results

Our first test was carried out on the prediction system selecting different LEO satellites to verify the correct orientation of the antenna. This was done using the control algorithm. The angle was measured with an inclinometer, which matched with the values obtained from the Orbitron prediction software. The physical azimuth and elevation values that we obtained were exact, and suffered no modification in the algorithm software and gear system. The velocity and strength of the two rotors was sufficient and it was possible to support more weight from the antenna to increase the gain in the reception audio from LEO satellites.

The CJU antenna was capable of receiving incoming signals from a satellite during 70% of its orbit when the minimum elevation angle was 10 degrees. The only adjustment done in the CJU antenna was in its polarity. The first test included mounting the antenna with a vertical polarization that received some fading signals from the some LEO satellites. To solve this, we changed the antenna to horizontal polarization, but we obtained more fading. Upon orienting the antenna manually, the polarization can

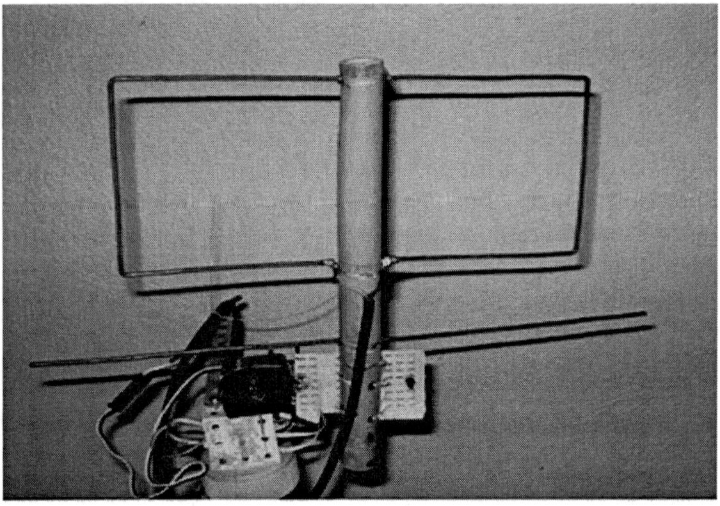

**Figure 8.** CJU antenna.

be changed manually between vertical and horizontal to correct o decrease fading. In this case, the best audio reception was when the antenna was mounted between the two polarizations (45 degree clockwise). That position best received an acceptable audio signal from the satellite, covering 70% of the satellite orbit.

The embedded system and the Slackware Linux worked acceptably with memory and CPU use was really low, with sufficient excess capacity to support other processes or applications in the future. The 2 GB Flash drive was enough for the applications and software that we developed; the response time for all the processes let the embedded system be quickly pointed at and follow the LEO satellites, comparable to commercial solutions.

## Conclusion

Our embedded system makes the tracking of LEO satellites easier to perform than manually pointing an antenna, thus significantly reducing orientation errors and enhancing the quality of audio received from an amateur satellite.

The modular approach used in our system allows anyone to change procedures to control the azimuth or elevation rotors without affecting the rest of the system. The use of an embedded system reduced the energy consumption. We obtained the same results as a commercial solution while using only a minimal hardware setup.

Most of the Linux control daemons (pieces of software) where written in PERL language and only the parallel port control software was written in C. This can be seen as an advantage because the software development time can be dramatically reduced, compared to other programming languages (Love 2007).

GNU/Linux can be compiled as a reduced version, which includes only the modules and programs that we only needed for our particular satellite tracking application. This allowed more control of the embedded architecture and reduced the memory, storage, and CPU usage. In addition, GNU/Linux is a quick and inexpensive way to develop and control embedded systems and it has many technical advantages.

A number of videos of our satellite tracking system prototype can be downloaded from this link: http://xe1azj.sat-xe.org/video.html

# REFERENCES


Alvarez, O., M.G. Mayoral, M.A. Garcia-Ruiz and H.E. Preciado. 2010. Antenas IOio y Moxon para satélites LEO. Onda Corta, No. **424:**7–10.

AMSAT. 2011. Professional Tracking Software for Purchase. Available from http://www.amsat.org/amsat-new/tools/software.php.

ARRL. 2007. The ARRL Antenna Book: The Ultimate Reference for Amateur Radio Antennas, Transmission Lines and Propagation. 21st edition. American Radio Relay League, Newington, CT.

ARRL. 2010. The ARRL Handbook for Radio Communications: The Comprehensive RF Engineering Reference, 88th Edition. American Radio Relay League (ARRL), Newington, CT.

ARRL. 2011. The ARRL Handbook. Eighty-Eighth/One Edition. The American Radio Relay League, (ARRL). Newington, CT.

Britian, K. 2009. Cheap Antenna Designs for the AMSAT LEOs. Kindle Edition. Wilhite Write. San Francisco, CA.

Cakaj, S. 2009. Practical horizon plane and communication duration for low earth orbiting (LEO) satellite ground stations. In Proceedings of the 8th WSEAS international conference on Telecommunications and informatics (TELE-INFO'09), 62–67.

Chou, C.H., T.H. Yang, S.C. Tsao and Y.D. Lin. 2007. Standard Operating Procedures for Embedded Linux Systems Linux. Journal, Volume 2007, Issue. **160:**10.

Davidoff, M.R. 2009. The ARRL Satellite Handbook. American Radio Relay League (ARRL), Newington, CT.

FCC. 2011. Federal Communications Commission. Amateur radio service. Available from: http://wireless.fcc.gov/services/index.htm?job=service_home&id=amateur

ITU. 2008. Amateur and Amateur-satellite Services. Radiocommunication Bureau, International Telecommunication Union, Geneva, Switzerland.

Leick, A. 2004. GPS Satellite Surveying, Third Edition. John Wiley and Sons, Hobokewn, NJ.

Love, R. 2007. Linux System Programming: Talking Directly to the Kernel and C Library. O'Reilly Media, Sebastopol, CA.

Newport, J. (n/d). Amateur Satellites as a Vehicle for Satellite Communication Education. Available from http://www.oosa.unvienna.org/pdf/sap/centres/satcomE.pdf.

NOAA. 2011. National Oceanic and Atmospheric Administration website. Available from: http://www.noaa.gov/satellites.html

Roddy, D. 2006. Satellite Communications, Fourth Edition. McGraw-Hill Professional, Berkely, CA.

Roll, A. 2008. PICServo controller. Available from http://www.digitalnemesis.com/info/projects/picservo/.

Sally, G. 2009. Pro Linux Embedded Systems. 1st edition. Apress, Berkely, CA.

Surhone, L., M. Timpledon and S. Marseken. 2010. *QEMU*. Betascript Publishing, Mauritius.

Taggart, R. 1996. Weather Satellite Handbook, 5th edition. American Radio Relay League (ARRL), Newington, CT.

Tomasi, W. 1996. Electronic Communications Systems. Pearson Education, Upper Saddle River, NJ.

Yaghmour, K., J. Masters and G. Ben-Yossef. 2008. Building embedded Linux systems. 2nd Edition. O'Reilly & Associates, Sebastopol, CA.

15

# A VANETs Routing Protocol Adaptation to WSNs

*M. Gonzalez-Gutierrez*

**ABSTRACT**

With the development of *ad hoc* networks, research effort focusing on the benefits of using low rate devices that can communicate in an autonomous and distributed manner have significantly increased. Advantages of wireless sensor networks include the small size of nodes, inexpensive implementation, and their ability to function in hostile terrain or dangerous environments for humans without extra management. However, wireless sensor networks need to deal with challenges about routing, power supply, memory and processing capacity. Early routing techniques were not suitable because IP routing protocols used in wireless networks excessively consumed energy resources. Multiple efforts have combined to develop solutions that optimize resources to ensure high network performance.

This chapter outlines a wireless sensor network protocol that was developed to ensure the reliable delivery of data and avoid routing traffic flooding, while providing a distributed topology that ensures an optimal operation in changing environments. LORA-CBF, a location routing algorithm was originally created for vehicular *ad hoc* networks. This work presents an adaptation to wireless sensor networks and a summary of the key concepts concerning them.

## Introduction

Wireless Sensor Networks (WSNs) consist of a number of autonomous nodes called motes. A function of node in a WSN is to perform monitoring tasks about phenomena occurring in the environment with the goal of making decisions for a specific application, according to the information collected. Every node, or mote, has at least one sensor. A sensor is a device that converts an input signal to another kind of output signal is a transducer.

Sensors have been vastly improved as new technologies become available. These technologies have contributed to making sensors smaller, more precise, and less expensive to produce and purchase, among others. This has resulted in a significant increase of their application in a wide variety of areas like health, environment, industry, agriculture, security, and military, among others.

WSNs have properties that make them a better solution than conventional networks for covering large areas, to work in hostile terrain areas, or in places that present danger to humans. In addition, WSNs do not require special attention to configure or manage after their nodes are deployed. However, despite the large number of advantages of WSNs, there remain many challenges that motivate researchers to develop solutions, including the optimizing energy, memory and packet routing.

Routing in WSNs is not as simple as it is for *ad hoc* networks. It is challenging because of the large number of nodes in WSNs. Consequently global routing with IP Protocols does not result feasible because of the high energy and processing costs. Hara et al. (2010) mention that routing protocols must deal with a large number of motes, a high probability of information redundancy, as well as significant energy, processing, and storage restraints.

Routing protocols in WSNs are classified in two ways: according to the operation and their destination location strategy, and, of course, according to the network structure.

According to the protocol functions, WSN protocols can be classified as proactive, reactive and hybrid. In a proactive network protocol, sensor nodes are in sleep state for a pre-defined period; then, they sense their environment and identify changes that must be transmitted to neighbors. In a reactive network protocol, motes are continuously monitored and when a change occurs, they immediately transmit the data captured by sensors to a designed mote or a server (Kim et al., 2011). Finally, hybrid protocols combine characteristics of proactive and reactive types.

According to the network structure, routing protocols are classified into flat, hierarchical, and location-based. In flat-based protocols, every node in the field has the same characteristics and functions. In hierarchical protocols, nodes perform different roles. Nodes can function as cluster heads that are responsible for performing the function of generating routing packets, thus helping them to optimize energy. Location-based protocols use position information of the network nodes and rout packets to specific network zones (Lazakidou et al., 2010).

This chapter introduces a new wireless sensor networks protocol that was originally developed for vehicular *ad hoc* networks, or VANETs. LORA-CFB is a geographic location routing algorithm with cluster-based flooding, which reduces control traffic, ensures the delivery of packages, and performs a dynamic routing.

## State of the Art of WSNs

### History of Wireless Sensor Networks

The beginning of WSNs dates back to the Cold War, with extensive efforts to detect submarines during the World War I, and the improvement of sonar technologies used in World War II. The U.S. Navy developed a system to face the threat of submarine warfare, which comprises an array of sensors or hydrophones. This system, called Sound Surveillance System or SOSUS, was installed in the Atlantic and the Pacific Oceans. The signal received by the hydrophones was transmitted to stations called "Naval Facilities" or NAVFACs and analyzed by the spectral frequency which resulted from the acoustic signals produced by enemy submarines. From 1958 onwards, SOSUS identified several Soviet submarines and discovered the location of submarine remains belonging to the ill-fated USS Scorpion (Whitman 2005).

Experimentation and development of *ad hoc* networks began in 1973. The Defense Advanced Research Projects Agency of the United States (DARPA), a military department responsible for the development of new military technologies, began a program called PRENET (Packet Radio Network), whose goal was to improve node communication, propose novel routing techniques, and ensure energy use (Witt 2008). Years later, in 1978, DARPA organized the Distributed Sensor Nets Workshop to help solve problems regarding the limits and challenges of sensor networks, including networking technologies, signal processing techniques, and distributed algorithms (Dargie and Poellabauer 2010). In 1980, the agency began a program called DSN (Distributed Sensor Networks) to research how the autonomous operation and communication features of Arpanet could be extended to wireless sensor networks (Hong 2011).

During the following years, the improvement and development of new technologies contributed to new features being adapted to sensors. Some remarkable achievements that contributed to improving WSNs were, the definition of Wireless Integrated Network Sensors (WINS) proposed by the Rockwell Science Center in cooperation with the University of California, which led to the development of the Low Power Wireless Integrated Microsensor (LWIM), in 1996, (Dargie and Poellabauer 2010). Another effort, the Sensor Information Technology program (SensIT), was created in collaboration with DARPA, had the goal of developing new networking techniques for dynamic *ad hoc* environments. This was an important line of research because sensors needed to be rapidly deployed in the battle field and the current solutions for voice and data handling were insufficient. Furthermore, they wanted to improve data processing to ensure data reliability (Chong and Kumar 2003). The interest of research in the area of wireless sensor networks began to increase again in 2002, with the Smart Dust project, whose aim was to design a small node, or mote, to demonstrate that a complete sensor system the size of a grain of sand or a dust particle could be produced (Dargie and Poellabauer 2010).

In 2003, the Institute of Electrical and Electronics Engineers (IEEE), released the IEEE 802.15.4-2003 standard for low rate Wireless Personal Area Networks (WPAN), followed by a revised version in 2006. A year later, a group of research formed by companies engaged in the development of control devices and monitoring called "ZigBee Alliance", published the ZigBee Standard for wireless sensor networks based on the IEEE 802.15.4 standard.

Currently, there are several developers of applications and platforms for wireless sensor networks with micro-electro-mechanical systems (MEMS), devices that are characterized by their high sensitivity. Companies such as Crossbow, Ember Corporation, Sensys, Dust Networks, Sensicast, Ant Wireless, among others, offer a wide range of devices for different kinds of applications.

## Hardware Structure of Sensor Node

The sensor node is the basic element of a wireless sensor network. This device is responsible for analyzing its environment, processing the information, and transmitting data to neighbor nodes. Network features depend on node sensor attributes; therefore, selection of sensors to meet the specific system requirements is a critical factor.

Figure 1 shows the basic components of a sensor, which consists of five principle elements: controller, memory, communication devices, sensors, and power supply.

The central processing unit of a sensor node is the controller. Controllers perform the task of collecting information through their sensors. Then the information is processed and a decision is made regarding its transmission as the controller simultaneously executes protocols and applications programs (Karl and Willig 2007). The common controllers for wireless sensor networks are specially designed for embedded systems, which are also called microcontrollers. A microcontroller is a tiny integrated circuit composed of memory, output and input interfaces, a clock generator, digital analog converter, and a serial interface (Dargie and Poellabauer 2010).

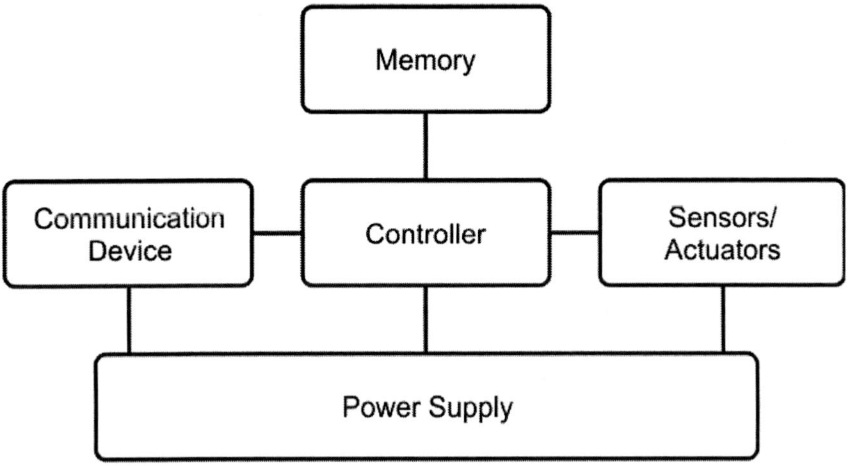


**Figure 1.** Hardware Sensor Components (Karl and Willig 2007).

Other kinds of controllers are also used besides microcontrollers, including Digital Signal Processors (DSPs), Application-Specific Integrated Circuits (ASIC), or devices like the Field Programmable Gate Array (FPGA). It is important to add that Field Programmable Gate Array is used for specific purposes because it offers a more efficient use of resources, although at a higher cost at the hardware end.

DSP processors are geared towards processing discrete signals through digital filters. These powerful processors, which can take millions of samples per second in a real time, are recommended for multimedia applications that involve audio and video signals and for applications in hostile environments where there is significant noise that can cause interference (Dargie and Poellabauer 2010). ASIC processors are specialized processors that are designed according specific application requirements. Finally, FPGA processors can be reconfigured during network operation to add new features. However, reconfigured FPGA is not convenient because of the require cost of time and energy. (Karl and Willig 2007).

Sensor node memory is generally RAM (Random Access Memory), ROM (Read Only Memory) or EEPROM (Electrical Erasable Programmable Read-Only Memory). In selecting the type of memory to be used, one should take into account variables such as the time required access memory and the quantity of energy needed to carry out the required operations.

To transmit information, sensor nodes can use radio frequency, optical transmission, ultrasound, and magnetic inductance. Radio frequency (RF) is the most common means of transmission as it can more easily be adapted to different types of applications, providing them with a high data rate, low error rates, and efficient power consumption; furthermore, radio frequency does not need line of sight between the transmitter and the receiver (Karl and Willig 2007). Every node is composed of a transmitter and receiver, with most novel platforms offering both in the same device called a transceiver.

Sensors are the core of nodes. Gravilovska et al. (2010) define sensors as the interpreters of physical reality into comprehensible and processable representations. Table 1 summarizes the main kinds of sensors presented by Dargie and Poellabuer (2010) and Gravilovska et al. (2010).

The power supply of a sensor node is another important element to consider. The power supply will define the lifetime of the entire network. Operations such as data transmission and reception, storage, etc., all depend on energy availability and careful calculations must be made taking into account the energy consumption of network to determine its life.

Hardware platforms that consider all of the above mentioned elements are developed by important technology manufactures. Most common platforms include: Btnode developed by ETH Zurich, EyesIFX by Infineon, Mica by Crossbow, Kmote by TinyOS, Tmote by Moteiv, and SensiNet by Sensicast Systems.

## Software Basics in WSN

Sensor node software must ensure that application objectives are met at the same time it overcomes the limitations imposed by the hardware (energy use, memory, processing capacity, etc.). The software components are

**Table 1. Summary of sensors used in wireless sensor networks.**

| | *Sensors* | *Characteristics* |
|---|---|---|
| Temperature Sensors | Electromechanical Temperature Sensors | Composed of Bi-metal thermostats that respond to the contraction or expansion of motion when they undergo temperature change. |
| | Electronic Temperature Sensors | Based on a pair of metals that generate a flow of current when they are subjected to temperature change. |
| | Silicon sensors | Operate through a bulk electrical resistance of semiconductor materials |
| | Resistive Temperature Sensors | Also known as Thermistors, use ceramic resistors that change according to the temperature. |
| | Resistive Temperature Detectors | Use a metal resistor and handle higher temperature ranges. |

| | | |
|---|---|---|
| Humidity Sensors | Capacitive RH Sensors | Based on three layers (thermose polymer, platinum electrodes and a silicon chip), react to relative humidity in the environment. |
| | Resistive Humidity Sensors | Use resistor changes to measure humidity. |
| Chemical Sensors | Interdigital Transducer Sensors | Composed of a layer used as the dielectric between two electrodes that change with different substances. |
| | Conductivity Sensors | Use chemical substances to measure conductivity. |
| | Optical Chemical Sensors | Detect chemical reactions through optical waveguides. |
| | Piezoelectric Chemical Sensors | Measure electric charges to identify chemical composition changes. |
| Mechanical Sensors | Pressure Sensors | These sensors measure the applied force over something. |
| | Magnetic Position Sensors | Measure the intensity, direction, rotation or variation of a magnetic field. |
| | Ultrasonic Position Sensors | Similar to radar, these identify objectives by means of high frequency sound waves. |
| | Accelerometers | Can identify volcanic activity, structure stiffness or rigidity of human bones and human joints. |
| Medical Sensors | Oximeters | Measure blood oxygenation by determining the hemoglobin of the patient. |
| | Blood Flow Sensors | Employ Doppler shift to reflect an ultrasonic wave to measure blood flow. |
| | ECG (electrocardiogram) | Measures heart rate. |
| | EMG (electromyography) | Measures muscular activity. |
| | Glucose monitor | Detects blood glucose levels. |
| Other Sensors | Piezoelectric cylinders | Measure gas velocities. |
| | Passive infrared sensors | Detect infrared radiation to detect movement. |
| | Seismic Sensors | Detect and warn of possible earthquakes by measuring seismic waves and vibrations. |
| | Acoustic emission sensors | Identify micro-structural changes or displacement. |

commonly composed of three layers. The operating system, which is at a single node level, is located in the first layer. The second and third layers are located in the entire network level, and are comprised of middleware and applications, respectively. The software components of WSNs are shown in Figure 2.

The operating system is the means by which a node accesses the hardware. It is responsible for managing energy and memory by using commands, besides controlling processes required by the application.

Operating Systems can be classified by their architecture (monolithic, modular or virtual machine), by the type of programming (dynamic or static), and by the mode of execution (event-based, multi-threaded or hybrid).

Monolithic architectures have of a single system image with the necessary components of the operating system and application. In modular architectures, the operating system and application consist of several modules which interact with one another. Virtual machine architectures, on the other hand, have of a single system image plus the application, static and dynamic components (Reddy et al., 2009).

In static operating systems resources are assigned at the time of design by the type of programming. In dynamic operating systems, resources are allocated and deallocate at run-time, making them the most flexible and suitable systems for changing environments (Dong et al., 2010).

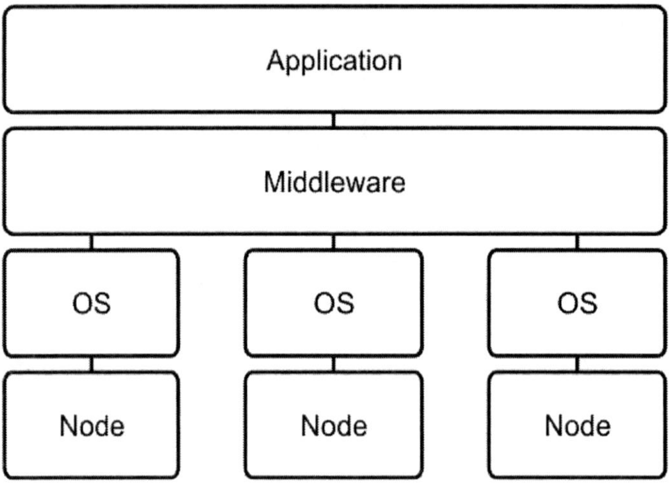


**Figure 2.** Software Layers of a sensor node (Reddy et al., 2009).

For event-based operating systems, programmers maintain the application state manually, which makes them suitable for sensitive applications with limited resources. Multithreaded systems can run multiple events at the same time so they are simpler and less complex than event-based operating systems (Reddy et al., 2009).

When selecting an operating system for sensor networks, it is important to take into account several factors. First of all, one should consider the type of data to be managed because the use of resources depends on the complexity of data collected from the sensors. Secondly, the tasks to be scheduled need to well-defined as the operating system must have the capacity to organize and execute the process. Finally, how the operating system stores information must be considered in order to provide efficient access to hardware components.

Listed below are the major operating systems for wireless sensor networks:

- TinyOS: is one of the most popular operating systems because it is adaptable to energy and memory constraints. TinyOS is a monolithic operating system that uses a component model, meaning that a single image is produced and components can be integrated according to application requirements.
- SOS: is a modular operating system in which every module represents a specific function. Contrary to TinyOS, SOS allows dynamic memory association (Dargie and Poellabauer 2010).
- Contiki: is an open source modular operating system that is portable, event-driven, lightweight, and can multitask. Contiki features include a graphic user interface, a browser and a personal web server (Farooq and Kunz 2011).
- MagnetOS: reduces energy consumption by using divided code that is distributed throughout the entire network. However, its main characteristic is that it has a single-system similar to Java virtual machine (García-Hernando et al., 2008).
- LiteOS: is a thread-based operating system that produces a single image which uses dynamic memory association and can support multiple applications.
- NanoRK: is a monolithic multitasking operating system which supports real-time applications and programming and is useful in multiple hop networks (Farooq and Kunz 2011).

On the second layer in a network level, middleware acts as an intermediary between operating systems and applications. Common middleware is not convenient for WSNs because of sensor node constraints (i.e., their size, hardware limitations, and limited resources).

Middleware can be programmed according two classes. On one hand there is the programming support class which provides services and run-time mechanisms. On the other hand, it provides programming abstraction regarding ideas and concepts about sensor nodes and sensor data (García-Hernando et al., 2008).

In the application layer, algorithms define the specific parameters to be read by sensors. With the wide range of sensors available it is possible to collect information from multiple environmental factors. Current wireless sensor network applications cover a vast array of areas which can be classified into three main areas: security, healthcare, and environment monitoring.

In the area of security, military applications have advanced considerably over other areas due to the greater investment in research and development. Additionally, WSNs had their genesis as military projects and military researchers continue to study the rapid deployment of WSNs, along with fault tolerance and inexpensive devices that do not represent significant losses if lost in the battle field.

Today, the main military applications include monitoring allied forces, battlefield conditions, equipment, and ammunition, as well as terrain reconnaissance, enemy-force identification, guidance systems and battlefield damage assessment (Akyildiz et al., 2002).

Surveillance applications can also be classified as security applications. Garcia-Hernando et al. (2008) categorized surveillance applications into outdoor and indoor applications due to the different requirements of both kinds of environments. Indoor surveillance applications are found in public buildings like hospitals, government offices, schools, etc., while outdoor surveillance applications provide perimeter security as in an airport. A common indoor application is for fire detection in the form of smoke and temperature sensors. Nauman et al. (2011) propose an algorithm to detect a fire as it simultaneously identifies secure routes to assist evacuation. Narendra et al. (2011) install a monitoring and firefighting system in a hospital that employs a wireless sensor network and a LCD monitor that displays the current temperature. When exposed to high temperatures, if a preset threshold is crossed, a sensor node activates a voice message indicating that extinguishers will be automatically activated, without further warning.

WSNs are also used in the area of intrusion detection. By using thermal and volumetric sensors that employ infrared radiation (García-Hernando et al., 2008) create a highly sensible detection field. In addition, these systems can be improved with video sensors. For example, Pham (2011) proposes the using video sensors to minimized the intrusion detection time.

To guarantee outdoor security, wireless sensor networks are used to detect abnormalities in public spaces to detect possible criminal or terrorist activities. Sensors can identify potential threats including the presence of toxic gases, radioactivity, biohazards, or chemical threats. WSNs are also used for biometrics, including facial identification (García-Hernando et al., 2008).

For healthcare, WSNs are used to manage hospitals and medical personnel, monitor patients, identify possible allergies and symptoms, and for emergency systems. Gravrilovska et al. (2010) describe four possible scenarios for healthcare applications: pre-hospital care, in-hospital care, post-hospital care, and continuous care. Pre-hospital care refers to emergency situations, disasters, or massive accidents where emergency services need information about patient condition. Ahmad et al. (2011) present a detection model for survivors where sensor nodes exchange information before sending it to an emergency station to help in immediate rescue.

In-hospital care, on the other hand, involves the continuous monitoring of patients. Harun et al. (2011) present a prototype heartbeat monitor bracelet that provides patient blood oxygen saturation levels in the blood along with applications to monitor different organs, as well as physiological parameters such as blood glucose levels (García-Hernando et al., 2008).

Post-hospital care and continuous care implies constant patient monitoring during recovery, including in-home diagnostic or chronic health monitoring for the ill and elderly. Mileo et al. (2008) present a WSN prototype to collect information about patients and their surrounding environment called SINDI (Secure and Independent Living), which consists of sensors deployed inside a house with inertial measurement devices to locate and detect motion.

Healthcare projects have been developed to improve patient conditions and treatment. For example, the Code Blue project at Harvard University that uses a wireless sensor network to offer response to emergency situations, provide patient care, and rehabilitation (Iyengar et al., 2010).

The goal of environmental monitoring is to preserve natural resources and ecosystems. Khedo et al. (2010) present a system to monitor air pollution called WAMPS (Wireless Sensor Network Air Pollution Monitoring System), which consists of a set of sensor nodes that collect information about pollutants including ozone, nitrogen dioxide, carbon monoxide, and sulfur dioxide, among others. Importantly, however, sensors can also detect chemicals waste dumped into waters by factories (García-Hernando et al., 2008). In addition to monitoring the condition of water, air or soil, WSNs have applications that assist preserve animal and plant life. For example, in Los Angeles, California, there is a network that monitors the habitat which also provides data used for to study weather patterns, identify the presence and spread of diseases, and controls the animal population (Chong and Kumar 2003).

Environmental monitoring applications include precision agriculture, which requires information about humidity, temperature, light conditions, and soil, among others. This information is needed to control irrigation systems, measure fertilizer and pesticide application, increase crop yield, reduce labor costs, and control and reduce greenhouse gas emissions (Chaudhary et al., 2011).

Other applications have meteorological research purposes; WSNs can collect information about the rain, wind speed and direction, temperature, humidity, solar radiation, etc., to predict and mitigate possible natural disasters such as hurricanes or floods. Yawut and Kilaso (2011) present a disaster prevention system that employs a WSN to issuing warnings.

WSNs in industry help achieve a high level of precision, improve production processes, and ensure low operating costs. The IEEE and the NIST (National Institute of Standards and Technology) cosponsored the P1451 Smart Transducer Interface Standard to allow plug-and-play of sensors and networks in industrial environments (Chong and Kumar 2003).

Some WSN applications are intended to access places or environments that are too hostile or dangerous for humans; a clear example is the use of wireless sensor networks in greenhouse and mining. For example, miners descend underground under dangerous conditions, often characterized by poor ventilation, gas explosions, and cave-ins. Dargie and Poellabuer (2010) present detection task for WSN such as locating people in normal and abnormal conditions, location of landslides, earthquakes, measuring and forecasting within the mine and the concentration of gases including methane, oxygen, or carbon dioxide.

## IEEE 802.15.4

The IEEE 802.15.4 is a standard that defines low rate interconnections of wireless personal area networks or LR-WPANs, which have low data rates, low power consumption and a limited radio transmission. The most important goals of LR-WPANs are to provide easily installation, reliable data transfer, short-range operation, efficient energy consumption, and low cost—all the while employing a protocol that is as flexible and simple as possible (IEEE 2006).

The standard defines the lower layers of the OSI model: the physical layer (PHY) and medium access layer (MAC). The PHY is responsible for of modulating and demodulating signals received through a radio transceiver, data reception and transmission and frequency selection; the MAC layer manages access to the media, performs beacon management and association and dissociation tasks.

Radio-propagation behavior depends on the appropriate selection of the carrier frequency and its available capacity. A single frequency does not support the required capacity so a frequency band that is a finite portion of the electromagnetic spectrum is used (Karl and Willig 2007).

Different frequency bands are allowed under certified or unlicensed operation. Devices functioning under the 802.15.4 standard can transmit on 4 different frequency bands: an 868/915 MHz direct sequence spread spectrum (DSSS) PHY with binary phase-shift keying (BPSK) modulation, an 868/915 MHz DSSS PHY with quadrature phase-shift keying (O-QPSK) modulation, an 868/915 MHz parallel sequence spread spectrum (PSSS) PHY with BPSK and amplitude shift keying (ASK) modulation, and a 2450 MHz DSSS PHY employing O-QPSK modulation (IEEE 2006) whose characteristics are defined in Table 2.

**Table 2. Frequency bands and data rates (IEEE 2006).**

| *PHY (MHz)* | *Frequency band (MHz)* | *Spreading parameters* | | *Data parameters* | | |
|---|---|---|---|---|---|---|
| | | *Chip rate (kchip/s)* | *Modulation* | *Bit rate (kb/s)* | *Symbol rate (ksymbol/s)* | *Symbols* |
| 868/915 | 868-868.6 | 300 | BPSK | 20 | 20 | Binary |
| | 902-928 | 600 | BPSK | 40 | 40 | Binary |
| 868/915 (optional) | 868-868.6 | 400 | ASK | 250 | 12.5 | 20-bit PSSS |
| | 902-928 | 1600 | ASK | 250 | 50 | 5-bit PSS |
| 868/915 (optional) | 868-868.6 | 400 | O-QPSK | 100 | 25 | 16-ary Orthogonal |
| | 902-928 | 1000 | O-QPSK | 250 | 62.5 | 16-ary Orthogonal |
| 2450 | 2400-2483.5 | 2000 | O-QPSK | 250 | 62.5 | 16-ary Orthogona |

To select a frequency band, it is necessary to take into account important system design factors. Sensor nodes need to coexist with other systems using the same frequency. The effective power transmission, also called the antenna efficiency, can be affected by the small antenna size employed by sensor nodes in WSNs because wavelength decreases in direct proportion to antenna size and if antenna efficiency decreases, more energy must be expended to obtain the transmission power necessary to compensate (Karl and Willig 2007).

To access to the physical layer, the second layer of the OSI model, the MAC Layer, employs mechanisms to coordinate and synchronize the network, and offer channel access. The most common mechanisms are the Carrier Sense Multiple Access (CSMA) and the Multiple Access with Collision Avoidance (CSMA/CA), and their variations. The CSMA mechanism has two modes: non-persistent mode and persistent mode.

In CSMA non-persistent mode, nodes are allowed to transmit when the medium is released. If the medium is busy, a backoff period occurs and a temporizer starts to retry sending the transmission. In CSMA persistent mode, nodes analyze the medium to ensure it is free and then transmit. If a collision is detected, transmitting nodes wait a specified time to retransmit again.

CSMA/CA improves collision avoidance when the medium is released to transmit. If nodes do not transmit immediately, they start a waiting period. If two or more nodes are waiting, the node with the shortest waiting period transmits.

## Routing Protocols for WSN

Routing in wireless sensor networks is very challenging because methods for sending and receiving data must deal with situations that require the constant transmission and reception of data, originating from different regions in the network. Some nodes can sense the same information and create redundant information which can waste resources, thus compromising network performance in other areas such as such as processing capacity, energy and memory, all of which are extremely harmful for networks which must handle rapidly changing topologies. The main purpose of routing protocols is to optimize available resources, including energy, mobility, self organization, location, information management and exchange, among others.

In WSNs, sensor nodes transmit information to a specific node called a sink. According to the specific routing algorithm used, WSN protocols can be classified in data-centric, hierarchical, and location-based protocols.

In data-centric protocols, routing functions are based on queries. The sink node broadcasts a query and waits for data from different regions of the network. Hierarchical protocols use clusters and special nodes known as cluster heads. A cluster head allows efficient use of resources because routing occurs between cluster heads, thereby eliminating the need for flooding. Finally, location-based protocols are based on node position. In order to save energy, nodes are generally in a sleep state and only activate when receiving a hello message (Hara et al., 2010).

The main routing protocols for WNS are shown below in Table 3, a summary of routing protocols by Aquino Santos et al. (2007), Garcia Villalba et al., 2009; Garcia-Hernando et al., 2008; and Hara et al., 2010.

It becomes evident that routing in wireless sensor networks is very complex. Many research projects aim to provide algorithms that are suited to application needs but have limited hardware resources.

## Proposal

The algorithm presented in this chapter was originally created for vehicular *ad hoc* networks, or VANETs. Location Routing Algorithm with Cluster-Based Flooding (LORA-CBF) is a protocol that is designed to reduce redundant information and ensures a high performance for dynamic and distributed topologies.

Aquino Santos et al. (2005) describe LORA-CFB as a protocol that minimizes communication costs and builds a solid link to share information. Validated through simulations in which the performance of LORA-CBF was compared with protocols, including AODV (*Ad hoc* On-Demand Distance Vector), DSR (Dynamic Source Routing Protocol), and GPSR (Greedy Perimeter Stateless Routing). LORA-CBF was evaluated and validated as a functional and predictive algorithm for vehicular scenarios under characteristics such as high mobility and large numbers of nodes.

Vehicular *ad hoc* networks are a variant of mobile *ad hoc* networks or MANETs (Mobile *Ad Hoc* Networks). VANETs provide a communication system between vehicles or vehicles to base stations placed on the road. VANET applications aim to improve traffic flow or to provide vehicle safety. Data transmission is of critical importance in VANETs and extremely efficient routing algorithms are required for this type of network to maintain network integrity. This is due to the large number of nodes and the dynamic topology of VANETs, which are characterized a highly dynamic topology that contains nodes that are constantly entering and exiting the network, making it difficult to maintain efficient communication system in order to ensure delivery data. These networks are characterized by a complex topology due to rapid changes in links, possible fragmentation, and inconsistency in network density (Piran et al., 2011).

However, it is important to note that VANETs differ from WSNs in important aspects such as energy consumption because vehicular sensors can use energy provided by the automobile batter whereas wireless sensor networks face severe power restrictions. VANET sensor movements can be organized in certain ways (platoon behavior, merging behavior, etc.). On

**Table 3. Routing Protocols in Wireless Sensor Networks.**

| *Routing Mode* | *Routing Protocol* | *Description* | *Goals* |
|---|---|---|---|
| Data-centric | *Flooding* | Is the simplest algorithm. Packets are consecutively retransmitted to find the destination node. Does not require expensive maintenance or route discovery algorithms. However, can present implosion caused by redundant data received by nodes. | Implements a simple algorithm to makes packets arrive at a defined destination node |
| | Gossiping | Sends packets to a randomly selected neighbor instead of all neighbors. | Improves a simple algorithm such as flooding routing but with a simple technique added in order to avoid implosion problem of flooding |
| | Sensor Protocol for Information (SPIN) | Uses the "shortest route" strategy and routing packets as ADV (advertising) packets, which allows a sensor to advertise new data to its neighbors, REQ (request) packets to request data, and DATA packets that is the information. | Extended lifetime, exchanges metadata to reduce the number of messages |
| Data-centric | Directed Diffusion | Sensor nodes use a naming scheme for data. A node that wants to transmit generates an interest consisting of an attribute-value based scheme. The node establishes a reply link called a gradient, which is based on the received message. | Fault tolerance in n-way of communication. |
| | Rumor Routing | Is a direct diffusion variation that uses paths to identify events | Reduces queries in network |
| | Gradient-Based | Is another variation of direct diffusion. This protocol memorizes the number of hops when the interest is diffused through the network. | Data delivery through minimal number of hops |
| | Active Query Forwarding in Sensor Networks (ACQUIRE) | As a distributed database, ACQUIRE uses complex queries that can be divided into sub-queries. If nodes do not have enough information to define the path, sub-queries are added through the network in order to complete it. | Optimization of queries in network |

| | | | |
|---|---|---|---|
| Hierarchical | Low-Energy Adaptive Clustering Hierarchy (LEACH) | Minimizes energy dissipation with randomly selection nodes as cluster heads. | Lifetime through distributed clusters. |
| | Power-Efficient Gathering in Sensor Information System (PEGASIS) | Sensor nodes communicate only with their nearest neighbors using a shift strategy to communicate with the sink node. | Lifetime and bandwidth optimization |
| | Threshold-sensitive Energy Efficient protocol (TEEN) | Nodes constantly analyze the medium; however, data is transmitted less frequently. | Real-time and Lifetime |
| | Adaptive Periodic TEEN (APTEEN) | Changes the threshold value used in the TEEN protocol according to user needs and the type of application. | Real-time and Lifetime |
| Location | Minimum Energy Communication Network (MECN) | Minimizes energy consumption through GPS coordinates to identify a subnet with fewer nodes that requires less power for transmission. | Lifetime and self-reconfiguration |
| | Small Minimum Energy Communication Network (SMECN) | Is an extension of MECN, but with the advantage that the potential obstacles between selected nodes, are taken into account. | Lifetime and self-reconfiguration |
| | Geographic Adaptive Fidelity (GAF) | Divides the network area in order to form a grid. Nodes in the same cell are equivalent. Equivalent nodes can communicate with nodes in adjacent cells and these negotiate which one is to transmit while the others remain in wait. | Lifetime with a large number of nodes |

the other hand, WSN sensor nodes are deployed in a more random manner. VANETs also share some important features with WSNs which make their algorithms possibly adaptable to WSNs. In this case we consider adapting LORA_CBF, an algorithm designed for VANETs to WSNs.

WSNs and VANETs are infrastructureless and the routing is done by nodes that must self organize. It is also important for WSNs to have a protocol that can provide solutions to topological changes and ensures scalability, collection, and data delivery.

LORA-CBF is a location-based protocol. This kind of algorithm possesses lightweight routing in comparison to other protocols, mainly due its packet forwarding strategy that sends packets based on information gathered from its neighbors, thereby reducing control traffic and preventing flooding of redundant information.

LORA-CBF is a protocol that defines different states for VANET nodes, with each state being responsible for a different routing task. According to manner in which sensor nodes exchange routing packets, they can be change their status to reflect the different states listed below:

- **Undecided:** is the basic state of a node when the network is deployed.
- **Member:** is formed when a sensor node listens to a cluster head that makes it change its state to member node. If a member node stops receiving information from its cluster head, it become in an undecided node and the cluster formation restarts.
- **Cluster Head:** is responsible for controlling neighbor nodes.
- **Gateway:** is formed when a sensor node is a member of two or more cluster heads. Gateway nodes minimize flooding of broadcast messages reducing retransmissions in a same region.

Figure 3 shows the LORA-CBF algorithm. To send information, sensor nodes first check if they have knowledge of the destination's location, and if it is within range in order to determine whether or not to begin transmitting data or to employ the destination prediction algorithm.

At the beginning of network deployment, sensor nodes start to transmit periodic small packets called "hello packets". Hello packets contain basic information about the node, including its address, status, cluster head id, etc.; this information is used to identify other nodes and to make itself known. In the cluster formation mechanism, a new state is assigned to sensor nodes.

As an undecided node, a sensor node broadcasts a hello message and waits for a reply from a neighbor node. When the hello message is transmitted, the node starts a timer. If the time limit is reached and the sensor node does not receive an answer, it changes its status to cluster head. Immediately after this, the sensor node transmits a new hello packet and restarts the timer and waits for a new reply packet.

If a new sensor node receives a hello message from a cluster head node, it changes its status to member, changes its status from undecided to member, and broadcast a new hello message with its information updated.

When a sensor node receives information about a new neighbor or updated information, it stores the information in a neighbor table. Each node builds a neighbor table that is updated when changes occur when neighboring nodes change their states.

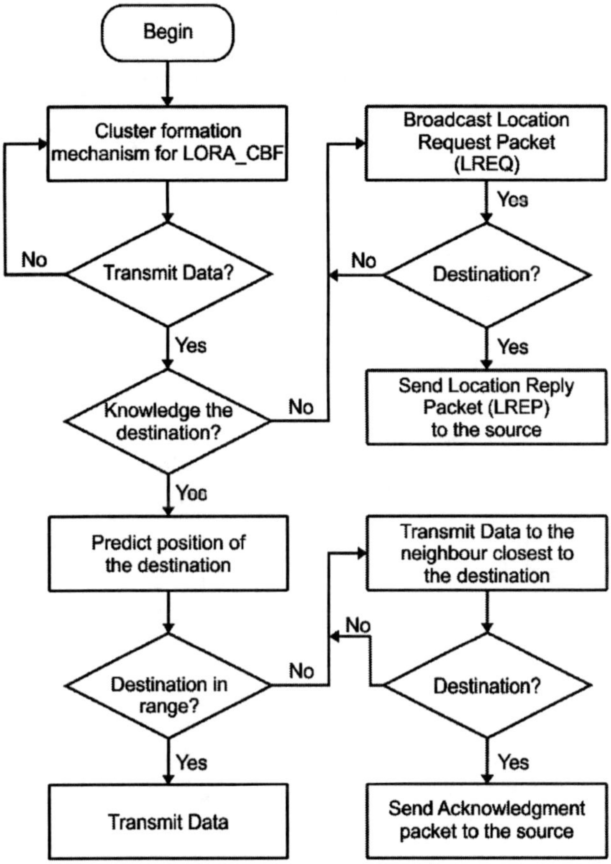


**Figure 3.** Flow diagram for LORA-CBF (Santos et al., 2007).

If a member node receives a hello message from a cluster head which posses a different id to the one stored in its neighbor table, the status of the member node changes to gateway; however, if the new cluster head id in the neighbor list does not correspond to its cluster head it will be replaced with the new cluster head id if the hello timer expires.

Figure 4 shows the procedures employed by the cluster formation mechanism of LORA-CBF. This diagram is an adapted version of the VANET cluster formation mechanism presented by Santos et al. (2005). Changes made on the original diagram were minimal.

In order to avoid duplication of neighbor information, when new information is received, nodes compare the newest information to the information already stored. If two cluster heads are in the same region, one of them must change its state based on the size of their neighbor tables.

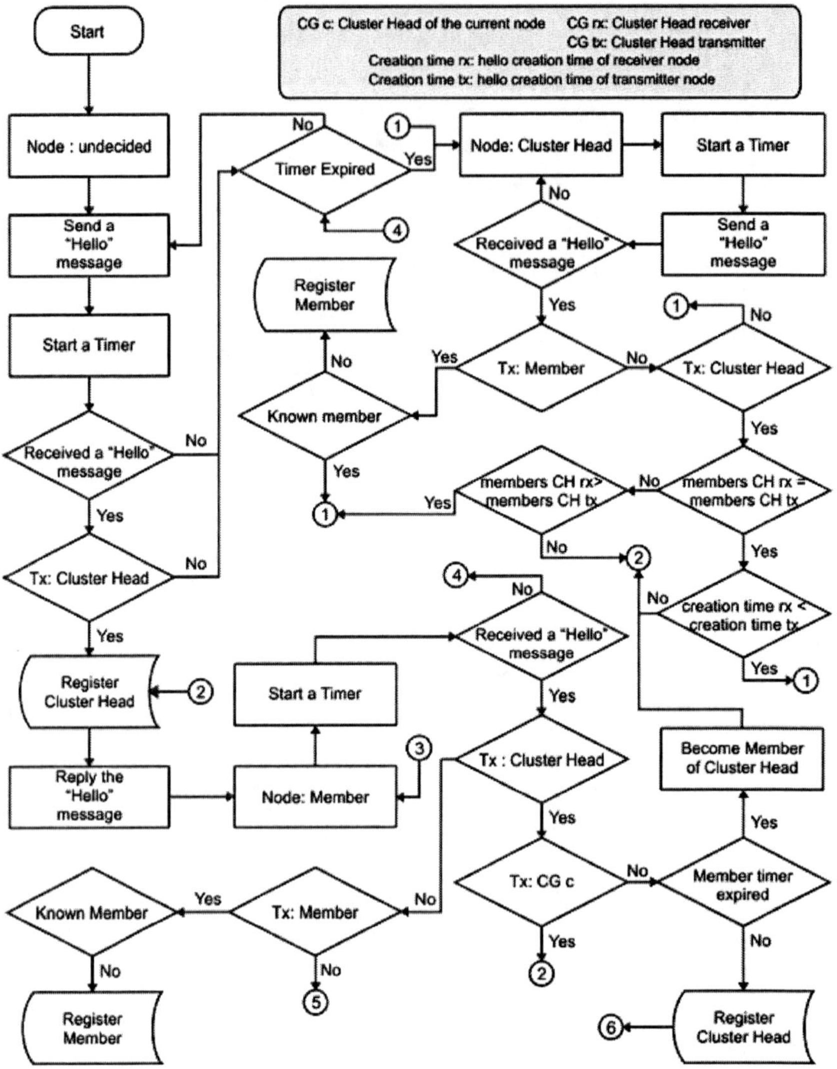


*fig. contd. on next page*

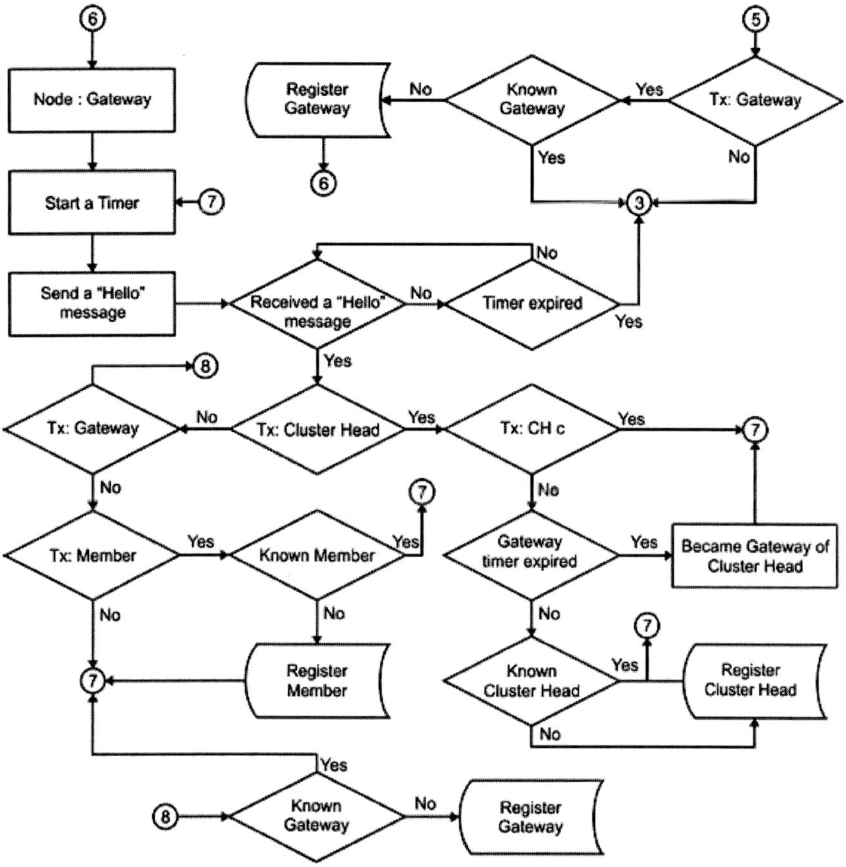


**Figure 4.** Flow Diagram of the Cluster Formation Mechanism for LORA-CBF.

The way which nodes return to their previous state is also different. In the original algorithm, a node sends a hello message before it changes its status; only when a response is received will the node changes its status. In the modified LORA-CBF algorithm, a node changes its state and sends a hello message with the current information. In addition, it temporarily stores the previous address. If the next hello message comes from a different node, the node transmits another hello packet to ensure a fast cluster formation.

Once cluster formation is complete, data transmission can start. When a node wants to transmit information, it searches the destination address in its neighbor table. If the address exists, the node will send the data without problems. However, if the address does not exist in its neighbor list, the node starts the destination location prediction.

In the prediction of position of destination, LORA-CBF uses two types of packets to predict the position of the destination: Routing Request Packet (RREQ) and Routing Reply Packet (RREP).

A RREQ packet provides information about the source node, destination node, previous and next hop, number of hops, and an identifier to avoid duplicating messages.

When a sensor node starts to predict the position of the destination node, it sends an RREQ message in broadcast to locate the cluster head node. In LORA-CBF only cluster head nodes can retransmit a RREQ among themselves by means of gateway nodes that serve as antennas. Gateway nodes transmit messages to other gateways as long as they belong to different cluster heads.

When a cluster head receives a RREQ packet, it searches for the destination address in its neighbor table. If the destination address does not exist, the cluster head stores the information about the source node and rebroadcasts the message. However, if the cluster head finds the destination address, it sends a RREP packet to the source node. As the RREQ packet travels through the network, nodes receiving a RREQ packet. Know the closest neighbor to the source node and can therefore determine a path for RREP packets.

The RREP packet will then arrive at the source node and the transmission of information will start. To ensure data delivery, if the source node does not receive an RREP before the timer has expired, it will send a new RREQ.

If the source node knows a route to the destination node, packet data transmission starts. When a node receives a data packet destined to it as the next hop, but not as the final destination, the node will search its routing table to transmit the location to the nearest node and then retransmit it. If it is a data packet for the sensor node, it keeps the relevant information and generates an acknowledgment packet to the source node.

## Results

We used a simulation in OPNET modeler to evaluate the performance of LORA-CBF in wireless sensor networks. Our protocol was developed for sensor nodes under the IEEE 802.15.4 standard at the network layer. Several scenarios were created to test scalability, robustness, and multicast methods.

Scalability is a critical factor in WSNs; this allows us to increase network size without compromising network performance. Wireless sensor networks are commonly composed of hundreds of nodes. Therefore, a WSN routing protocol must offer high performance in small or large-scale networks. The term "network robustness" refers to how a network ensures high performance when conditions such as the disappearance of a node or saturation of the transmission channel occur. Finally, multicast transmission provides greater scope to the protocol. This type of transmission is necessary when a sensor node wants to transmit to more than one destination. Although broadcast transmissions could be the solution, it unfortunately means that much more control traffic is necessary, which greatly increases the use of limited resources.

LORA-CBF performance was evaluated by the following metrics: end to end delay, routing overhead, overhead, and packet delivery rate.

End to end delay represents the time a data package requires to be transmitted to the destination node. In the evaluated scenarios LORA-CBF offers a reasonable delay. Scenarios with 10, 9, 8 and 6 hops were tested and the end-to-end delay obtained was not greater than one second.

One of most important factors of LORA-CBF in VANETs is the low routing overhead. Routing overhead represents all packets used during the simulation time that are not data packets. In WSNs, LORA-CBF guaranteed a low number of routing packets. Results show that routing overhead is related to the length of the routing path. However, LORA-CBF routing overhead was low.

Overhead is the representation of the number of routing packets required to transmit data packets to the destination node. In communications systems, it is directly related to network performance. A protocol that generates low levels of overhead ensures better network performance.

The LORA-CBF overhead obtained proves that this algorithm can optimize resources in WSNs. The number of routing packets used in the LORA-CBF simulations were minimal, indicating that it effectively ensures data transmission while preserving resources, while using fewer process.

The delivery rate of LORA-CBF proved the most promising. Due to the routing method of LORA-CBF, it knows the closest neighbor to destination node. Therefore, delivery rates obtained were over 90%.

LORA-CBF was also tested according to three important characteristics: scalability, robustness, and multicast transmission. The scalability evaluation shows that LORA-CBF performs well in both small and large networks. Robustness was tested with many transmitter sensors to determine channel performance when it is shared among more than one transmitter. In this evaluation, the delivery data rate was good. However, the loss of acknowledgment packages significantly increased. Finally, LORA-CBF includes a multicast transmission that we also tested. Results show that LORA-CBF possesses a high delivery rate.

## Conclusions

WSNs are an effective and efficient means of analysis and monitoring of the phenomena in their environment, features of WSNs enable them deployed in a wide variety of terrains: hostile environments, risk situations, hard-to-access areas, etc., reasons why their use has continued to show a substantial growth rate in certain areas, such as industry, healthcare, surveillance and agriculture, among others.

With regard to the advantages over conventional networks as that WSNs offer about not need physical infrastructure and an inexpensive cost, these also have hardware constrains as lack of energy, memory and process capacity, mainly because of the size of sensor nodes. This has led a research effort to develop new methods to optimize sensor features and to expand the reach of wireless sensor networks.

This work presents an adaptation and evaluation of a routing protocol: LORA-CBF, this protocol was created to satisfied necessities of vehicular *ad hoc* networks; we propose to expand the benefits of the algorithm to wireless sensor networks. Although VANETs are characterized by high mobility and other factors specific to the vehicle, also share characteristics that distinguish to WSNs such as a complex topology, no infrastructure and a large number of elements.

LORA-CBF was successfully adapted to WSNs through modeling software, which also made an evaluation performance.

## REFERENCES

5th Kuala Lumpur International Conference on Biomedical Engineering 2011. In N.A.A. Osman, W.A.B.W. Abas, A.K.A. Wahab and H.-N. Ting [eds.], (Vol. 35, pp. 348–351): Springer Berlin Heidelberg.

Ahmad, N., N. Riaz and M. Hussain. 2011. *Ad hoc* wireless Sensor Network Architecture for Disaster Survivor Detection. *International Journal of Advanced Science and Technology,* Vol. 34. pp. 9–16.

Akyildiz, I.F., W. Su, Y. Sankarasubramaniam and E. Cayirci. 2002. Agosto. A Survey on Sensor Networks. IEEE Communications Magazine, Vol. 40, pp. 102–114.

Aquino Santos, R., A. Edwards, R. Edwards and L. Seed. 2005. Performance evaluation of routing protocols in vehicular *ad hoc* networks. *International Journal of Ad Hoc and Ubiquitous Computing,* Vol. 1, 80-91. doi: 10.1504/IJAHUC.2005.008022

Aquino Santos, R., L. Villaseñor González, J. Sánchez and J. Rosario Gallardo. 2007. Routing Strategies for Wireless Sensor Networks. *IFIP International Federation for Information Processing,* Vol. 248, pp. 191–192. doi: 10.1007/978-0-387-74899-3_17

Chaudhary, D.D., S.P. Nayse and L.M. Waghmare. 2011. Application of Wireless Sensor Networks for Greehouse Parameter Control in Precision Agriculture. *International Journal of Wireless & Mobile Networks (IJWMN)*, Vol. **3(1)**:140–149. doi::10.5121/ijwmn.2011.3113

Ching Hong, L. 2011. Wireless sensor network for precision agriculture. Bachelor of Information Technology (Hons) Communication and Networking, University Tunku Abdul Rahman. Retrieved from http://eprints.utar.edu.my/226/

Chong, C.-Y. and S.P. Kumar. 2003. Sensor Networks: Evolution, Opportunities, and Challenges. Proceedings of the IEEE, Vol. 91. doi: 10.1109/JPROC.2003.814918

Dargie, W. and C. Poellabauer. 2010. Fundamentals of Wireless Sensor Networks Theroy and Practice J.W.A.S. Ltd. (Ed.)

Dong, W., C. Chen, X. Liu and J. Bu. 2010. Providing OS Support for Wireless Sensor Networks: Challenges and Approaches. IEEE Communications Surveys & Tutorials. **12(4)**:519–530. doi: 10.1109/SURV.2010.032610.00045

Farooq, M.O. and T. Kunz. 2011. Operating Systems for Wireless Sensor Networks: A Survey. *Sensors*. doi: 10.3390/s110605900

García-Hernando, A.-B., J.-F. Martínez-Ortega, J.-M. López-Navarro, A. Prayati, and L. Redondo-López. 2008. Problem Solving for Wireless Sensor Networks Computer Communications and Networks.

García Villalba, L.J., A.L. Sandoval Orozco, A. Triviño Cabrera and C.J. Barenco Abbasm. 2009. Routing Protocols in Wireless Sensor Networks. *Sensors, 9*, 8399–8421. doi: 10.3390/s91108399

Gavrilovska, L., S. Krco, V. Milutinovic, I. Stojmenovic and R. Trobec. 2010. Application and Multidisciplinary Aspects of Wireless Sensor Networks: Concepts, Integration and Case Studies: Springer.

Hara, T., V.I. Zadorozhny and E. Buchmann. 2010. Wireless Sensor Network Technologies for the Information Explosion Era Studies in Computational Intelligence, Vol. 278. doi:10.1007/978-3-642-13965-9

Harun, F.K.C., N. Zulkarnain, M.F.A. Aziz, N.H. Mahmood, M.F. Kamarudin and A. Linoby. 2011. Pulse Oximetry Color Coded Heart Rate Monitoring System Using ZigBee.

IEEE. 2006. Part 15.4: Wireless Medium Access Control (MAC) and Physical Layer (PHY) Specifications for Low-Rate WirelessPersonal Area Networks (WPANs) (pp. 0_1–305 ). New York, NY. USA.

Iyengar, S. S., N. Parameshwaran, V.V. Phoha, N. Balakrishnan and C.D. Okoye. 2010. Fundamentals of Sensor Network Programming: Applications and Technology: John Wiley & Sons.

Jalil Piran, M., G.R. Murthy and G.P. Babu. 2011. Vehicular *Ad Hoc* and Sensor Networks; Principles and Challenges. *International Journal of Ad hoc, Sensor & Ubiquitous Computing*. **2(2)**:38–49. doi: 10.5121/ijasuc.2011.2204

Karl, H. and A. Willig. 2007. Protocols and Architectures for Wireless Sensor Networks: John Wiley & Sons.

Khedo, K.K., R. Perseedoss and A. Mungur. 2010. A Wireless Sensor Network Air Pollution Monitoring System. *International Journal of Wireless & Mobile Networks (IJWMN)*, Vol. **2(2)**:31–45. doi: 10.5121/ijwmn.2010.2203

Kim, T., H. Adeli, W.C. Fang, T. Vasilakos, A. Stoica, C.Z. Patrikakis, G. Zhao, J. G. Villalba and Y. Xiao. 2011. Communication and Networking: International Conference, FGCN 2011, Held as Part of the Future Generation Information Technology Conference, FGIT 2011, Jeju Island, Korea, December 8–10, 2011. Proceedings: Springer-Verlag New York Inc.

Lazakidou, A., K. Siassiakos and K. Ioannou. 2010. Wireless Technologies for Ambient Assisted Living and Healthcare: Systems and Applications: IGI Global.

Logic Programming. In M. Garcia de la Banda and E. Pontelli [eds.], (Vol. 5366, pp. 145–159): Springer Berlin/Heidelberg.

Mileo, A., D. Merico and R. Bisiani. 2008. A Logic Programming Approach to Home Monitoring for Risk Prevention in Assisted Living.

Narendra, R., Basarkod and S.S. Manvi. 2011. Wireless Sensor Network based Fire Monitoring and Extinguishing System in Real Time Environment. Advanced Networking and Applications. Vol. **3(02):**1070–1075.

Nauman, Z., S. Iqbal, M.I. Khan and M. Tahir. 2011. WSN-Based Fire Detection and Escape System with Multi-modal Feedback Multimedia Communications, Services and Security. In A. Dziech and A. Czyżewski [eds.], (Vol. 149, pp. 251–260): Springer Berlin Heidelberg.

Pham, C. 2011. Scheduling Randomly-Deployed Heterogeneous Video Sensor Nodes for Reduced Intrusion Detection Time Distributed Computing and Networking. In M. Aguilera, H. Yu, N. Vaidya, V. Srinivasan and R. Choudhury [eds.], (Vol. 6522, pp. 303–314): Springer Berlin/Heidelberg.

Reddy, A.M.V., A.V.U.P. Kumar, D. Janakiram and G.A. Kumar. 2009. Operating Systems for Wireless Sensor Networks: A Survey. *International Journal of Sensor Networks.* Vol 5 (No. 4). doi: 10.1504/IJSNET.2009.027631

Whitman, E.C. 2005. SOSUS The "Secret Weapon" of Undersea Surveillance. *Undearsea Warfare,* Vol. 7.

Witt, M. 2008. Robust and Low-Communication Geographic Routing for Wireless *Ad Hoc* Networks: Cuvillier.

Yawut, C. and S. Kilaso. 2011. A Wireless Sensor Network for Weather and Disaster Alarm Systems Paper presented at the International Conference on Information and Electronics Engineering Singapore.

# Index

**N**



**P**



**R**



**S**



**T**



**U**



**V**



**W**

# Color Plate Section

## Chapter 5

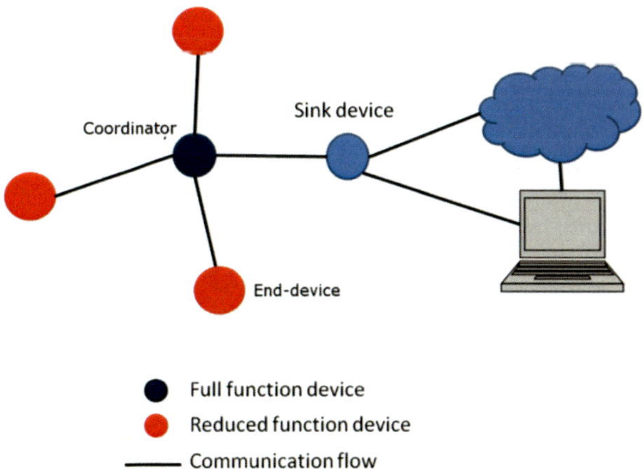


**Figure 1.** The WSN represented as a tiered structure.

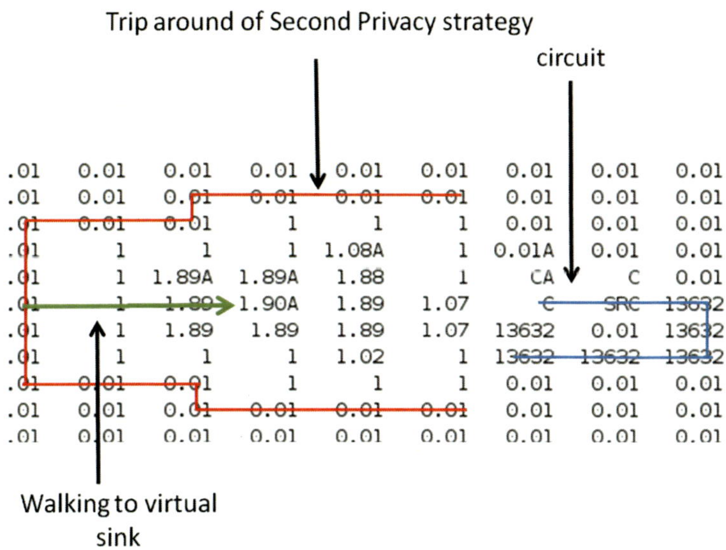


**Figure 8.** Privacy strategies by phase on transmissions.

## Chapter 6

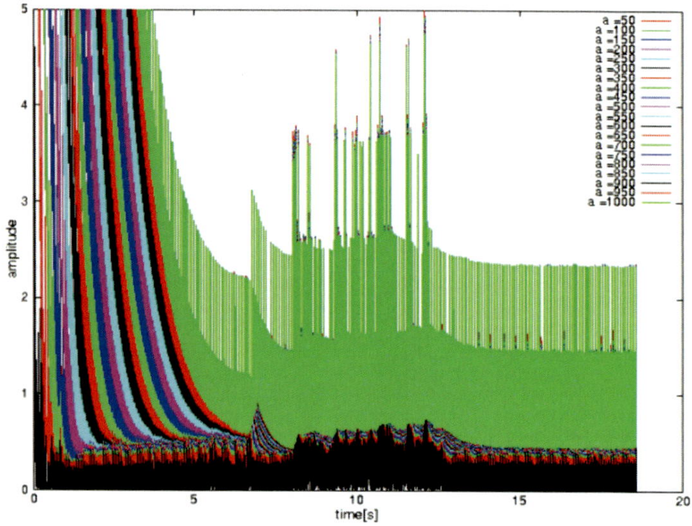


**Figure 9**

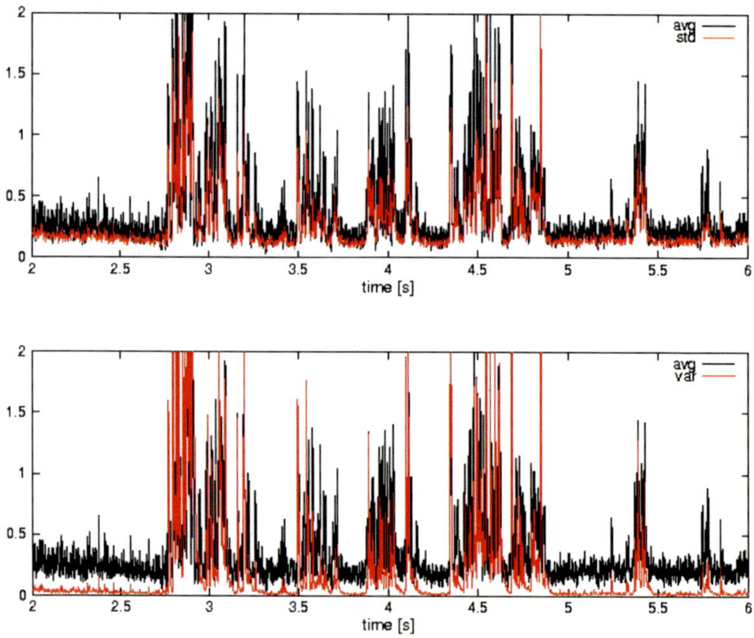


**Figure 10**

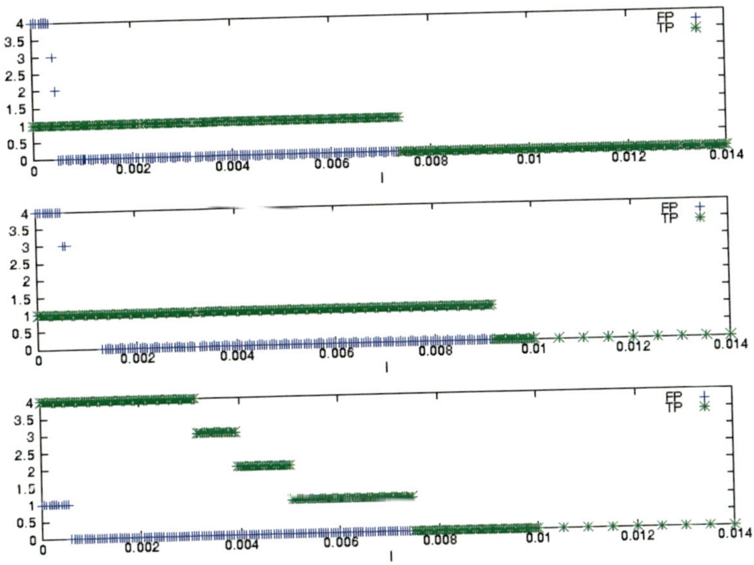


Figure 11

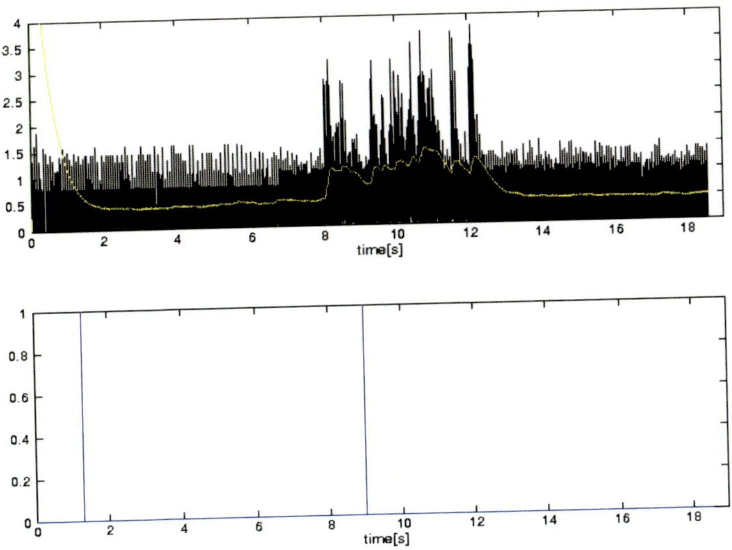


Figure 12

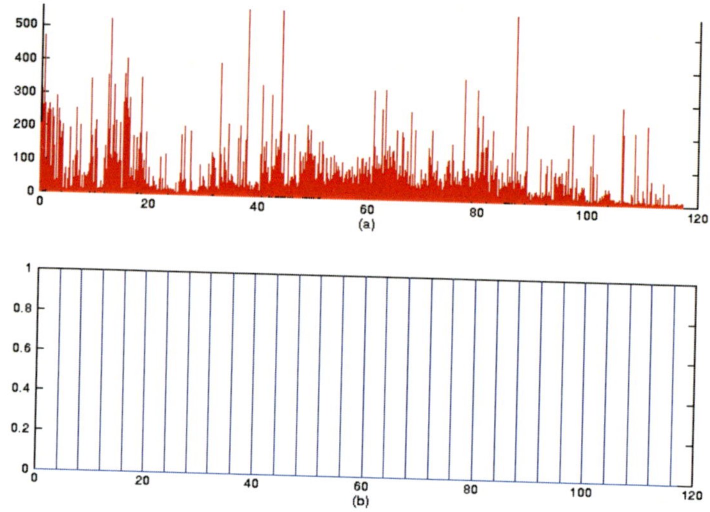


Figure 13

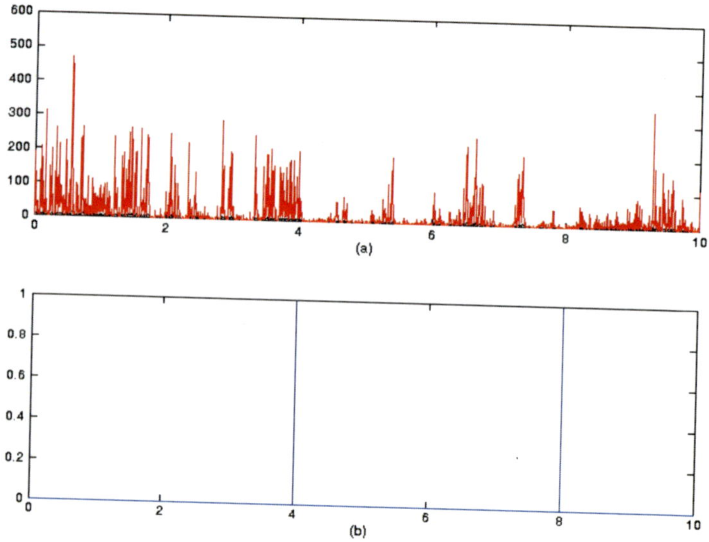


Figure 14

## Chapter 8

Figure 12

## Chapter 9

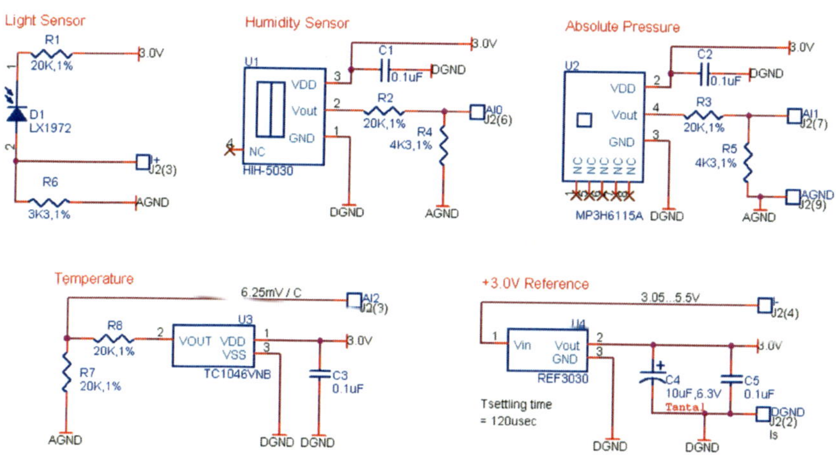


Figure 7

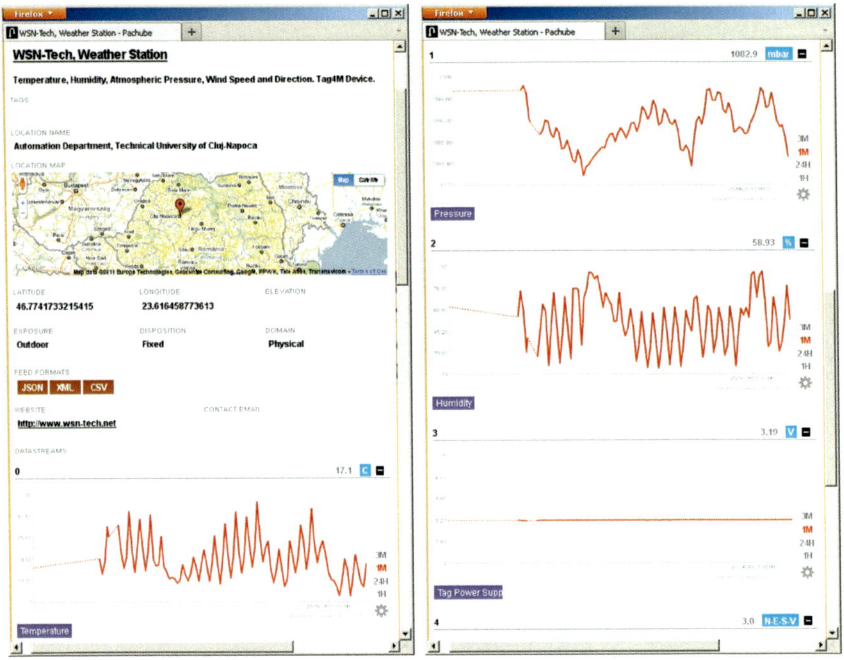


Figure 10

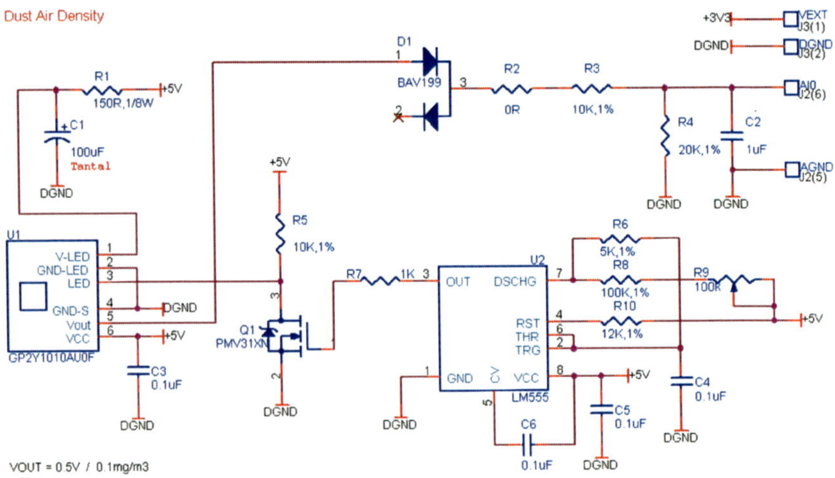


Figure 15

Figure 25

## Chapter 13

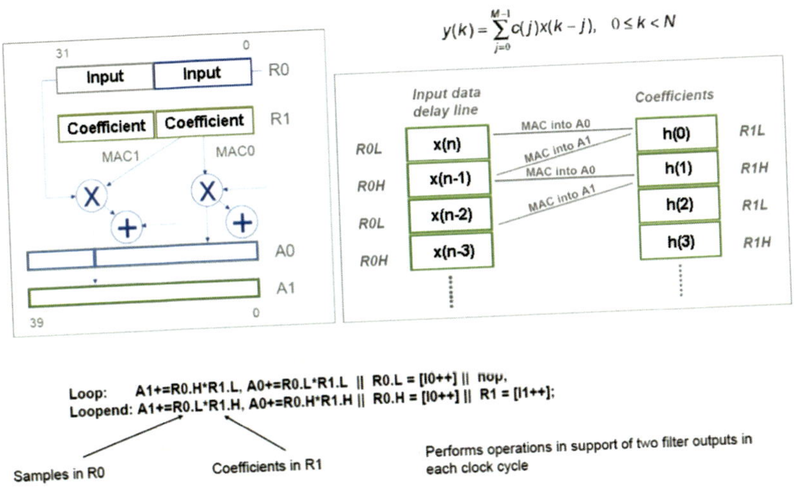


Figure 19

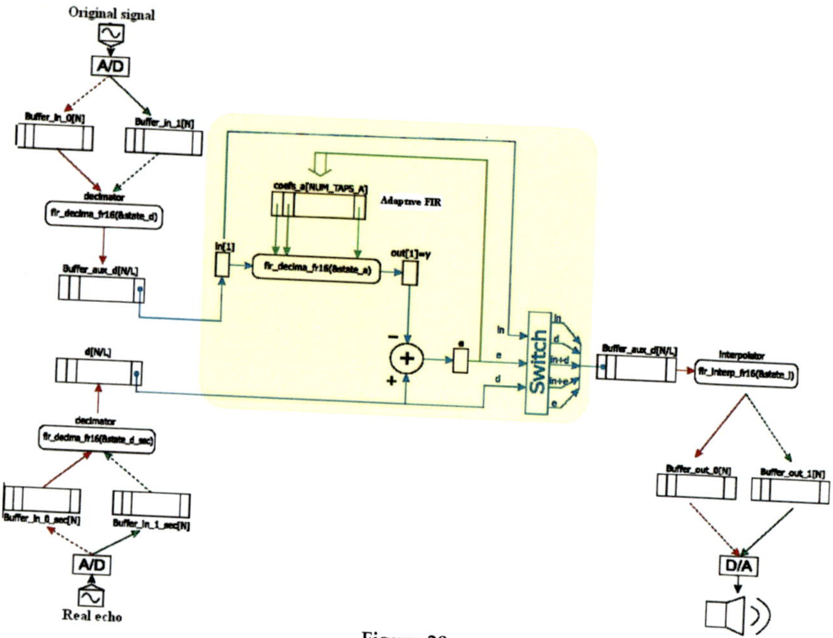


Figure 20

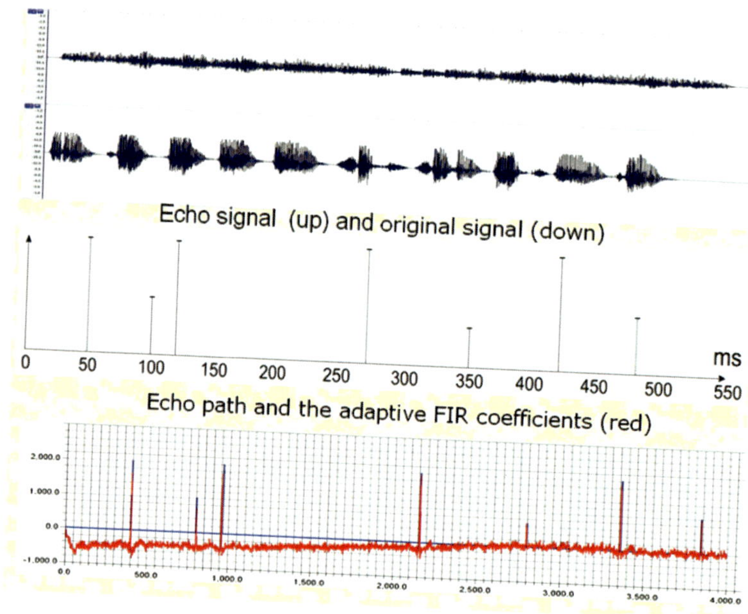


Figure 22